Lorenz Hart

Lorenz Hart

A POET ON BROADWAY

Frederick Nolan

New York   Oxford
OXFORD UNIVERSITY PRESS
1994

Oxford University Press

Oxford   New York   Toronto
Delhi   Bombay   Calcutta   Madras   Karachi
Kuala Lumpur   Singapore   Hong Kong   Tokyo
Nairobi   Dar es Salaam   Cape Town
Melbourne   Auckland   Madrid

and associated companies in
Berlin   Ibadan

Copyright © 1994 by Frederick Nolan

Published by Oxford University Press, Inc.
200 Madison Avenue, New York, New York 10016

Oxford is a registered trademark of Oxford University Press

Grateful acknowledgment for permission to quote from the following
works is made to:

Random House, Inc. for

"*Mr. Abbott,*" by George Abbott. Random House, Inc., 1963.
Getting to Know Him: A Biography of Oscar Hammerstein II, by
Hugh Fordin. Random House, Inc., 1977.

Carter, Ledyard & Milburn, Attorneys at Law, as Trustee for United
States Trust Company of New York, and Random House, Inc., for

Pal Joey, by John O'Hara. Random House, Inc., 1952.

Library of Congress Cataloging-in-Publication Data
Nolan, Frederick.
Lorenz Hart: a poet on Broadway /
Frederick Nolan.
p. cm.   Includes bibliographical references and index.
ISBN 0-19-506837-8
1. Hart, Lorenz, 1895–1943.
2. Lyricists—United States—Biography. I. Title.
ML423.H32N6 1994
782.1'4'092—dc20
[B]   92-41968

1 3 5 7 9 8 6 4 2
Printed in the United States of America
on acid free paper

*This book is for
Lisa and Rory,
Emma, Jody, and Courtney,
and
Max and Jessye,
for whom the joys of Rodgers and Hart still lie ahead.*

To the Reader

I had, of course, intended to examine many of Larry Hart's
lyrics—including some twenty that have never been published
before—at length and in detail; but, unaccountably, the copyright holders denied me permission to use any of them. I can
only apologize and say that I cannot imagine for one moment
that this would have been the case had Hart himself still been
alive.

ACKNOWLEDGMENTS

This biography is based upon three principal areas of information. The first is a series of interviews, some written but mostly oral, conducted by the author (plus some by Samuel Marx and Jan Clayton, asterisked) with, alphabetically, George Abbott, Larry Adler, Muriel Angelus*, Desi Arnaz, Irving Berlin, Leighton Brill*, Sammy Cahn, Irving Caesar, Jean Casto, Jan Clayton, Marc Connelly, Howard and Lucinda Ballard Dietz, Irving Eisman*, David Ewen, Dorothy Fields, George Ford, Helen Ford, Bennett Green*, John Green, Nanette Guilford, Hildegarde Halliday, Dorothy Hammerstein, James Hammerstein, William Hammerstein, Dorothy Hart, June Havoc, Yvonne Kalmán, Gene Kelly, Philip Leavitt, Alan Jay Lerner, Joshua Logan, Rouben Mamoulian, Samuel Marx, Jessie Matthews, Edith Meiser, Mabel Mercer, Constance Moore, Henry Myers, Milton Pascal*, Irving Pincus*, Richard Rodgers, Dorothy Rodgers, Gene Rodman*, Morris Ryskind, Arthur Schwartz, Vivienne Segal, Leonard Spigelgass*, Benay Venuta, Harry Warren, Alec Wilder, and Max Wilk. To every one of them I extend my heartfelt gratitude for so unstintingly sharing their personal remembrances of Larry Hart.

The second major source is the Richard Rodgers scrapbooks in the Music Division of the Performing Arts Research Center, New York Public Library. And the third, of course, is the books listed, by way of acknowledgment, in the bibliography.

I am especially indebted to two good friends: the late Samuel Marx, who gave me (as I had earlier given him to mine) unrestricted access to his research materials on Rodgers and Hart; and Dennis Moore, who equally open-handedly shared his considerable knowledge of their lives and music. I owe a further debt of gratitude for advice and sound judgment to my editor and publisher, Sheldon Meyer, and I also wish to thank Joellyn Ausanka.

Many others have significantly helped bring this work into being, and I gratefully acknowledge the contributions of Ben Bagley, Painted Smiles Records, New York; Bill Barrow; George E. Boziwick, Music Division, Performing Arts Research Center, New York Public Library at Lincoln Center; Gerald R. Bordman; Seymour Britchky; Rita M. Chambers; Theodore S. Chapin, Rodgers & Hammerstein, Inc., New York; Michael Colby; Desmond Elliott; Theodore C. Fetter and William Richards, Theatre and Music Collection, Library of the City of New York; Ian Marshall Fisher; James Fisher, American Society of Composers, Authors and Publishers,

London; Gary Ford, American Society of Composers, Authors and Publishers, New York; Dr. Philip Furia, University of Minnesota, Minneapolis; Robert Gersten, Brant Lake Camp; Conrad Goulden; Robert Grafton; Benny Green; Stanley Green; Hollee Haswell, Curator of Columbiana, Columbia University, New York; Dr. Hart Isaacs Jr. and his wife Patricia; Richard C. Kestler; Prof. Joseph Kissane; Aletha Kowitz, American Dental Association, New York; Kristine Krueger, National Film Information Service, Academy of Motion Picture Arts and Sciences, Los Angeles; Phyllis Lerner; June Levant; James T. Maher; Billy Matthews; Bob Medina, the *New York Times*; Geoffrey Millais; Stanley Musgrove; Arthur and Harriette Pine; Leslie Pound, Director of Publicity, Paramount Pictures, Los Angeles and London; Mrs. Alice Regensburg; Paul R. Rogers, Low Memorial Library, Columbia University, New York; Ned Sherrin; Mrs. Paulette Silverberg, Mount Zion Cemetery, New York; Stephen Sondheim; Cathy Surowiøc, British Film Institute, London; Marc Wanamaker.

F.N.

CONTENTS

1. *Max and Frieda* 3
2. *Love at First Sight* 12
3. *A Lonely Romeo* 20
4. *A Couple of College Kids* 28
5. *Doldrums* 35
6. *The Blond Beast* 43
7. *Campfire Days* 48
8. *The Brief Career of Herbert Richard Lorenz* 54
9. *Gilding the Guild* 58
10. *Crest of a Wave* 67
11. *More Gaieties* 77
12. *The Great Ziegfeld* 85
13. *"One Dam Thing After Another"* 95
14. *A Great Big Beautiful Hit* 104
15. *A Willing Ham for Dillingham* 110
16. *Oscar Hammerstein Was Right* 115
17. *Makers of Melody* 122
18. *Wall Street Lays an Egg* 129
19. *Ten Cents a Dance* 136
20. *Hard Times on Broadway* 147
21. *Hollywood Bound* 154
22. *A Jolson Story* 167
23. *Goldwyn's Folly* 178
24. *Night Madness* 186
25. *Yesterday's Men* 195

26. *Billy Rose's* Jumbo   202

27. *"The Saddest Man I Ever Knew"*   210

28. *Twice in a Lifetime*   223

29. *"The Biggest Opening Since the Grand Canyon"*   229

30. *A Special After-dark Existence*   236

31. *Musical Comedy Meets Its Masters*   243

32. *"If It's Good Enough for Shakespeare . . ."*   251

33. *The French Have a Word for It*   264

34. *Bothered and Bewildered*   271

35. *I Could Have Been a Genius*   284

36. *Nobody's Hart*   294

37. *To Keep My Love Alive*   300

38. *What Have I Lived For?*   308

    Coda   313

    APPENDIX I: *Lyrics by Hart: A Show-by-show Listing*   319

    APPENDIX II: *Lyrics by Hart: An Alphabetical Listing*   347

    BIBLIOGRAPHY   359

    INDEX I: *Rodgers and Hart Shows and Movies*   363

    INDEX II: *Rodgers and Hart Songs*   365

    INDEX III: *General*   370

Lorenz Hart

LARRY HART, LED AS A SONG WRITER

Lyricist Who Had Combined His Talents for 20 Years With Richard Rodgers Dies

Lorenz (Larry) Hart, one of America's foremost song writers died last night in Doctors' Hospital after a short illness of pneumonia. He was 47 years old.

Lorenz Hart was Broadway's "Laureate of Lyrics." For more than 20 years he and Richard Rodgers combined their talents to write distinguished musical comedies. Mr. Hart's last contribution to the Broadway musical-comedy scene was the elaborate revival of the Fields-Hart-Rodgers hit of 1927, "A Connecticut Yankee," which included not only five of the popular numbers from the original version, such as "My Heart Stood Still," but several new numbers.

Born in this city, the son of Max M. and Frieda Hart, he prepared for college at the Columbia Grammar School and was graduated from Columbia University.

Messrs. Hart and Rodgers met at the University on Morningside. Rodgers, 16 years of age at the time, had just entered as a freshman, and Hart was doing post-graduate work. The student body requested that they join their talents in producing the Columbia Varsity Show of 1920. For Hart, the opportunity was golden. It was a decided change from translating German plays for the Shuberts. The first effort of the two had the audience cheering. So came into being the team of Rodgers and Hart.

From 1920 to 1925 Hart and Rodgers provided such hits as "The Poor Little Ritz Girl," "Garrick Gaieties," "Dearest Enemy," "Peggy-Ann," "Betsy," "The Girl Friend." They wrote "The Greenwich Village Follies," which ran for 211 performances. Then followed such shows as "Chee Chee," "Spring Is Here," "Evergreen," "America's Sweetheart," "Jumbo," "On Your Toes," "Babes in Arms," "I'd Rather Be Right," "I Married an Angel," "The Boys from Syracuse," "Pal Joey."

Mr. Hart wrote the lyrics for such numbers as "With a Song in My Heart," "Isn't It Romantic?," "Soon," "It's Easy to Remember but So Hard to Forget."

Rodgers and Hart never waited for the muse to strike. When a project was given them, they began work at 11 A.M. and continued with amazing dispatch. For instance, they wrote the entire score of "I'd Rather Be Right" in three weeks.

Both wrote lyrics and music for more than a half-dozen motion pictures.

Mr. Hart was not married.

1

Max and Frieda

*W*ith that final, familiar journalistic euphemism, the *New York Times* correctly identified Larry Hart's personal tragedy; but practically all of the other "facts" which preceded it were wrong. He never graduated from Columbia, nor did he and Rodgers meet there: Rodgers was still in high school, Hart a long way past postgraduate work. The student body did not invite them to write the 1920 Varsity Show: they wrote one of five that were submitted that year. Hart was not yet translating shows, and when he did, it was not for the Shuberts, but United Plays. Rodgers and Hart did not write *The Greenwich Village Follies*, although a couple of their songs were used in the touring version of one edition . . . and so on.

It is time to set the record straight. The place to begin is at the beginning: with Max and Frieda.

Max Meyer Hertz and Frieda Eisenberg were married on November 6, 1886, just nine days after city-wide celebrations surrounding the dedication of the Statue of Liberty by President Grover Cleveland. Max was twenty, she eighteen, only a year over from Hamburg, Germany, and with, by her own account, less than twenty cents in her purse that blustery Saturday. Frieda was a tiny woman, well under five feet in height; handsome, portly Max was himself only a few inches taller.

Both of them were from large immigrant families. Danish-born Meyer Hertz and his German wife Taubchen (Heyne) Hertz (her first name translates as "Little Pigeon," but the family called her Tessié) had arrived from Hamburg in 1878, when Max was twelve. There were nine children in all. The Eisenbergs numbered ten. They all lived on Allen Street, in the heart of the Fourth Ward, the noisy, teeming, dirty Lower East Side. In those days, half of New York's population of a million lived below 14th Street. There were only sixty or seventy thousand houses for all of them, if you didn't count the tenements. There were a lot of those, especially in Allen Street.

The whole area hadn't changed much since 1872, when an observer noted that the Fourth Ward contained five hundred tenement houses, three hundred of which were in "bad sanitary condition"—a euphemism for disgusting. The other two hundred, we may assume, were those occupied by what the same writer called "well-to-do working people" and described as "immense, but spruce looking structures" kept cleaner than the other kind, but in all cases "suffering from the evils incident to and inseparable from such close packing."

Max Hertz was an ambitious man, a *Macher*, a pusher. He wasn't the type to sit in somebody else's office and wait for seniority to get him to the top. He never had that kind of job. He was always on the street, on the move, putting together deals, fixing, arranging, promoting, buying, selling, bringing together people who ought to know each other. Max got to know the ropes fast and bought drinks in the right saloons for the right people. He worked his way up into Tammany circles, among the other fixers and gofers and ward heelers running errands for the heirs of "Honest John" Kelly, who had died just a few months before Max and Frieda married, and who had been succeeded by another Tammany "boss" as venal as any who had gone before: Richard Croker.

Max and Frieda's firstborn, James Hart—Max had wasted no time Americanizing his name—was born in 1892. It was a difficult birth, and baby James died in infancy. Accurate genealogy is somewhat difficult to come by. The "authorized" version by Teddy Hart's widow, Dorothy Hart, names only three children: James, Lorenz Milton, and Theodore van Wyck, born in that order. Lorenz Hart's birth certificate indicates two children had been born prior to his birth, one of whom was still alive, although this may have been no more than misunderstanding of the registrar's questions by Frieda, whose command of English was not good, or more likely a simple clerical error.

By the time his second son was born—another difficult birth for the diminutive Frieda—Max Hart had moved uptown in both senses of the word, to an apartment at 173 East 111th Street, between Lexington and Third. He gave his occupation at this time as "Commission Merchant," which suggests he was making a dollar any way—literally *any* way—he could. It is no exaggeration to say Max Hart made a lifetime career out of making a dollar any way he could. If God in His wisdom sent Max an opportunity, Max didn't just take it, he grabbed it with both hands and ran. Exactly how is illustrated in a story that Max, with others present—including his son Larry—once told Philip Leavitt:

I had the job of Assistant Coroner. The Coroner was a man named Messmer; he was always drunk, so I did the work. Well, one Sunday the Coroner's office got a call to come up to the Vanderbilt mansion at 59th Street, so up there we went. When we walked into the big front parlor, there was the whole family, headed by old man [Cornelius] Vanderbilt; and lying dead on the floor with a bandage around his head was his son, heir to the family fortune.

Vanderbilt said he was sorry to disturb us on Sunday, but his son had unexpectedly had a stroke, and as no doctor had been present, and as the law stated that in that event the Coroner had to certify as to the cause of death, would we please sign this—and he held out a release, all written out properly.

I said I would have to examine the body, and he asked: Was it necessary to go through all that formality? But I insisted, and whenever Messmer asked what was going on, I said, "Beer, beer," and he quieted down. I undid the bandage that was around the deceased's head, and as I expected, there was a bullet hole right in the middle of the forehead.

"There must be some way we can keep this quiet," Vanderbilt said, and I answered: "Ten thousand dollars." He sat down to write a check, but I said, "Cash." He objected that it was Sunday, and no banks were open on Sunday. And—a Vanderbilt check! I just kept repeating "Cash," so he sent out and got it and gave it to me, and I wrote out "Death from natural causes" and had Messmer sign it.

It's possible, of course, that this story is gospel truth, but it's equally possible that Max—whose persona was a cross somewhere between P. T. Barnum's and Baron Münchhausen's—told it for effect, to amaze, to shock, and because it was a hell of a good story. Larry Hart was not unlike his father. Indeed, he loved to tell people, not without a certain defiant pride, that his Old Man—"O.M.," as he called him—was a crook.

Max became many things in life: now an investor, again a banker, yet again a promoter. One year he would be in real estate, in railroads, stocks, bonds; later this, still later that. Leavitt recalled that at one time O.M. was also a banker.

The bank's address, printed on the check, was somewhere over on Second Avenue, about midtown, and I went there. Sure enough, there it was, in the ground floor of a building, in a space that seemed to have been intended for a shop. The entire personnel consisted of one man, who sat within a cashier's cage; otherwise, there was no one—no executives, managers, clerks, guards, and at the moment at least, no customers except myself. I endorsed my check and presented it; [the cashier] immediately telephoned M. M. Hart, asked him if it was all right, then paid me the cash.

Max and Frieda decided on classical names for their new baby, born in the 111th Street apartment on Thursday, May 2, 1895: Lorenz Milton. He was cute as a button, a tiny little fellow with dark hair and deep brown eyes. The family language at home was German; Frieda never lost her accent, which is probably why Mr. Savage, the Recorder at 181 East 116th Street where Frieda reported her son's birth on July 25, wrote the boy's name as "Laurence"—the way Frieda pronounced it. Later, when fame came, they would say it a variety of ways: LorENZZ, LorENTS, and correctly, as Frieda did. At home, he was always Lorry. Until the day she died, Frieda never called him anything else.

Max continued to prosper, although there were plenty of people who shook their heads at his methods. New York directories for 1896–97 list him as a "builder" living at 220 East 105th Street, with offices at 147 East 125th Street; among the businesses he would become involved with at one time or another were property development and rental in Harlem, the Pittsburgh and Allegheny Coal Company, and—it was rumored—an interest in the Manhattan brothel run by the notorious Polly Adler. He was crude, coarse, vulgar, often outrageous, apparently unaware or at any rate unashamed of his huge laugh, his heavy accent, his lisp, or his reputation. Although he stood only a roly-poly five feet four inches small, you knew when Max Hart was in the room. Hell, they said: you knew when Max was in the neighborhood.

By the time the next baby came along, on September 25, 1897, Max had become brash enough to name the boy after the Vice-President of the United States, and friendly enough with the first Mayor of Greater New York to give the child the middle name van Wyck in His Honor's honor—a dubious one, as it turned out: Robert C. Van Wyck was as venal as the politicians who had fixed his election, and lasted but one term.

Max, however, was still upwardly mobile. Soon after Theodore van Wyck "Teddy" Hart's birth, the Harts moved even further uptown, as their affluent friends were doing, to welcome the new century in a new home just a couple of blocks south of the Harlem Courthouse: 59 West 119th Street.

Of course, Harlem was not then the ghetto through which New Yorkers drive now with all their car doors locked. At the turn of the century, it was a pleasant, spacious middle-class suburb where children played on streets rarely disturbed by traffic, and families sat out on their stoops on summer evenings. In this friendly neighborhood, nearly every parlor displayed the emblems of respectability: an aspidistra, a suite of mahogany-stained furniture upholstered in velveteen, an upright piano, and gilt-framed engravings or chromos on the wall. Like Brooklyn, Harlem rejoiced in its "small-town" ambiance.

West 119th was a street of solid brownstones with high stoops; the Hart house was on the north side of the street, just west of Third Avenue. A little further along was the Mount Zion reform synagogue. Across Mount Morris Park, Seventh (Lenox) Avenue, with its theatres and showplaces, was a boulevard of high style and fashion where rich "uptown Jews" promenaded, dressed in top hats and black coats, swinging silver-topped canes.

On West 125th Street, the main business and entertainment thoroughfare, were the Harlem Opera House, where sultry actresses like Fanny Davenport and Olga Nethersole appeared in "shocking" French plays; Hurtig & Seamon's Music Hall, where Sophie Tucker and Fanny Brice sometimes headed the bill; the New Orpheum Theatre; and cinema houses like Proctor's, the Orient, and Loew's Victoria. Here too, housed in solid, respectable, European-style sandstone buildings, were the banks, insurance companies, and department stores.

The wealthier families dined at Pabst's restaurant; poorer ones at any of the many bakeries, hash houses, and beaneries. At the intersection of Seventh Avenue and 125th Street (now Martin Luther King Jr. Boulevard) was Harlem's equivalent of Union Square, where suffragettes, stepladder Socialists, and Henry George single-taxers shouted themselves hoarse.

Money burned a hole in Max Hart's pocket: he made it to spend, and spent it to make. He was a hustler, a *hondler*, a *schmeikler*. He employed two housemaids and, when the automobile came in, a chauffeur and even a footman. The Harts lived a noisy, crowded family life, with uncles and aunts visiting, with Max's business cronies stopping by for drinks, with lavish parties that sometimes went on into the early hours, as vulgar and manners-be-damned as the man who was throwing them. When Max's business ventures were going well, it was champagne and caviar all the way. So Teddy and Larry Hart grew up accustomed to being petted and spoiled by the New York beau monde: movers and shakers, politicians and tycoons, newspaper proprietors, actors, actresses, and theatrical producers. Tonight, John Morrisey or Gustave Amberg, Jesse Lewisohn or Lillian Russell; tomorrow, Tony Pastor, Willie Hammerstein, and Diamond Jim Brady.

Of course, when things got tight, and there were more than a few times when they did, the small fortune in jewels Max had bought for his wife found its way to the pawnbrokers "until things looked up." Shameless, but it was impossible to shame Max Hart, who never changed. He remained to the end what he had been from the beginning: a noisy, outrageous chiseler. In genteel Harlem, Max's antics were considered shocking: in those days, decent people distrusted flamboyance, loose morals, ostentation. Max, on the other hand, wallowed in them, and even had he cared a fig for his neighbors' opinions, the last thing he would have done was show it. The opposite, in fact: Milton Pascal recalled an evening when Max came into Larry's bedroom on the third floor and cavalierly peed out of the window into the street.

As for the boys, they wanted for nothing. There was always an aunt or an uncle around to bring them presents, to take them for an ice cream, to go on

an outing. From the time he was six, Lorry was being taken to the German-
language Irving Place Theatre on 14th Street. Its stars came frequently to
the Hart house, where German was as fluently spoken as English. For a
while Willie Hammerstein—whose son Oscar, born two months after little
Lorry, was named after Willie's august brother, the famous impresario—
became Max's partner. Such partnerships were frequent and usually lasted
only as long as it took the partner to discover how badly he was being
cheated, but, for a while at least, the association meant a supply of "Annie
Oakleys"—complimentary tickets, so called because of the perforation that
looked like a bullet hole—for Hammerstein's Victoria. Precocious little
Lorry soaked it all up: by the time he was six or seven, he was writing poems
and verses, reciting them with the orotund, declamatory style of the born
ham.

The family tradition is that the boys were also taken to Hurtig & Seamon's Music Hall (later to become the renowned Apollo Theatre). Some
might have thought a burlesque house raw meat for such young teeth, but
both Lorry and Teddy reportedly loved it. In their early teens, they formed a
double act called the Hart Brothers, entertaining the family with skits based
on the sketches they had seen. Even then, Lorry loved to make people laugh.

When the time came for him to go to school, Lorry was sent to the best: the
De Witt Clinton High School at 59th Street and Tenth Avenue. His acceptance may be taken as further evidence of his precocity; the school's aca-
demic standards were high, and money alone was not enough to ensure
entrance. In 1908 the boys went to their first summer camp—the Weingart
Institute at Highmount in the Catskills, in upstate New York.

It was run more like a military institute than a camp; its owner, Sam
Weingart, believed bugle-call reveilles, room and bed inspections, cold
showers, and plenty of calisthenics kept young boys out of trouble. Of
course, so did a lot of parents in those days: the Bonwits of Bonwit Teller
fame, two of the Selznick brothers, Oscar Hammerstein (who hadn't yet
tagged "the Second" onto his name), and Herbert Sondheim (who hadn't yet
become the father of Stephen) all went to Weingart's at one time or another.

As far as can be ascertained, Lorry Hart's theatrical debut took place at
Weingart's on Sunday, July 19, 1908, when he appeared in a farce called *New
Brooms*, playing the third broom. A month later, he appeared in the big event
of the Institute's season, the minstrel show. As end man, Lorry got to sing,
which by all accounts was a mixed blessing (in later years, Richard Rodgers
said his partner sang like a raven; Hart himself described his voice as "low,

but disagreeable"). The song was a now-forgotten gem, probably composed by Larry, called "Pass It Along To Father."

In his second year at Weingart's, in 1909, Lorry became editor of the camp magazine, "The Weingart Review." Some fourteen-year-olds might have thought this quite an achievement, but Lorry appears to have been less than awed: he was wont to refer to the journal as the "Highmount Daily Dope Sheet," which hints at some of his other hobbies. He claimed to be writing a play, *Inky Ike, the Ashbarrel Detective*, and appeared in another called *A Warm Reception*, in which a reviewer gravely noted the performance of "our famous tragedian." A few of the things he wrote at Weingart's have survived to provide still further evidence of his precocity and of the facts that he talked non-stop ("five paragraphs a second") and, even then, was making defensive jokes about his height ("my heavy brain does not permit me to grow in a vertical direction").

It is easy to see that writing, to him, was a breeze. His casual brilliance is already apparent in the dual-language parody of Irving Berlin's "Alexander's Ragtime Band" he produced in 1911 for Max and Frieda's silver wedding anniversary party at Harasted Hall in the Bronx.

> So clink your glass, each lad and lass
> For Max and Frieda's wedding day.
> Put on a smile, make life worthwhile
> Let each wrinkle shout hooray!
> *Nur einmal ist es erlaubt*
> *Dass der Mensch gut leben kann.*
> *Die Zeit fliegt ja und sorg beraubt*
> *Alle Jugend von dem Mann!*

As Arthur Schwartz percipiently observed, Larry Hart never "grew." He had that effortless facility with words and rhyme right from the start—and to the very end.

In 1910, when he was twelve, Lorry's parents had sent him to a new and larger summer camp, at Lake Paradox, near Highmount in the Adirondacks. Getting to Camp Paradox was an initiative test in itself; it involved taking a boat from the 125th Street dock of the Hudson River Line and steaming up the river to Ticonderoga, where buckboard wagons met the boys and ferried them the twelve miles to the camp.

Every boy had a trunk containing everything he would need for his stay in the mountains. Well, not every boy. Teddy Hart's wife Dorothy recalled that when the coach driver who brought them from the boat tried to lift down Larry's trunk, he couldn't budge it. His tentmates—Eugene Zukor, the son of Adolph Zukor; Mel Shauer—were not unnaturally curious about what it contained. When he opened it, it turned out Larry had packed nothing but books: dictionaries, a thesaurus, and the complete works of Shakespeare. This, inevitably, landed him with the nickname "Shakespeare" Hart.

Later on they would dub him "Dirtyneck" Hart, too. Larry had no sense of possession, especially clothing. He was frequently "porched"—that is, confined to the porch of the main building—for leaving clothing or a baseball glove or shoes lying wherever he had dropped them. Whenever he needed something—a shirt, sweater, socks—Larry would simply hit up one of his tentmates. "You've got two sweaters," he'd say to Gene Zukor, "let me have one." He'd grab socks from Mel, something else from Teddy; which is why in pictures taken at camp, he always looks like he is wearing the first thing he laid hands on. Apart from the—compulsory—two- or three-day hikes, he was really only interested in one thing: the camp shows.

Shauer, later to play an important part in the founding of Brant Lake Camp, was a tall, skinny, good-looking youngster who wanted to be a songwriter and later became a producer at Paramount. He recalled that one time Max Hart came up to visit his sons. He arrived bright and early, at five a.m. He had to climb a steep hill to get to the camp, and as he approached, the hills were alive with the sound of Max yelling, "Where are my boys? Where's Teddy? Where's Lorry?" The first thing he did on encountering them was bawl them out for not meeting him.

"Wherever Larry was there was excitement," Shauer recalled. "He was a laughmaker, a nonconformist." The Hart household was the same, bohemian and kind of shocking. Kids like Mel who came from conventional middle-class homes were simultaneously attracted and shocked by the Harts' "open house" policy. The door was always open, and even if Larry and Teddy were upstairs asleep, the kids were still welcome. They could do whatever they wanted.

"Larry loved it all," Shauer said. "He loved people around him. Loved to see them eat and drink. Everybody was welcome at all times." Sometimes Max would give Larry a hundred-dollar bill. Larry would gather a group around him and take them out on the town, paying for everything with crumpled bills from his back pocket—never a wallet—saying, "Here, I've got it! I've got it!"

From time to time, one or the other of the "gang" would protest feebly, but Larry would wave such protests aside with a lordly gesture. "Friends meant everything to him," Shauer said. "His generosity was unbelievable."

The "gang" loved to gather at Lorry's, because Frieda—Mel Shauer called her "the Cameo," and Larry referred to her as "the Cop"—didn't seem to mind them stripping her front parlor of furniture and turning the room into a sort of debating hall where politics, literature, poetry, and girls were hotly discussed until dawn. While the youngsters reveled in such freedom, most of their parents disapproved of the Hart household, Shauer recalled. They thought it "too loose"—and with reason. Sometimes when the discussions got heated—and noisy—Max "O.M." Hart would thunder down the stairs, the long flannel nightgown he always wore emphasizing his huge belly.

"Ain't you goddamn kidth got anyplathe elthe to go?" he would scream. "How the Chritht am I thuppoosed to thleep with all thith fucking racket going on?"

Lorry knew exactly how to handle the O.M. He'd solicit his opinion, invite Max to join the discussion. If there was a pretty girl or two in the gang, Max would get himself a drink and take over the meeting. If they didn't watch him closely, he'd pinch a few fannies while the going was good, too, laughing uproariously when the girls squealed. As playwright Morrie Ryskind once tartly remarked, if Larry had written dirty poems, Max would probably have sold them.

It was all good fun. Not good clean fun, of course, but that was the way young Lorry liked it. He was convinced—as he wrote in a 1913 school essay about his favorite fictional character, Falstaff—that he had "inherited some inborn rascality from a bacchanalian progenitor" which justified and excused his precocious predilection for hard liquor, spendthrift living, and "shocking" conduct.

Although, unlike Max, he was far too intelligent to have ever really believed it, Larry Hart managed to remain firmly attached to that proposition for as long as he lived.

2

Love at First Sight

*A*fter De Witt Clinton High School, Larry attended Columbia Grammar School; there he dashed off all sorts of pieces for the school newspaper, showing off that extraordinary ability to transmute what was going on around him into witty, even erudite concoctions like "The Rock of Refuge," which extolled the pleasures of the "ten-twenty show" in the "temple of Loew" (Loew's Victoria cinema on West 125th Street). And if he couldn't come up with an appropriate rhyme, he simply manufactured one.

In the fall of 1916, having heard the boy recite some of his poems at the Hart house, Gustave Amberg, an actor-turned-producer of the German-language shows at the Irving Place Theatre—begged by Larry, browbeaten by Max?—gave the youngster a break and introduced him to a producer named Rachmann who invited him to provide lyrics for a show called *Die Tolle Dolly*.

It was presented at the Deutsches Theatre in Yorkville with a cast that included Mizi Gizi, Rudi Rahe, Lieschen Schumann, and Ernst Naumann, and consisted of the usual knockabout German-language nonsense. Some of the songs were by Hans Kronert, others by Walter Kollo and one by Arthur Steinke. Whether Larry got paid for his adaptations—which were English translations for the published versions of the songs—is not recorded; probably not. They included such gems as "Ticky, Ticky, Tack" (a train song in which the wheels go "ticky ticky tack" on the track, or *Zicke Zucke Zack* in German), "Hubby Dances On A String," "Every Man Needs A Wife," "The Kiss Lesson," "Bummel, Bummel, Bummel" (a march, not an insult song), and the glorious—don't forget Max's middle name—"Meyer, Your Tights Are Tight."

> Meyer, your tights are tight,
> As tight as tights can be,
> It's very plain to see,
> They don't fit properly.
> Meyer, those tights were never
> Modelled for the sea,
> If you but try
> To wink an eye
> You'll lose your dignity.

Alas, *Crazy Dolly* wasn't even crazy enough to get itself a mention in the record books, much less launch Larry Hart into the world of musical theatre.

In 1915 Larry transferred from Columbia Grammar to the Columbia University School of Journalism. There he joined an extension class in dramatic technique conducted by Professor Hatcher Hughes, himself an aspiring playwright who would collaborate with Elmer Rice on the 1921 production *Wake Up, Jonathan* and win a Pulitzer Prize in 1924 with his play *Hell-Bent fer Heaven*. In the same class was a young fellow named Henry Myers who recalled that Larry was in love.

> He never spoke to her about it. He never spoke to her at all; she was just a fellow-student. He didn't dare—I'll tell you why in a moment; first I want to corroborate his taste: she was gorgeous. She was Miss Langevin (a French name, so pronounced).
>
> "Oh, that Miss Langevin!" Larry would sigh when we got talking about girls, and give a deep sigh and roll his eyes. "If he felt that deeply," the skeptic may say, "he'd have told her so. The worst he could get was a smack in the face."
>
> That's just the point. It's not the worst he could get, and it's not what he was afraid of.
>
> He was afraid she would laugh at him.

Larry apparently never had any serious intention of becoming a journalist, but his interest in literature, poetry, and the theatre soon led him, along with many of his classmates, to submit smart little verses and prose items to the famous "Conning Tower" column written by Franklin P. Adams.

Born in Chicago in 1881, Adams—who was Jewish, in spite of the WASP surname—had started out in life as an insurance salesman. He became a journalist on the *Chicago Journal* before moving to New York in 1903, where he started a column in the *Mail* called "Always in Good Humor." In 1914 he moved to the *New York World*, changing the name of his column to the better-remembered one, and giving himself the acronym F.P.A. He had a formidably sharp tongue; his witticisms were quoted in smart circles everywhere. One of his more celebrated inventions was a character named Dulcinea whose malapropisms, known as "Dulcyisms," received wide circulation throughout the twenties; she even became the central character of a hit play.

The "Conning Tower"—subtitled "The Diary of Our Own Samuel Pepys"—appeared Saturday mornings. It was considered to be just about the best written column in America, perhaps even the world, and contributors—who were not paid for their submissions—vied furiously to get something of their "wit and wisdom in prose and verse" published. Adams was a tough but unerring judge of talent. "Famous initials and names spattered the diary like a translucent Milky Way," wrote Moss Hart. "G.S.K. and Beatrice—A.W. and Harpo—Alice Duer Miller and Smeed—Benchley and Dottie—Bob Sherwood and Marc—I. Berlin and J. Kern—H. Ross and Sullivan—H.B.S. and Maggie. The initiate knew that G.S.K. was George S. Kaufman, Dottie was Dorothy Parker, H.B.S. Herbert Bayard Swope, and so on ad infinitum." Among the writers who made early appearances in print in F.P.A.'s column were Sinclair Lewis, Edna St. Vincent Millay, Edna Ferber, and Ring Lardner.

How much Adams appreciated Larry Hart's submissions is not known, but we do know that his teachers, notably Professor Walter B. Pitkin, thought highly of his work. There were exceptions, however. In one instance, Hart had turned in a profile of the Lew Fields family; by all accounts it was some distance away from journalism.

"Mr. Hart, this isn't good newspaper prose," Prof. Pitkin remonstrated. "In fact, I don't know what it is." From across the room came Hart's high-pitched query: "Poetry, maybe?"

In fact, as he himself said many times, what Lorenz Hart majored in at Columbia was neither literature nor journalism but Varsity Shows. Conceived as fund-raisers for the varsity athletic teams, these end-of-year musical extravaganzas had been a tradition at Columbia since the turn of the century. While any connoisseur of the genre would have insisted that Columbia's efforts didn't hold a candle to such paragons of production as Princeton's Triangle Show or Harvard's Hasty Pudding, the college was nonetheless able to command some notable talent: contributors over the years included Howard Dietz, Herman Mankiewicz, I.A.L. Diamond, Richard Rodgers, Herman Wouk, William deMille, Terrence McNally, Oscar Hammerstein II, Ed Kleban, Corey Ford, and Morrie Ryskind.

It was a necessary tradition in these shows that all the female parts were played by men. Diminutive Larry Hart's first association with the Columbia

Varsity Show was in just such a role. For the 1915 show, *On Your Way,* he wrote a skit satirizing Mary Pickford, in which he also played the part of the movie queen. Recalling the performance, Oscar Hammerstein II wrote:

> Imitating the way movie ingenues were chased around trees by playful but pure-hearted heroes, Larry skipped and bounced around the stage like an electrified gnome. I think of him all the time as skipping and bouncing. In all the time I knew him, I never saw him walk slowly. I never saw his face in repose. I never heard him chuckle quietly. He laughed loudly and easily at other people's jokes and at his own, too. His large eyes danced, and his head would wag. He was alert and dynamic and fun to be with.

The following year, the "electrified gnome" was seen as Mrs. Rockyford, a maid in *The Peace Pirates.* For the first time in four years, undergraduates were the authors; book and lyrics were written by Herman J. Mankiewicz. Oscar Hammerstein II won plaudits for his contributions, which included a blackface routine, an impersonation of Nijinsky, and a sketch involving a long-haired poet à la Bunthorne. Hammerstein stayed on to write and direct two more Varsity Shows, but Hart quit Columbia without a degree, impatient to get started in the theatre. His immediate ambition was to become a producer-director; the profession of lyricist was by no means his first choice.

That summer of 1917, he worked as counselor in charge of weekend entertainment, mounting summer shows at Brant Lake Camp, a new resort in the Adirondacks established the preceding year by Bob Gerstenzang, previously at Camp Paradox. At summer's end, needing a job, he turned for help to Gustave Amberg, who was associated through his company, United Plays, with the Shuberts, the largest producing organization in New York.

The Shubert Organization was run by two brothers, Lee and J. J. (Jake) Shubert, who cordially hated each other and had only one thing in common—the desire to make as much money as they could out of each other, or failing that, out of their artistes and the public who paid to watch them. Amberg's deal with "Mr. Lee" was simple: whenever he hit upon a German play or operetta that might lend itself to an English adaptation, he would offer it first to the Shuberts. If they passed on it, Amberg was then free to show it to other producers. The deal was also typical of the Shuberts: they got first crack at anything Amberg found, but had to give him no commitment unless he brought them something they could use.

To do this effectively, Amberg needed someone to translate the plays from German to English. He gave Larry Hart the job. In the course of his work, Hart often took the liberty of inserting dialogue to suit American audiences. Amberg minded this not at all: it made the plays that much easier to sell. For his efforts, Hart got a flat salary, fifty dollars a week: no royalties, no percentage. He didn't give a damn. Money meant nothing to him; he probably would have done the job for nothing.

World events shaped what happened next. A month after America entered the Great War, Lorenz Hart turned twenty-two, and in due course he received greetings from the draft board. Intensely patriotic, he was desperately keen to serve and was humiliated when he was rejected because of his height: how did you tell your pals you'd been turned down for the army because you were *too small?*

His physical appearance was a cross Hart would find increasingly difficult to bear as the years went by. As a full-grown adult, he was barely five feet tall. Although his head was of normal proportions, his body, hands, and feet were not much larger than those of a child. "What's the shortest distance from A to B?" they wisecracked. "Larry Hart." Or, "What's Larry Hart up to these days?" "Oh, about my belt buckle."

However, Larry hid his true feelings then, as he would hide them always, behind a shrug, a grin, a joke, a cloud of cigar smoke. He was full of ideas, full of life, full of hope. Okay, if he couldn't go over to Europe and win the war, he'd make a big name for himself on Broad Way, say it in two distinct syllables like a New Yorker. All he needed was a chance, one break that would enable him to show the world what he could do.

Its pointed references to the bandit Pancho Villa—("He is a dog!")—suggest it was around this time that Larry, as "L.M. Hart," collaborated with a J. N. Robbins on a one-act play called *Mexico*. Its four characters were James Clemens, his wife Anne, their friend Tom Sterrett, and a dark and passionate Mexican named Fernando. Whoever the authors had in mind for the part of Fernando, they wrote it with all the melodramatic stops pulled out.

Tom and Anne are lovers. Fernando overhears them planning to go away together. Intensely loyal to her husband, who once saved his life, he has been watching her on James's orders. But he, too, loves Anne and begs her to forget Sterrett and come with him to his hacienda in Mexico. She spurns him. Enter her husband. He knows all. Confrontation. He produces a revolver. Before he can kill his wife, Fernando shoots her. "Señor, you can call the police," he says. "You saved my life once."

The partnership of Robbins and Hart sank without trace. What happened to J. N. Robbins is not known—could he have possibly been a young Jack Robbins, on his way to becoming a songplugger at Remicks and later a hugely successful music publisher? As for Larry, he went on slogging away at his translations for Amberg and waited for his big break to come along.

Enter Philip Leavitt.

The youngest student in the Columbia University class of '17, Phil Leavitt was newly returned from France. Having played the role of conquering hero as long as he could get away with it, Leavitt joined his father's paint business, meanwhile pursuing his dream of writing, acting, or directing in the theatre. He also joined a small group called the Akron Club, a bunch of gregarious kids who had banded together "to add up five for basketball in the winter, nine for baseball in the spring, and eleven for football in the fall."

The cost of the equipment and the all-important uniforms was raised by raffles, dances, and the like. "In order to bait the hook to get the tickets sold," Leavitt recalled, "a portion of the net profit was given to charity." Anxious also to do their patriotic bit for the American Expeditionary Force, the members of the Akron Club decided to put on a show whose proceeds would go to buy cigarettes for the troops via the *New York Sun* Tobacco Fund.

They called it *One Minute, Please* and presented it (during a howling blizzard) at the Plaza Hotel, New York, on the evening of December 29, 1917. The book and most of the lyrics were written by Ralph Englesman, a club member. The music was by the fifteen-year-old kid brother of another member, Mortimer Rodgers.

> Richard Rodgers, composer, was born right then and there. His music had a lilt. The real difficulty was to find someone who could match words to it. A few club members took a crack at that, even Dr. Bill Rodgers, his father. The members realized they had an unusual find in Dick, but in order to round out his original score, it was necessary to find a suitable lyricist.

They tried; but by the following spring, they were no nearer finding one. For their new show, *Up Stage and Down*, which opened on March 8, 1919, all but three of the lyrics had been written by composer Rodgers himself. True, there was a new contributor, recently graduated from Columbia University: Oscar Hammerstein—now with the II after his name—who was universally known as "Ockie." No special significance was attached to the collaboration, although for the history books it would mean that the partnership of Rodgers and Hammerstein was actually begun a year before that of Rodgers and Hart. And there was certainly a promise of things to come in the first three Rodgers and Hammerstein songs: "Weaknesses," "Can It," and notably, "There's Always Room For One More," a cute number between the girls and the hero, Jimmy, that bears more than a passing resemblance in construction to "Tell Me, Pretty Maiden" from *Florodora*.

With the success of this show, and another in the works for the following year, the Akron Club's need for a regular lyricist became acute. It was at this point that Leavitt remembered Larry Hart, whom he'd met at Columbia and heard at the West 119th Street house expounding on the musical comedy and occasionally reciting lyrics he'd written for German melodies.

> We realized that Dick needed someone who could write lyrics that approached his music, and I suggested to Dick that he do a tie-up with Lorry because Lorry had done some translating from German [Max] Reinhardt shows abroad, and I suggested he meet Lorry.
>
> Lorry had a habit of sitting in front of an old Victor phonograph and listening to the Kern, Wodehouse, and Bolton operettas, and he loved the lyricisms that were in there, and he developed a style all of his own. . . . He had a very quick and accurate mind, [and] he had a different objective in what he saw and heard both in music and lyrics.

A few days later, Leavitt took young Richard Rodgers around to the Hart house. Larry Hart met them at the door. They were a study in opposites. Dick was fresh, tanned, athletic, handsome, a high school champion swimmer and tennis player. Larry was unshaven—according to Rodgers, Hart invariably looked like he needed a shave five minutes after he'd had one—and wearing a bathrobe over an evening shirt and trousers, with carpet slippers on his feet. He was already talking—and doubtless puffing a cigar and rubbing his hands together as he invariably did—as his visitors climbed the steps, and kept going non-stop as he took them back to the overstuffed library, where there was a piano. Bridget, the cat, strolled in, and Hart introduced her as an "old fencewalker." That broke the ice.

> It was all so simple. Dick sat down at the piano, and Lorry said, "What have you written?" and Dick played some of his music, and it was really love at first sight. All that had to be done was [for me to] sit, listen, and let nature take its course.

Puffing on the ever-present cigar, hands flying, brown eyes flashing with enthusiasm and energy, Hart expounded upon the craft of lyric writing, excoriating the intellectual poverty of simpleton writers who rhymed "slush" with "mush" while neglecting the possibilities inherent in double and triple rhymes, slant rhymes, fragmented rhymes, false rhymes, interior rhymes, feminine rhymes—but most of all, witty rhymes. Rodgers was captivated.

"I listened in rapt astonishment as he launched into a diatribe against songwriters who had small intellectual equipment and less courage, and who failed to take every opportunity to inch a little further into territory hitherto unexplored in lyric writing. I was enchanted," Rodgers recalled in what has become perhaps his most-quoted comment on his mercurial little partner. "Neither of us mentioned it, but we evidently knew we would work together, and I left Hart's house having acquired in one afternoon a career, a best friend, and a source of permanent irritation."

Elsewhere Rodgers has been quoted—more accurately, one feels—as saying he left the Hart house bubbling over with excitement, repeating over and over to himself, "I have a lyricist, I have a lyricist!" Ever since he had first heard Jerome Kern's music, Dick had only wanted to do one thing: write songs. It takes no imagination to picture him walking down 119th Street that Sunday evening, sixteen years of age, elated, exhilarated, full of hope and anticipation.

Of Hart's feelings there can equally be no doubt. As Philip Leavitt put it with such unwitting percipience, it was love at first sight.

"Poor Larry," one of his close friends said. "What a shame he had to fall in love with Dick."

3

A Lonely Romeo

So Richard Rodgers, the would-be composer, had a lyricist. And Lorenz Hart, the as yet unpublished lyricist, had a composer. Not everyone thought this cause for celebration. Bennett "Beans" Cerf, a friend of Rodgers's brother Mortimer, and later a famous publisher and raconteur, warned Dick that Hart was shiftless, disorganized, and undependable, and that he would never amount to anything. As was often the case, Cerf was about half right. More important, Dr. William Rodgers and his wife did not share Cerf's doubts; according to Rodgers, they never expressed any reservations about Hart. If he was the partner their son wanted to work with, that was all right with them.

Although it is apparent that most of Larry's friends (apart from Dick, who throughout his life protested, perhaps far too much, that he was never aware of it) recognized and accepted his sexual ambivalence—probably born either at summer camp or, more likely, while he was a counselor at Brant Lake—no one seems to have thought the relationship between the teenaged Rodgers and the twenty-four-year-old Hart remarkable. Today, perhaps, it would be different; but this was a more innocent age, and they were perceived as nothing more than a couple of well-off college kids trying to get into show business. New York was full of those.

The war years had wrought enormous social changes in America. With their men away at war, women had shimmied off the prim morality of 1914. The skirts that had swept the pre-war floor had climbed six inches off the ground. The long black stockings that in olden days a glimpse of was looked on as something shocking had been replaced by sheer silk hose that exposed the leg. Why, some girls even went so far as to roll their stockings down below the knees when jazz bands started playing new dances like the fox trot! More and more young women were beginning to wear bobbed hair, lipstick and rouge, silk underwear, and pumps that showed off well-turned ankles.

Like the returning soldiers marching in the grand victory parades up Fifth Avenue, young Richard Rodgers also had some adjusting to do. Coming from the sober surroundings of the Rodgers household to the chaos of West 119th Street was like being thrown out of a church into Bedlam; Dick's people were as conventional as Larry's were not. His father, William Abraham Rodgers, was of Alsatian extraction, born William Abrams in Holden, Missouri. A graduate of the College of the City of New York, he was a struggling general practitioner when he met and married Mamie Levy, the second of the three children of Jacob and Rachel (Lewine) Levy, on November 24, 1896.

Will and Mamie went to live with her parents at their Lexington Avenue home. It was a fraught and stormy relationship. Mamie was dominated by her strong-willed parents. While he could get along with Jacob, Will disliked his mother in law intensely; there were periods of weeks when he would not even deign to speak to her. Their first son, Mortimer, was born January 13, 1898, at the family home; the second, Richard Charles, on June 28, 1902, at a summer house the Rodgerses had bought on Brandreth Avenue in Arverne, Long Island.

The following year the Levys sold their town house and moved to a five-story brownstone at 3 West 120th Street, near Mount Morris Park, where Dr. Rodgers conducted his practice from a ground-floor office. On the second floor were the dining quarters and a spacious living room with a Steinway piano. Red-haired and strikingly handsome, Dr. Rodgers loved music and was a devoted theatregoer; he and Mamie attended all the leading musical productions.

The first musical Dick saw was probably Victor Herbert's *Little Nemo*, a children's entertainment which opened at the New Amsterdam in October 1908. It is remembered mostly because it featured a mythical beast called a Whiffenpoof; this apparently captivated some Yale students who saw the show to such an extent they named their glee club after it. Next came *The Pied Piper*, starring De Wolf Hopper, which opened December 3 at the Majestic on 125th Street. But it was not until Dick was thirteen that a musical came along that fired him with the desire to write songs. The show was *Very Good Eddie*, and its composer was Jerome Kern. He became and remained always Rodgers's idol.

By this time the family was living in a large fifth-floor apartment at 161 West 86th Street, which would remain Dick's home for the next decade or so; Dr. Rodgers's office and waiting rooms were on the ground floor. He had a successful upper-class practice; among his patients was the movie star

Norma Talmadge. Mortimer, in temperament and personality like his father, aimed toward medicine via Columbia University. Dick, no scholar, could not cope with the intensive curriculum of Townsend Harris Hall, and transferred after a year to the less demanding but no less prestigious De Witt Clinton.

He was his mother's son: quiet, self-contained, and a complete conformist: punctual, organized, sober, precise, diligent and ambitious—just about everything Lorry Hart was not. Right from the start, Dick had to get used to the fact that his new partner didn't care to—wouldn't—work in the mornings because more often than not, he was getting over the night before. The way he adapted to Larry's undisciplined ways reveals an aspect of Rodgers's character that everyone who knew him remarked upon: he knew, always, what was right for him. And once he had identified what it was, he would never let go of it.

Dick Rodgers knew all about Larry Hart's bohemian habits. Perhaps, as the curate's son is fascinated by sin, he was even attracted by them. He knew Larry drank too much, smoked too much, forgot to eat, spent money he didn't have, stayed up till dawn, slept till noon—and he could live with all that. It wasn't important. What mattered was the man's talent. He was dazzled by it. Much later, he rationalized his feelings by saying he knew

> that his kind of lyric writing could work in the theatre. It was revolutionary. Nobody had ever written that way before . . . I think Larry and I were very sensitive to each other immediately, and yes, had a genuine liking for each other. He was about as different from me as you could possibly get. He was badly disorganized, and I was not. His whole family life was not only disorganized, it was *un*organized, and mine was quite the opposite. . . . I think the interesting thing from a technical point of view that obtained all through our years of working together [was] the fact that he would not work without me in the room with him. He rarely wrote anything without my being there. On the other hand I could write without him, and frequently did.

It is a truism that survivors write history. In a hundred similar interviews over the years following Hart's death, Rodgers fashioned the legend of his partner's unreliability, his reluctance to work, his drinking, his dislike of the business end of their business, his inability to write without Rodgers there to make him. In the end, he probably believed it himself. It needs only a moment's reflection to see that—Hart's precipitant decline during the last few years of their partnership apart—what Rodgers said was not altogether true. At the beginning of their partnership, and for a long time afterwards, Rodgers was the acolyte, not the other way around.

After all, Hart was twenty four, vastly better educated, a wealthy young man with a wide circle of friends and cronies. He had seen his first play when he was seven, entertained his friends with one-man routines in his living room, constructed scenarios on the toy theatre he was given for a childhood birthday present. He was translating European operettas for Gustave Amberg while Rodgers was still in high school. *He* was the one with the experience, the contacts, the know-how. It was from his fertile mind, not Rodgers's, that the titles for their songs, the sketches for their revues, the stories for their shows sprang. Rodgers was a talented melodist, and would become a supremely good one. Later, he would display intelligent daring in much that he did. But not when he was seventeen, nor even twenty-seven. At the beginning, Hart was the dominant partner.

Late that spring, before going up to Brant Lake Camp as a summer counselor, Larry Hart directed a "revival" of *Up Stage and Down* (now called *Twinkling Eyes*) at the 44th Street Theatre. Four of the original sixteen songs were dropped, and a new one, "I'm So Shy," was added. Because it was a benefit show—presented by the Brooklyn YMHA for the Soldiers and Sailors Welfare Fund—it ran only one performance, on May 18, so the score had little opportunity to make any great impression. The new team was still waiting for a break. It came sooner than anyone expected.

Phil Leavitt's parents had rented a house next door to the summer home of the producer Lew Fields. In short order, Leavitt got acquainted with Lew's dark-eyed daughter, Dorothy—he and Larry Hart had gone to camp with her brother Herbert—and prevailed upon her to take a part in the next year's Akron Club show. He always maintained that it was he who persuaded Dorothy Fields to get her father to listen to a song Dick and Larry had written called "Venus." To everyone's excited delight, she managed to set up a meeting at the Fields house, which Leavitt remembered as being on Franklin Avenue in Far Rockaway.

"In his nervous, jerky way," Leavitt said, "Larry drew up a strategy for the meeting. He was sure that 'Venus' would inspire Fields enough to invite Dick to play their whole repertoire. He would sing while Dick was at the piano, and throw in a bit of acting for good measure. But it didn't quite work out that way."

Perhaps that was because, according to Rodgers, Hart wasn't there: claiming a splitting headache, he had begged off. Whenever it came to selling their stuff or negotiating their contracts, Rodgers said, Hart would always go missing, and that sweltering summer Sunday afternoon was no exception. Dorothy and Herbert Fields concurred with his memory, putting the Leavitt version under some strain, quite apart from the fact that the Fields summer home seems to have been not at Far Rockaway but at Allenhurst, New Jersey. However, it should not be forgotten that Larry had written a profile of the Fields family while he was still at Columbia, and was co-producing camp shows with Herb Fields at Camp Paradox in 1910; or that later, Dick and Herb also later wrote camp show songs there, including "The Land Where The Camp Songs Go," a parody of a Kern song from *Miss 1917*. So perhaps Rodgers didn't go in quite as cold as his account indicates.

When he got there, he found the entire Fields clan assembled to hear him perform—paterfamilias Lew, his wife, his eldest son Joe, Joe's handsome, wavy-haired twenty-two-year-old brother Herbert, and the two girls, Dorothy and Frances. It was a big moment for the young composer: Fields was one of the most celebrated performers on the American stage, renowned as the comic half of the famous vaudeville comedy team of Weber and Fields.

Cultured and wealthy, an accomplished raconteur, always immaculately dressed, Fields in his middle forties was something of an eccentric. He was a fastidious, autocratic, charming man who scowled habitually, smiled infrequently, rarely slept longer than three hours, took long runs in all kinds of weather half a century before anyone had heard the word *jogging*, did handstands to clear his head, and seldom remembered anyone's name.

"Pop" Fields and his multi-talented family lived and breathed—breakfast to bedtime, seven days a week—in a non-stop showbiz atmosphere of skits, blackout lines, and funny routines. Everybody in the family was expected to collect jokes for Papa's act, entering them in an enormous ledger into which the old man would delve when he was writing a show. The children had regular boxes for Saturday matinees at all Fields shows, but were nonetheless regularly admonished not to consider the stage as a career. "You children must be extra polite to strangers," Mrs. Fields told them, with endearingly Dickensian logic, "because your father is an actor."

The career of Weber and Fields is better documented than that of Fields's talented family. Boyhood friends Joe Weber (his first name was actually Morris) and Lew Fields (Lewis Schoenfeld) were born seven months apart in 1867. Both were the children of Polish immigrants. They first met in the Allen Street public school—P.S. 42—when they were ten years old. According to Lew Fields, both had a fancy for clog dancing, and it was this that drew them together:

> During recess and after school hours we would practice dancing, and every time we could, we would sneak off to the London Theatre and gaze with awe upon the performers from our seats in the gallery. We resolved to be actors, too, and it was not long before we made our first appearance on the stage. . . .
>
> From the pennies we could scrape together we each bought a pair of green knickerbockers, a white waist, black stockings, dancing clogs, and a derby hat. Then we were ready to make our debut as dancers and singers. We had a song,

words of our own composition, music cribbed, called "The Land of the Shamrock Green."

Well, Joe and I made our first appearance at this benefit, and we received such praise from a very slim audience that we felt we were cut out for actors. At that benefit we decided on our vocation, and we vowed that neither one of us would do any other sort [of work] except dance and sing. The Lord knows we didn't look like the two Irish boys our song told about, but our first audience didn't care whether we came from Ireland or Jerusalem.

We played three or four benefits after this, and kept on going to school. Then we were bold and struck out for a job. Morris and Hickman's East Side Museum was then on Chatham Square. We demanded an engagement there, and we were hired for one week at $3 each. We went to school in the morning and played hookey in the afternoon while we were billed at the museum. . . . It wasn't long before we had another engagement. This time it was at the New York, another Bowery museum. . . . Well, we played at the New York nine weeks, and when we ended our stay there we were getting $12.50 each for a week's work. By that time nothing could drive us away from our set purpose of continuing on the stage. We left school and let our parents in on our game.

From the dime museums of the Bowery, where they played nine shows a day, six days a week for five years, Weber and Fields went on to play Harry Miner's Bowery Theatre at $30 a week and the Comique in Philadelphia at $80 a week, after which they joined Carcross's Minstrels as the first non-blackface act Carcross had ever featured. By 1895, they were getting $750 a week at Hammerstein's Olympia. It was there, Fields said, that they first got the idea of a burlesque company of their own. They got the money together and bought the Imperial Music Hall on Broadway at 29th Street, changing its name to theirs.

For nine years they produced hugely successful burlesques of stage hits (the names of their skits—"Cyrano de Bricabrac," "Quo Vass Iss?"—clearly indicate their content) with variety acts and their own now-distinctive ethnic comedy. The tall, skinny Fields played Meyer, the aggressive hustler who cons and bullies the short, bulging-bellied Mike, played by Weber. Both men used their own brand of fractured German-Yiddish-English dialect. In their famous pool hall sketch, Mike tells Meyer he doesn't know how to play, and Meyer reassures him by saying, "Don't vorry, votever I don't know, I teach you." Meyer's pool rules are simple: "De von dot gets de money vins de game." Their act always ended with an emotional farewell scene. "I hop you'll alvays look beck on me as de heppiest moment off your life," Fields would say. "End I vish to express you—charges collect—my uppermost abbreciation off de dishonor you hev informed on me," Weber would reply.

Weber and Fields were among the biggest names in vaudeville, but toward the end of their career dissent drove them apart. During the run of *Hoity-Toity*, their 1901 production, Weber's displeasure with his part in a sketch called "The Man from Mars" was so great that he stopped speaking to

Fields. While for over a year the team concealed the break, it was to prove irreparable, and they went their separate ways as producers.

Fields had many successful productions, but Weber's star as a producer shone ever less brightly; and in 1912, he agreed to appear in a new show with his old partner. *Roly Poly* was to be the last of their great burlesques. It opened at the new Weber and Fields Music Hall, which the Shuberts had built on 44th Street, and followed the duo's traditional two-parter format. While the critics and the public loved it, things were less than happy backstage. The old feud was still smoldering, and Fields, in particular, had found lucrative avenues for his talents elsewhere. The eight-week run was the only time they played at the theatre named for them; after a decent interval, the Shuberts renamed it the 44th Street Theatre (the site is now occupied by the *New York Times*). For generations who never saw Weber and Fields at work, the nearest approximation of it can be seen in a play perhaps suggested by their lives—Neil Simon's *The Sunshine Boys.*

Fields returned to producing (*All Aboard; A Glimpse of the Great White Way*) and starring in (*Step This Way; Miss 1917*) Broadway musicals. After four weeks at the Shubert Theatre, his newest show, *A Lonely Romeo*, had moved to the Casino shortly before Phil Leavitt paved the way for Dick Rodgers's audition. How much of all this Rodgers knew, he did not mention. What he said he remembered most about that afternoon were the dark eyes of fourteen-year-old Dorothy Fields: he found himself trying much harder to impress her than her father.

When Dick was through, Lew Fields astonished him by offering to buy one of the songs he had played, "Any Old Place With You," and interpolate it into *A Lonely Romeo*. Rodgers floated back to the city, unable to believe what had happened: sixteen years old, and he already had a song in a top Broadway show!

It seems a shame to explode such a romantic tale, but the truth—as is so often the case—appears to have been considerably more mundane. Asked to contribute some memories to its "Flashbacks" column on December 5, 1929, Herb Fields told the *Evening Graphic* that one of his favorite memories was of himself, Larry, and Dick "serenading Pop Fields with 'Any Old Place With You' nightly outside his dressing room until in disgust, to keep us quiet, he put it in a show."

No matter: regardless of how it got there, the fledgling team had their first song in a Broadway show. We may fairly assume that both composer and lyricist made a point of being in the audience at the Casino when, on August 26, 1919, at a Wednesday matinee, Eve Lynn and Alan Hale (a burly blond lad who would later be seen alongside Errol Flynn in many a Warner Bros. costume epic) launched "Any Old Place With You," and with it the professional Broadway partnership of Rodgers and Hart.

Well, almost.

A Lonely Romeo ran only seven more weeks before the actors' strike of 1919 closed the show. "Any Old Place With You" was cute enough—Larry had shamelessly rhymed "Syria" and "Siberia," "Virginia" and "Abyssinia," "court you, gal" and "Portugal," "corner ya" and "California," and finally "go to hell for ya" and "Philadelphia"—but the song attracted no more attention than the other one Fields had interpolated, "I Guess I'm More Like Mother Than Like Father," by Richard Egan and Richard Whiting. It certainly didn't start producers beating a path to Rodgers and Hart's door. Apart from that single, unremarkable Broadway credit, they were still, very definitely, amateurs.

4

A Couple of College Kids

*1*919. It was the kind of summer the newspapers had someone fry
eggs on the sidewalk as a stunt. Jack Dempsey beat Jess Willard by a
technical knockout after three rounds in the heavyweight championship bout at Toledo on Independence Day. Daily airmail service
between New York and Chicago was inaugurated. People tried to make light
of the fact that the prosperity of the pre-war years was drying up faster than
the ink on the Treaty of Versailles. "Every time we tell someone to cheer up,
things could be worse, we run away for fear we might be asked to specify
how," cracked F.P.A. in "The Conning Tower."

On June 25, Al Jolson's wife Henrietta Keller filed for divorce after twelve
years of marriage. Jolson said he couldn't understand it. "Outside of my liking for wine, women, and racehorses," he told reporters, "I'm a regular husband." The next day, William A. Brady produced a hot-weather thriller by
Owen Davis called *At 9.45*. The long-run champion was *Lightnin'*, featuring
Frank Bacon, at the Gaiety. Vaudeville was flourishing at the Palace, burlesque at the Columbia, silent movies at the Rivoli and the Strand. Other
early theatrical offerings were *The Five Million*, a post-war comedy; Holbrook Blinn in Eugene Walter's *The Challenge*; and a new play by Thomas
Dixon, author of *The Clansman*, called *Red Dawn*. On August 1, just as the
Broadway season was about to begin, impresario Oscar Hammerstein died.
In October, George Gershwin and Irving Caesar got together on a little
number called "Swanee" that took about ten minutes to write and would
make both of them rich young men.

With no part to play on the Broadway scene, Richard Rodgers enrolled at
Columbia University. He became Class of '23 hero almost as soon as he got

there by writing both words and music for this deathless ditty, with which his freshman class won a singing contest:

> C boys, it's '23 boys, and
> O we'll give them
> L.
> U boys, be true, boys, and
> Mmm we'll make 'em yell.
> B boys, forever steadfast
> I will for
> Aye
> (All shout) Columbia!
> '23—hooray!

Very collegiate, but hardly the stuff of which theatrical careers were being made that year; not much in tune with the times, either. Strikes paralyzed the nation: steel, meat packing, railroads, the garment industry, building, even the Boston police force. On the Atlantic seaboard, the longshoremen shut down the ports. In Pennsylvania, coal and iron police bent iron bars over the heads of strikers.

The biggest theatrical event of 1919 turned out to be the actors' strike. It was called by Equity, the actors' union, over the most contentious of all differences between artistes and management: the absence of a standard contract. It was more than that, though: backstage conditions on Broadway were usually dreadful, and managers held the whip hand. Many contracts contained a "satisfaction" clause, which meant the performer could be fired on the spot if—for any reason—the manager decided he or she was not "satisfactory." There was no limit to what was called free rehearsal—a straight play might need ten weeks, a musical eighteen: nobody got paid till opening night. Artistes had to provide modern costumes, sometimes period ones as well; often a player would invest money in costumes only to be dismissed as not satisfactory, or be left holding the bag when the show folded before opening. They had to pay their own transportation, so out-of-town strandings were frequent. No bonds were posted to guarantee salaries in the event of a show failing or going broke.

Unprecedented in the annals of American theatre, the actors' strike began on August 7 and settled in for a long run. It was an ongoing spectacle of picketing stars and chorus dancers, speeches and parades, fights, lawsuits, arrests, and closing theatres. A number of famous movie stars—Douglas Fairbanks, Dustin Farnum, Francis X. Bushman—took the side of the actors. Ranged against them were some equally big guns: Sam Harris, president of the Producing Managers' Association; Bruce McRae; Donald Brian; L. Lawrence Weber; Marc Klaw; Lee Shubert, who resorted to some very spectacular lawsuits; and William A. Brady.

Strangely enough, one of the most prominent players in the free-for-all with which the show business fraternity regaled the astonished late-summer metropolis was that Yankee Doodle Boy himself, actor-producer-playwright-composer George Michael Cohan, one of the biggest stars on Broadway. To everyone's surprise, Cohan took an anti-Equity stand, while insisting he was fighting for "the little people." This was arrant nonsense and no one can have known it better than Cohan, but he persisted, more often than not making himself look ridiculous in the process.

"If Equity wins," Cohan said for the record, "I'll quit the theatrical business and run an elevator." The Broadway mob chortled with glee when next day, Equity's supporters hung a banner from their window that read WANTED: ELEVATOR OPERATOR. GEORGE M. COHAN PREFERRED.

The strike spread across the country as stagehands and musicians came out in sympathy. Cohan and his fellow managers fought bitterly on, forming a new organization, the Actors' Fidelity League, with Cohan as its president. So intense was Cohan's feeling that he wrote a check for one hundred thousand dollars to the new League, which wisely refused it and turned it over to the Actors' Fund. Even with Mrs. Fiske, David Warfield, Otis Skinner, Laura Hope Crews, Fay Bainter, Alan Dinehart, Holbrook Blinn, and William Collier backing him, Cohan's cause was doomed. When 412 stagehands failed to appear for the evening performance of *Happy Days* at the Hippodrome, making that famous old landmark the twenty-third Broadway house to close, it was generally conceded that the managers had no chance.

On September 6, peace came with complete capitulation. The producers and managers signed an agreement with Equity fully recognizing the organization and binding both the Producing Managers' Association and Equity to the settlement terms for a period of five years. Lawsuits were dropped, actors went back to work, theatres reopened, and the lights came on again on Broadway. For George M. Cohan, it was a defeat from which he never recovered. Although he rejoined the Friars Club the following year, he hated Equity, as the Devil hates holy water, until the day he died.

———————

Larry Hart had a new companion: Milton G. "Doc" Bender. According to Teddy Hart's widow, Bender "inveigled himself into the family fold, playing the piano for Frieda, keeping Larry laughing with very bad puns." Tall and bespectacled (strangely enough, nearly everyone remembered him as short, squat, ugly, and pop-eyed) and incurably stagestruck, he "looked like a stooge." Almost as voluble as Larry, he fancied himself a bon vivant, but his real talent was as an inseparable hanger-on, insisting he "represented" Larry and encouraging him—not that Larry needed much encouragement—in all his excesses. Bender, who lived on West 87th Street near West End Avenue, was studying at the New York College of Dentistry on First Avenue. They had met the preceding summer at Brant Lake. Amiable, humorous, and unashamedly homosexual, Bender would have an increasingly pernicious effect on Larry's life.

As winter drew in, the boys from Morningside Heights started to put together a new Akron Club musical, *You'd Be Surprised*. Billed as "an atrocious musical comedy," the book was by Bender, the music by Rodgers, and the lyrics by Hart, Bender, Robert Simon, Oscar Hammerstein II (whose own first musical show, *Always You*, would open January 5 at the Central Theatre), and Herbert L. Fields. Bender directed, with "professional assistance" from Lew Fields; the cast included Dorothy Fields.

By this time, Rodgers and Hart had hammered out a working pattern which was to persist throughout their partnership. If Rodgers had a pretty melodic idea he would play it to Hart, who would listen for the dominant melodic strain and construct his lyric around it. In their alternative mode, Hart would come up with an idea, a title, or even a complete lyric, and Rodgers would improvise on the piano until he had a melodic line "not too sour, not too sweet," as Hart once put it, that pleased them both.

"We had rehearsals wherever we could," Phil Leavitt said, reminiscing about the Akron Club show. "While the book was being rehearsed, Dick would improvise at the piano. Whenever Lorry heard something that pleased him, he'd rush over to the piano, ask Dick to repeat what he'd just heard, and then make a lead sheet on a piece of paper. He'd take that home, and come in next day with the completed song. He was just fantastic."

You'd Be Surprised played a one-night stand at the Plaza Hotel's Grand Ballroom on March 6, 1920. Exactly eighteen days later, Rodgers and Hart attended the premiere of their second show that year. This time, however, it wasn't just another amateur production. This was the Varsity Show of 1920, *Fly with Me*. To them, it was the next best thing to being on Broadway, the sole reason both of them had gone to Columbia in the first place. So even while they were writing and rehearsing the fifteen songs for *You'd Be Surprised*, the boys were simultaneously at work on another dozen for *Fly with Me*.

The final credits—"book by Milton Kroopf and Philip Leavitt, adapted by Lorenz M. Hart"—indicate how the libretto reached the form which persuaded the judging committee of the Columbia Players—Richard Conried, Ray Perkins, and president Oscar Hammerstein II—to select it.

Twenty-three-year-old Herbert Fields, who'd been trying his hand at acting (in dialect roles not too different from those his father played), was brought in as choreographer, although his ideas rarely ranged much further than the one-two-three-kick variety. Ralph Bunker, a professional actor, directed. The plot, set far off in the future (1970!) was a nonsense about an island (called Manhattan) off the coast of America ruled by the Soviets.

If nothing else it was topical: the "Red Menace" was headline news all across America, and nowhere more than in New York. The Bolshevik uprising in Russia had stimulated anti-Red hysteria: when the *Boston Herald* ran a story headlined BOLSHEVIST PLAN FOR CONQUEST OF AMERICA, Woodrow Wilson's attorney-general, A. Mitchell Palmer, "the Fighting Quaker," authorized mass arrests of "Communists" among the strikers.

The show opened in the Love Laboratory of Bolsheviki and galloped off simultaneously in all directions, with plenty of double takes, coincidences, gags, puns, and double entendres. With the exception of "Always Room For One More" and "Weaknesses," lifted from *Up Stage and Down*, all the lyrics but one ("Twinkling Eyes," for which Rodgers wrote both lyric and melody) were by Larry Hart. They show him at the very start of his career to have been as inventive, witty, and spontaneous—how blithe, then, that virtuosity!—as he was to be until the very end.

The opening number, "Gone Are The Days," sets the pace, managing to rhyme "ecclesiastics,' "bombastics," "iconoclastics," and "gymnastics" in one stanza, and "freaky," "leaky," "sneaky," and "Bolsheviki" in the next. The lyrical fireworks flash throughout, in "Don't Love Me Like Othello," "Peek In Pekin," "Working For The Government," and "Inspiration." Other songs included "Dreaming True," a pretty waltz; a neat pastiche number, "Another Melody In F"; and the rousing choral finale "A College On Broadway," a hymn to Columbia.

Fly with Me opened—the first of four performances—in the Grand Ballroom of the Astor Hotel on Times Square on March 24, 1920, and got some rave notices ("music that sparkled from the rise of the curtain to its last descent," said the *New York Times*). Much more important, however, it convinced Lew Fields that Rodgers and Hart had real talent. To their further astonishment, he told them he was so impressed with their work he wanted them to write the score of his next Broadway show.

What he didn't tell them was that he was in deep trouble. The new show, which was to be called *Poor Little Ritz Girl*, was booked to open its Boston tryout on May 28, and Fields had already ditched one score. He shrewdly figured that Rodgers and Hart—just a couple of college kids, after all— would be so thrilled at the idea of writing a Broadway score they wouldn't read the small print in the contract; and he was right. They were far too eager to do the show to worry about things like "no interpolation" clauses, even if they had known about them. And Pop Fields clearly felt under no compulsion to enlighten them.

The partners excitedly went to work in the old Central Theatre at 47th and Broadway, producing seventeen songs in a two-week burst of creativity. "Collaboration is a good thing for everybody [involved in writing the show], [but] is merciless to everybody else, and tenderly nurtured ideas are only permitted to live when they are worthy," Larry told an interviewer who asked how they went about writing the songs.

In fact, several of them were reworkings of earlier numbers: "Don't Love Me Like Othello" became "You Can't Fool Your Dreams;" "Dreaming True" became "Love Will Call"; "Peek In Pekin" became "Love's Intense In Tents"; and "Princess of the Willow Tree," which had been in *You'd Be Surprised*, became "Will You Forgive Me?" Even Herb Fields got into the act—it was his father's show, after all—contributing lyrics for a clever little ditty about "poor bisected, disconnected Mary, Queen of Scots."

The story was simple, not to say simple-minded. A chorus girl rents the apartment of a wealthy young bachelor while he is out of town. He comes back unexpectedly and . . . well, that was the kind of show they were writing in 1920. Rodgers and Hart were invited up to Boston, then an overnight train ride from New York, to attend the premiere of the show on Friday, May 28. It nearly opened without Hart: the porter who was supposed to awaken him didn't see the diminutive form under the bundle of blankets in the upper berth of the sleeper, so Hart got no wake-up call. When he awoke, the train was parked way out in the railroad yards. He barely made it to the Shubert-Wilbur Theatre in time.

"I walked the streets all night waiting for the papers and the reviews of our show," Larry reported. "I drank about forty cups of coffee and telephoned the others at the hotel every time a paper came out."

The reviews for *Poor Little Ritz Girl*, which featured Victor Morley, Lulu McConnell, and Roy Atwell, were encouraging, and everyone was optimistic about its chances on Broadway. Before that momentous day, however, it still had to play six weeks in Boston, with further short tryouts in Stamford (Connecticut) and Atlantic City. Since there was nothing for them to do for the moment, the boys took summer counseling jobs: Larry Hart as usual at Brant Lake, and Rodgers and Herb Fields at Camp Paradox, Hart's old hangout.

They came down from the Adirondacks for the July 27 New York opening to one of the most stunning disappointments of their lives. The show was utterly, completely changed. Charles Purcell and Andrew Tombes had replaced Morley and Atwell; Eleanor Griffith had taken the lead played in Boston by Aileen Poe; Lew Fields had substantially rewritten the book (and taken a credit as co-author) and even brought in a new musical director. Worst of all, however, ten of their songs had been cut and replaced with new ones by Sigmund Romberg and Alex Gerber, leaving only seven of the original seventeen the boys had written. More than fifty years later, Rodgers still vividly remembered his frustrated anger and disappointment. As for Larry Hart, not only was he as angry and disappointed as his partner, he suffered additional humiliation when the critics ignored his songs and singled out most for praise the lyrics of Herb Fields's "Mary, Queen Of Scots."

5

Doldrums

hat was probably the most significant event of 1921 occurred at 12:01 a.m. on January 16, when the Volstead Act, the Eighteenth Amendment to the Constitution of the United States, went into effect. This Act, prohibiting the manufacture or sale of alcoholic beverages, finally slammed a door that had been closing slowly for decades. What Herbert Hoover called "the Noble Experiment," and everyone else called "Prohibition," had begun.

The cabaret phase of New York nightlife was over. Unable to operate profitably, the big old restaurants closed down one by one; the wreckers leveled such famous watering holes as Shanley's in Longacre Square, Louis Martin's, Reisenwerber's, and Maxim's. A new phenomenon appeared: where once there had been fifteen thousand legal saloons, there were soon more than thirty thousand illegal joints called speakeasies. The Roaring Twenties were off to a fast start.

Like it or loathe it, for Larry Hart it was back to the amateur grind or nothing. Just three weeks before the inauguration of the new President, Warren G. Harding, who had won election on the slogan "Back to Normalcy," the Akron Club unveiled its new showpiece, *Say Mama*. Originally conceived as *First Love*, it opened on February 10, 1921, in Brooklyn, with another performance the following night in the more palatial surroundings of the Plaza Hotel in New York. It was directed by Herb Fields and featured his sister Dorothy.

Rodgers and Hart came up with a largely fresh score for this one, which was in aid of the Oppenheim-Collins Mutual Aid Association and the *New*

York Mail Save-a-Home Fund. Only "I Surrender," one of the songs jetti-soned from *Poor Little Ritz Girl*, had been heard before. The other eight numbers were all new. The boys had high hopes for the knowing "Chorus Girl Blues," which they would use in two more amateur shows before the year was out.

That same February, Larry teamed up with Mel Shauer to write a couple of songs. "My Cameo Girl" may have been inspired by Mel's nickname for Frieda Hart; it featured Larry at his most outrageous, rhyming "cameo" with "mammy-oh" and, worse, "chamois-oh." The second song, "Fool Me With Kisses" was much better, an early example of the simpler, more lyrical style Larry would eventually perfect.

Cute enough, but nothing more, both songs disappeared into that peculiar limbo reserved for melodies that never made it. Larry—only friends from earlier days called him Lorry now—went back to working on a translation of yet another operetta, *Lady in Ermine*, for the Shuberts from which they would profit mightily and he not at all. He was still spending most of his time translating, still being paid a flat fee—with no accreditation—for each play. During this period, it is said, he also wrote many of the lyrics for a suc-cessful production of Offenbach's *La Belle Hélène*. It's also said that Benjamin F. "Barney" Glazer, a so-so Broadway actor, writer, and boulevardier, paid him two hundred dollars to do an adaptation of Ferenc Molnar's *Liliom*, which Glazer then sold to the Theatre Guild, whose production of the play was a distinguished hit. If this old Broadway legend is true—and Larry's sis-ter in law Dorothy Hart claims Joseph Schildkraut, who starred in the origi-nal production of the play, told her categorically that it was—how remark-able that neither Rodgers nor Oscar Hammerstein II, who might have been expected to know, never commented upon the irony when, a quarter of a century later, they transformed that selfsame play into *Carousel*.

Apart from their songwriting, Hart and Rodgers didn't socialize much. Out-side of the theatre, they had practically nothing in common. Rodgers was what today they call upwardly mobile: he wanted to be part of the smart set, in society, one of the Four Hundred. His father had movie stars as patients. His brother Morty was a member of Pi Lambda Phi, along with Bennett Cerf and Oscar Hammerstein II. Larry Hart couldn't have cared less about all that. So Rodgers went one way—home to West 86th Street, his parents, his studies, and his girlfriends—and Hart went the other, back to Harlem and his merry round of booze and cigars and talk and hangovers.

For all his straitlaced ways, Rodgers would always be a ladies' man. "I had my first date at eight," he once said, "and I've been in love ever since." Right now, he was living especially dangerously. The new love of his life was dark, diminutive, beautiful Helen Ford, a former vaudeville performer who was about to make her Broadway mark in *Always You*, the first Broadway musical for which twenty-five-year-old Oscar Hammerstein II had written book and lyrics. With a score by Herbert Stothart, what had originally been *Joan of Arkansaw* turned into *Toinette* and then *Always You* during its week at Poli's Theatre in Washington, a venue famous for having (but only by dint of a very long covered walkway) an entrance on Pennsylvania Avenue.

Always You opened at the Central Theatre in New York on January 5, 1920, and it was at this juncture that Rodgers and Helen Ford met: he not yet nineteen, she only a little more than twenty. The daughter of a Troy, New York, manufacturer, Helen Isabel Barnett was that town's musical prodigy throughout childhood. In her early teens, while studying piano and singing at the local conservatory of music, she was selected by the manager of a Lyceum Company to appear in a New York production of *The Heart of Annie Wood*, which opened at the Palace on Armistice Day 1918 and does not appear to have lasted long enough to get into the record books. It was either while she was appearing in this, or when she replaced Francine Larrimore in *Sometime* at the Shubert later in the year, that Arthur Hammerstein saw Helen and signed her up.

How she met Rodgers is not recorded; more than likely, it was through Larry. They liked each other immediately: both were passionate about music, the theatre, their careers. There was just one problem: Helen was married. It was a somewhat unconventional marriage. Her husband George, a nephew of the Ford who built the theatre in which Abraham Lincoln was assassinated, was a producer of touring Shakespearean festivals. Considerably older than his wife (he was nearly forty), George was away a great deal of the time. In fact, he had gone on the road two weeks after their marriage on August 9, 1918, and bride and groom hadn't seen each other for six months thereafter.

Right now, Ford was touring coast to coast—as he would for the next four years—with a repertory company featuring Fritz Leiber; in his absence, the romance flourished. Referring to her only as "Helen" in his autobiography, Rodgers said, "She was many things to me: teacher, mother, confidante and companion." His guarded references to their affair make it sound like a comparatively brief relationship, but it was more than that; and while it lasted they were inseparable. Helen Ford was to play a vital part in launching the careers of Rodgers and Hart, but that was still some years in the future.

Meanwhile, Larry and Dick buckled down to writing the Varsity Show of 1921, a musical comedy with a book by Herman Axelrod, father of playwright George (*The Seven Year Itch*), and Henry W. Hanemann. Directed by

Oscar Hammerstein II, it was called *You'll Never Know* and ran for four nights, April 21–23, 1921, at the Grand Ballroom of the Astor Hotel. Of its eleven songs, four were from earlier shows. The new ones, not including the opening prologue and chorus, were unmemorable; "When I Go On The Stage" would be rescued seven years later for a revue called *She's My Baby*.

Encouraged by his sweetheart, Rodgers decided to quit Columbia and transfer to the prestigious Institute of Musical Art at Claremont Avenue and 122nd Street, just a block over from the University. Founded by Frank Damrosch, the IMA—later the nucleus of the Juilliard School of Music—taught music theory, ear training, and harmony. "At the IMA," Rodgers said, in his typically understated way, "I started to learn for the first time what it was I had been doing right."

His harmony class was conducted by Percy Goetschius, who had an unbreakable rule that he would never teach more than five students at any one time. From Goetschius, Rodgers learned the technique that so characterizes his music: the avoidance at all costs of any musical phrase ending in a tonic chord (that is to say, the first, third, or fifth step of any scale).

No surprise, then, to find him volunteering immediately to write the score for the IMA's end-of-year show, the musical fantasy *Say It with Jazz*. Subtitled "A Coq d'Orian Fantasy" (it was loosely based on Rimsky-Korsakov's *Coq d'Or*), it had a book by Dorothy Crowther, Maurice Lieberman, and Frank Hunter, who also wrote some of the lyrics. Eight of the songs were earlier Rodgers and Hart numbers, including "Working For The Government," reconstituted as "Working For The Institute." The only new ones were "Just Remember Coq d'Or," "See The Golden Rooster" and "Oh, Harold!" Rodgers's best friend at IMA, a young man named Gerald Warburg (son of the banker Felix Warburg, whose house at Fifth Avenue and 92nd Street is now the Jewish Museum), contributed "If I Knew"; and William Kroll (who later founded the Kroll String Quartet) wrote "Hindoo Moon." The show was presented at the IMA on Wednesday and Thursday, June 1–2, 1921.

Then, since there was nothing doing on Broadway in the summer anyway, Larry left for his usual break at Brant Lake. He had been going up there since his intimate friend Bob Gerstenzang had opened the place in 1916. When it was no longer possible for him to be a counselor, he stayed at a nearby hotel. That he loved working on the "shows" and would have gone for that reason alone is self-evident, but his insistence on having Bender along paints a darker tone into the picture.

Someone said that in these early days with Bender, Larry was like a non-swimmer sitting poolside with everyone shouting, "Come on in, the water's lovely!" The analogy points up his continuing ambivalence. He was twenty-five years old, but he had never had a steady girlfriend. He never would. When someone suggested he get married, he replied bitterly, "Oh, yeah, I could always buy a stepladder, I suppose." Alan Jay Lerner defined the inevitable result of this predicament. "Because of his size, the opposite sex was denied him," he said, "so he was forced to find relief in the only other sex left."

It is hard to escape the impression that in early 1921 the partnership of Rodgers and Hart was very much in the doldrums: on one hand, Hart putting lyrics to Mel Shauer's songs; on the other, Rodgers working with a different lyricist. Clearly inspired by the huge success of the new Kern-Wodehouse-Bolton hit *Sally*, which had opened at Ziegfeld's New Amsterdam a few days before Christmas 1920, and which featured Marilyn Miller singing "Look For The Silver Lining," Rodgers concocted a tune to lyrics by Frank Hunter, who had worked with him on *Say It with Jazz*. He managed to persuade Lew Fields to interpolate it into a new revue called *Snapshots of 1921*, which starred Fields, Nora Bayes, and Gilda Gray. The song was called "Every Girlie Wants To Be A Sally," and if neither the song nor the show (it ran six weeks) was a big hit, the venture presented Rodgers with another opportunity for personal advancement.

The Shuberts, still Broadway's leading impresarios, were anxious to set up their own vaudeville circuit to compete with, and hopefully break the dominance of, the one operated by the B. F. Keith organization. Instead of booking only vaudeville acts, however, the Shuberts planned to send entire shows on tour. The first half of each show—the customary vaudeville routines—would be followed by a shortened, or "tab," version of a recent Broadway revue.

One of the shows they selected was the Lew Fields starrer, and Fields invited Rodgers to be the replacement musical director when the show had been on tour for two months. Rodgers made his debut as a professional conductor at the Shuberts' Detroit Opera House on December 4, 1921, and stayed with the show until it closed on the road the following spring. Oddly enough, neither the formidable *Fact Book* that lists his achievements nor his autobiography mentions the song that brought all this about.

Larry Hart next became deeply involved in co-producing, with Irving Strouse, a play by his friend Henry Myers called *The First Fifty Years*. Perhaps as thrilling to him as anything else was that the play was to be presented at the Princess Theatre, home of the early Kern-Wodehouse-Bolton musicals he adored. His involvement came about when Myers—whom Larry had dubbed "Morbid," and who was now a member of the literarily inclined group that met at the Hart household—persuaded him to stage a two-character play he had written called *The Blond Beast*, using the repertory company of a West 55th Street association, the MacDowell Club. Hart agreed, and became so taken with the play—what he was doing right now was always the best thing he had ever been involved with—that he convinced everyone it was good enough for Broadway.

Together with Milton Wynn, a lawyer, and Edwin Justus Mayer, another budding playwright, Hart came up with a startlingly new way to scare up a producer. They would give a single performance using professional actors, and put it on, just as Hart had staged it at 108 West 55th Street, in a Broadway theatre, inviting all the successful producers to come and see it live, rather than just read a script.

Meanwhile, Henry Myers had sent another of his two-handers to British-born character actress–director Margaret Wycherly (remembered now always as James Cagney's mother in *White Heat*—"Top of the world, Ma!"). When Hart heard this, he called on the actress in her Greenwich Village apartment, grandly informed her that he would personally produce the play, and invited her to star in it. Although Wycherly was in rehearsal for the Theatre Guild's production of Elmer Rice's *The Adding Machine*, she promised to read the play and invited Hart to bring the playwright to dinner.

When they got there the following Sunday evening, they found that Wycherly had definite ideas of her own and had all but rewritten the play. Myers was flummoxed, but Hart rose to the occasion, charmingly thanking the actress for the benefit of her judgment and experience, and suggesting that he and Myers take the play away and study it. Outside, his manner changed. "Screw her!" he said. "I'll get the actors myself. And a top director, too!"

Using as seed money a fifteen-hundred-dollar loan he wheedled out of his O.M.—Max must have been having one of his flush periods—Hart talked six of his more "persuadable friends" (to use Henry Myers's perfect phrase) into putting up the rest of the five thousand dollars they would need. Determined to make sure they got value for their money, the backers—Herbert Schloss, Mannie Diamond, and Irving Strouse among them—went everywhere Hart went: to interview directors, to audition actors, to rent scenery or costumes. Amused rather than offended by their lack of trust, Hart impishly registered the production company as Caravan, Inc.

After innumerable turndowns, he managed to persuade Livingston Platt to direct. Tom Powers, a journeyman actor who later scored in O'Neill's *Strange Interlude*, and Clare Eames, actress wife of playwright Sidney Howard, played the couple whose first half-century of marriage is seen at significant milestones. They went into rehearsals, and had one performance in Allentown, Pennsylvania. The play was enthusiastically received, and everybody was very optimistic—everyone, that is, except Hart, who pointed to Lent on the calendar. "That," he said gloomily, "is what is going to lick us. We should close now and wait until next season to open, but I can't wait any longer. My nerves can't stand it!"

The First Fifty Years opened with a flourish—Hart invited *le tout* Broadway—at the Princess on the evening of Monday, March 13, 1922. All the first-string critics were there: Alexander Woollcott of the *Times*, Gilbert Gabriel of the *Sun*, Henry Krehbiel of the *Herald*, Percy Hammond of the *Tribune*, Heywood Broun, Robert Benchley, Burns Mantle, and S. Jay Kaufman. While in main the critics liked the play, the public ignored it, and it folded after forty-six performances.

To rub salt into Hart's wounds, the Shuberts chose this moment to announce they would produce a new musical version of *The Lady in Ermine*, yet another of the operettas he had translated. It would open in October, with American interpolations by Al Goodman and what was to be a big hit for Sigmund Romberg, "When Hearts Are Young." Worse, now that Rodgers had left Columbia, the team couldn't even volunteer to write the score for the 1922 Varsity Show, *Steppe Around*, although they did contribute two songs, one of them, "When You're Asleep," notable for having only a five-line refrain.

The source of Larry's inspiration for the second song, "Kiki," was a David Belasco production adapted from the French play of that name by André Picard, which opened on November 29, 1921. It featured Lenore Ulric as the eponymous heroine, a Parisian chorus girl who lives platonically (but hopefully) in the apartment of a revue producer. With its "Belasco" and "tabasco," "sneaky," "cheeky," and "Bolsheviki" rhymes, "Kiki" might in more commercial times have been used to plug the show, but the Belasco play hardly needed a free plug; it ran for 580 performances, which was probably 576 more than Larry Hart's song got.

Yet another he wrote with Mel Shauer, called "Vixen," likewise got nowhere, maybe because, once again, Larry's rhymes—"vixen" with "fixin'" and "mixin'," "tricks in" and even "Old Nick's in"—tried just that bit too hard.

When Pop Fields told him about a new project reuniting Weber and Fields in vaudeville to be staged at the Palace Theatre, Hart got together with Herb Fields to write some special material for the show. They put words to a French song written in 1919 by a man named F. D. Bode, who used the pseudonym Clapson. It came out as "The Pelican," a new dance sensation which set absolutely nobody's toes tapping, in spite of its claim that "all the world of art/ in the gay Montmart'/ took it to its heart."

Much more appealing was Herb and Larry's collaboration with a some-time songwriter named Joe Trounstine. Yale graduate, ex-Army lieutenant, and now a twenty-eight-year-old Wall Street broker, Joseph H. Trounstine wrote songs as a hobby. Quite how he drifted into Hart's life is unknown; the lyricist's range of acquaintances was, to say the least of it, eclectic.

At any rate, perhaps in emulation of or as a salute to Lillian Russell, who had sung a "coon" number called "When Chloe Sings Her Song" in the turn-of-the-century Weber and Fields extravaganza *Whirl-i-gig*, they wrote a tune called "Chloe, Cling To Me." Published in October 1922, it deserves to be better remembered than many of its era which are, but the reuniting of Weber and Fields was—perhaps predictably—unsuccessful. The show—and the songs—sank without trace. No wonder at this stage of his life Hart com-plained bitterly to Henry Myers that bad luck stuck to him like a disease. "They won't let a new man in," he said. "But I'll show them! I'll show the bastards!"

While Hart was thus involved, Rodgers wrote another two amateur scores, the first for the Benjamin School for Girls, at which Dorothy Fields was a student. *Chinese Lanterns* contained four numbers for which Rodgers wrote both music and lyrics. The second was for the Institute of Musical Art: *Jazz à la Carte*, "A Musical Feast in Ten Courses," in which the lyrics for the new songs were all by Frank Hunter, and the rest a ragbag of Varsity Show num-bers and interpolations from Gerald Warburg, William Kroll, and Sigmund Krumgold. Prominent among them was "Every Girlie Wants To Be A Sally" as used in Lew Fields's *Snapshots of 1921*.

All of this would seem to dispose once and for all of the myth—propa-gated throughout his life by Rodgers—that once they met, Rodgers and Hart never worked with anyone else. At this juncture, the facts indicate the very opposite: that Larry Hart, older and infinitely better connected in show busi-ness circles, was growing impatient with the profitless grind of amateur shows, and continued to involve himself in them only out of that devoted loyalty he would always show to his partner, his best friend and—because Rodgers's problems with Hart were also Hart's totally different problems vis-à-vis Rodgers—his constant source of irritation.

6

The Blond Beast

*A*s far as musicals were concerned, the 1922–23 season—Broadway "seasons" run from summer to summer—was one of the worst of the decade. Out of forty-one new shows, only half a dozen were worthy of note, as producers played safe with barefaced cribs of earlier successes that were, to quote one authority, "lackluster, derivative, and banal."

In August 1922, Helen Ford was a big success in *The Gingham Girl*, a solid hit with songs by Albert von Tilzer. (One of them, "The Twinkle In Your Eye," predates a similar Larry Hart title by a decade and a half.) The season's biggest musical success was *Wildflower*, which opened on February 7, 1923, at the Casino with a score by Vincent Youmans and Herbert Stothart, book and lyrics by Oscar Hammerstein II. Oscar was having his usual mixed bag of luck: earlier in the season *Daffy Dill*, written with Guy Bolton and Herbert Stothart, had managed only nine weeks at the Apollo. All the signs indicated that the public wanted operettas. The public was going to get them in ever-increasing doses.

Still determined to break into the theatre on the production side, Larry Hart had meanwhile managed somehow to scrape together enough money to stage the long-deferred one-off production of Henry Myers's play, *The Blond Beast*. Myers's mother, a formidable lady known as "Muzzie" who was living on the tail end of a fortune at the Majestic Hotel, had ambitions for her son to become a concert pianist and composer. She asked Hart why he was wasting his time on her son's play instead of his music. "Henry's writing moves me," Hart told her flatly. "Henry's music does not."

Hart pulled absolutely no punches when criticizing someone's work, although he was rarely unkind in public. During rehearsals Henry Myers showed some resistance to aspects of Hart's staging, and asked whether there wasn't some way an author could stop production of his play if he didn't like what was being done with it. Myers recalled that Hart became furious, promising that if Henry ever again mentioned stopping the production, he would find some especially dreadful way to kill him. So with Arthur Holh and the floridly named Effingham A. Pinto in the only two roles, they got ready for their second Broadway show. The ads broke in the papers shortly after the opening of *Wildflower*.

PLYMOUTH THEATRE 45 WEST OF BROADWAY
Special Performance Friday Matinee 2 March 1923
Fourwalls, Inc., presents
THE BLOND BEAST
A Modern Comedy by Henry Myers
Staged by Lorenz M. Hart

Hart's plan to attract the attention of the leading entrepreneurs of Broadway worked beautifully. Every one of the major producers either attended or sent a representative. Alas for Fourwalls, Inc. (formerly Caravan, Inc.), just before the curtain went up, word arrived at the theatre that there might be another actors' strike (Equity was in the process of renegotiating the 1919 agreement), and every single one of the producers present left to attend an emergency meeting at David Belasco's office.

Dick Rodgers and Herb Fields now came up with another amateur show, a benefit for the Benjamin School for Girls. Hart got involved on his usual what-the-hell basis. Such writing brought in no fees, but it did provide a showcase to introduce their work to Broadway producers. Mrs. Benjamin grandly allowed them a budget of three thousand dollars—pretty good for an amateur show. And this time the idea was a good one. Looking back over his shoulder, as usual, Fields had remembered a 1901 production called *If I Were King*, which had starred the great Shakespearean actor E. H. Sothern. The play had its origins in a popular romantic novel by Justin Huntly McCarthy in which the poet François Villon saves Paris by becoming king for a day.

With Fields directing, and the cast drawn from the student body of the school (including Dorothy Fields, wearing a beard, as Villon), *If I Were King* was presented at the Maxine Elliott Theatre on 39th Street on Sunday evening, March 25, 1923.

Their brand-new score included a rousing opening chorus called "The Band of the Ne'er-Do-Wells," the title song, and (proof of Rodgers's growing musical confidence) both a march and a minuet. It was sufficiently good that after the show a producer named Russell Janney approached the boys to propose a Broadway adaptation. Not surprisingly, they agreed enthusiastically. Once more, Rodgers and Hart were teetering on the verge of a breakthrough; but despite Janney's genuine enthusiasm, he could find no one prepared to back a show by three unknowns. Two and a half years later, Janney did produce the story as a musical. With a score by Rudolf Friml, *The Vagabond King* became one of the most successful operettas of the decade.

Downhearted, but still determined, Fields and Rodgers next collaborated on the annual Institute of Musical Art show. Drawing its inspiration from recent archaeology and Mark Twain, the book was the work of Dorothy Crowther, who worked with Rodgers and Herb Fields on some of the lyrics. Larry Hart wrote only two new songs for *A Danish Yankee in King Tut's Court:* "If You're Single" and "Hermits." Around the same time, he and Rodgers produced a semi-topical song for the 1923 Varsity Show, *Half Moon Inn.* It was called "Babbitts In Love," doubtless inspired by the Sinclair Lewis novel published the previous year.

Somehow—probably through Helen Ford, who was its star—Larry Hart managed to get another of his songs interpolated into a Broadway show called *Helen of Troy, New York.* Even by the standards then applying on Broadway it had a speckled history. An agent-turned-producer named Rufus LeMaire invited playwrights George S. Kaufman and Marc Connelly to supply a book on any subject they liked, provided it fitted the title already chosen: perhaps it was not entirely coincidental that Helen Ford was a Troy, New York, girl. Kaufman and Connelly, who had written the big 1921 hit *Dulcy* (based on the famous "F.P.A." character), were keen to do a musical, but Connelly, who knew LeMaire's reputation for making deals for which he couldn't pay, demanded an advance of thirty-five hundred dollars.

LeMaire hedged (he didn't have the money) but stunned the writer by making a down payment with a thousand-dollar bill. Kaufman and Connelly learned afterwards that the first investors in the play were bootleggers (hence the big bills), who'd promised LeMaire fifty thousand dollars as soon as their next shipment came in. Nick the Greek, one of LeMaire's crapshooting pals, had also promised to make a big investment. Until they came across, LeMaire didn't have the wherewithal to pay either the librettists or the songwriters, Bert Kalmar and Harry Ruby.

LeMaire turned to a close friend, George Jessel, then an up-and-coming young nightclub comedian, who contrived a fake telegram using the name of Eugene V. Kahn (it sounded not dissimilar to that of the famous investment banker, Otto Kahn) promising to put up twenty-five thousand. The idea was to use this as a come-on to make the bootleggers and Nick the Greek spring for some cash. Unfortunately for LeMaire, the bootleggers' shipment got hijacked, and Nick the Greek had a bad run with the dice and ended up flat broke.

LeMaire somehow managed to convince Jessel to put up ten thousand dollars, his life savings, which was soon eaten up in advances to actors and chorus girls. Jessel went cap in hand to the garment district without success: the cloak-and-suiters were suffering a recession, and money was tight. He met LeMaire for breakfast at an Automat—all they could afford—and they were about to abandon the project when Jessel saw an item in a discarded copy of the *Times* which mentioned that Sidney Wilmer and Walter Vincent, owners of a string of theatres in Pennsylvania and Virginia, were going into the production end. LeMaire begged Jessel to go see them; Jessel agreed, not letting on that he had a couple of reasons of his own for doing so.

As it happened, Jessel had seen a play in the Village called *Dry Rot*. He knew he could acquire it cheaply if he could find backers to put it on. He decided to talk to Wilmer and Vincent about it. When he described it, however, they turned him down (a big mistake: when it came to Broadway under a new title, *White Cargo*, it was a million-dollar hit). As Jessel was leaving, they asked casually whether there were any musicals around. Jessel turned back and started talking, fast.

Helen of Troy, New York tried out in Fairmont, West Virginia, about as far away from Broadway as anyone had yet taken a show. The opening night was a complete disaster from which its creators fled in dismay. On the train going back to New York, Jessel and Kaufman found themselves sitting in opposite seats. With not a little trepidation, Jessel mustered enough courage to tell the great man that he thought the book disappointing and unsatisfying, because although it was supposed to be a love story, there wasn't one clinch in the entire play. He also felt the humor was too cerebral for the average audience. By the time he got through he was being pretty emphatic. Kaufman just peered at him over the top of his glasses and lapsed into unbroken silence; at which juncture Jessel realized—too late—how far he had overstepped the mark.

The two men sat uncomfortably looking out of the window and not at each other for what seemed like a year. Then, as the train stopped at a back-country crossing, Jessel spotted a ramshackle clapboard shack beside the tracks, the only sign of life a pair of flaming red ladies' drawers hanging on a clothesline. He leaned over and nudged Kaufman. "Let's get off the train," he suggested, "and go into that shack and . . . forget."

This was the show for which Larry Hart found his new collaborator, a strange duck named W. Franke Harling. Born in London, he had come to America as an infant and received his education at the Grace Church Choir

School in New York. After a stint as organist and choir director of the Church of the Resurrection in Brussels, he spent two years at West Point, where he wrote not only the Academy hymn, "The Corps," but also its official march, "West Point Forever." Harling was now thirty-six years of age, trying unsuccessfully to bridge the gap between classical works and the Broadway musical.

Exactly where, when, and how the Hart-Harlinge collaboration came about is no longer on the record; Marc Connelly recalled the show quite clearly but had no memory of Larry's involvement. Not that it matters: "Moonlight Lane," which was interpolated into the show during its shake-down tryout in Newark, was just another schmaltzy waltz with corny rhymes and turn-of-the-century sentimentality, and made no impression on any-body. Even with his name appearing incontrovertibly on the sheet music, it is still difficult to believe the lyric was written by Lorenz Hart.

Opening on June 19 at Broadway's Selwyn Theatre to gratifyingly good notices—the *Telegram* called it "a gem"; the *American* said "Don't miss it!"; and the *Herald* called it "the perfect musical comedy!"—*Helen of Troy, New York* settled in for a six-month run, although it was not to be Helen Ford that it lifted to stardom, but Queenie Smith, playing the second female lead. The Kalmar-Ruby songs "I Like A Big Town" and "It Was Meant To Be" had a run-of-the-play popularity, then also disappeared without trace. Larry Hart said the hell with it and took off for Brant Lake, where he was to meet, for the first time, yet another would-be songwriter, Arthur Schwartz.

7

Campfire Days

orn in Brooklyn on November 25, 1900, Arthur Schwartz left New York University in 1920 with a Bachelor of Arts degree, and received a Master of Arts degree from Columbia University the following year. He became a Doctor of Jurisprudence in 1923, the same year his poem "Damon and Pythias" was accorded the honor of filling the entire "Conning Tower" column, and would gain further distinctions in law in 1924. Despite having no formal musical training, Schwartz was a fine instinctive pianist, and was using his skill to help pay his way through college. Although his chosen discipline was the law, he had a hankering to write songs and work in the theater. The trouble was, he didn't know how to go about it.

> I was a camp counselor as a young fellow, and I met a man who was big in Larry Hart's life, Milton Bender, [who] was a counselor in another camp that I first was at, called Camp Kiowa in Pennsylvania. Bender knew a lot of the Rodgers and Hart songs before they were eversuccessful, and he would play these songs for me. And I thought, My God, what great lyrics! and he told me all about Larry Hart. He said, "If you want to meet Larry Hart one summer, don't go back to this camp, go to Brant Lake Camp, because that's where Larry Hart goes for vacations. He's not a counselor, but there's a little place near the camp for people to have holidays." So next year, I got myself a job at that camp in order to meet Larry Hart. It was a much better camp. I got much better money, much better food. And I got to work with Larry.

He also got to meet the family. "His father was a well-to-do businessman in businesses that were mainly crooked," Arthur remembered. "The way I know they were crooked was that Larry Hart said to me, when I was a lawyer and had not yet left [the law] for the theatre, 'If you need a client, my father always needs a lawyer.' And so I got involved with his father, Max Hart; and had I signed a certain paper that he wanted me to sign, I would have gone to jail.

"I used to go there for Sunday dinner; sometimes I would go on a Saturday night and sleep over, and Max Hart was like the father of the Katzenjammer Kids. He was a round-bellied, slightly bald man who swore furiously at the table all the time, in order to shock his wife, who wasn't shocked at all, but pretended to be."

⌁⚬⌁

Arthur Schwartz was at Brant Lake Camp when Billy Rose, one of the pushier and more obnoxious Broadway hustlers, came up and spent three weeks there "till he was thrown out because he was an objectionable person." The purpose of his visit was to work with Larry,

> and pay him a hundred dollars a day for collaboration on songs that Billy was working on. Their system was to go out in a boat at ten or eleven o'clock in the morning, take sandwiches with them, and BillyRose and Larry Hart would go over the melodies that Billy was working on, and at the end of the day Larry would show me a hundred dollar bill—every day. He never got any royalty, he never got any credit. This went on for three weeks, and I'm sure it went on later on in New York before Larry became established himself.

The degree of Billy Rose's obnoxiousness may be gauged from the fact that he was thrown out of Brant Lake, while Doc Bender, who everyone agreed was pretty vile, was not. Born in poverty on the Lower East Side, September 6, 1899, as William Samuel Rosenberg, Rose was already well on his way to becoming a Broadway legend. His unscrupulous pursuit of money, his ruthless disregard for the feelings of others, his unhesitating use of people, and his "borrowing" of their ideas would make him one of the most thoroughly hated men in show business.

Noting that he had been born on the night President McKinley was shot, Broadway wags said the assassin shot the wrong man. Others claimed that Rose hadn't been born at all: he was manufactured by press agents. But you couldn't insult him; the pushy, diminutive, aggressive Rose had a skin like a crocodile.

Rose was a genius at Gregg shorthand: by the time he was seventeen, he was winning contests and making good money from it. He could do 160 words a minute with 98.66 percent accuracy. He conned himself into a job as a speed stenographer with the War Industries Board in Washington at eighteen hundred dollars a year before he was eighteen, and then as a stenographer for Bernard Baruch. They say that if he'd won the national championship in 1919 (and the story of how he lost it reads like one of his own press agents' handouts) maybe he never would have drifted onto the Broadway scene. But he didn't, and he did.

Late at night he hung out around the Turf and Lindy's, where the song-writers and songpluggers gathered. He knew he had to get acquainted with them if he wanted to get into the business. What they didn't know, what nobody knew, was that Rose was taking everything down in that incredible shorthand of his. What Rose heard came out a few hours later as his own creation when he talked with somebody else and repeated some smart wise-crack, thought, phrase, or witticism that he had jotted down earlier.

All through 1920 and 1921 he was everywhere in Tin Pan Alley—the Leo Feist offices; the Bourne company; Remicks, Mills and Witmark; Shapiro, Bernstein & von Tilzer; and the most celebrated of all the professional par-lors, T. B. Harms, Inc., at 62 West 45th Street. Most of the professional composers and lyricists made it a habit to drop in there around noon to hear the latest gossip, shop talk, and songs: George Gershwin, Harry Ruby, Joe Meyer, Bert Kalmar, Buddy DeSylva, Ray Henderson, Phil Charig, and "court jester" Irving Caesar, who was famous for his impromptu parodies. Sometimes Max Dreyfus would appear and take a few of them to lunch in the Hunting Room of the Astor Hotel, where a special table was always reserved for him.

Rose was always around, peddling lyrics nobody wanted—yet. He later claimed he lived at this time on two packs of salted peanuts a day. It might have been true. You never knew with Rose. He made himself known to everyone: publishers, bandleaders, writers, songpluggers, vaudevillians, the determination hardening in him to find a way to use, to exploit, and to out-smart them all.

In 1921, shortening his real name to the one he would use for the rest of his life, he collaborated with lyricist Benny Ryan on a song called "I Hold Her Hand And She Holds Mine" with music by Irving Bilbo. That was to be Rose's big trick: collaborating. Get your name on the song, it didn't matter how, just so you got a share of the royalty. Then go out and hustle it: get someone to print it, someone to plug it, someone to sing it in a show. Then collect.

By 1922, Rose had his name on half a dozen songs; but it wasn't until 1923 that he scribbled a few lines about the cartoon character Barney Google and took them to songwriter Con ("Ma, He's Making Eyes At Me") Conrad, who presented Rose with a hit tune. Now Rose had money, and something else as well: he had the smell of success. "That Old Gang Of Mine" consoli-dated it. He rented a brownstone apartment in the mid-Fifties. He made connections, he gave parties, he put himself about. It worked: soon he was writing songs with top composers like Ray Henderson, Sammy Fain, Jimmy Monaco, Pete Wendling, Lou Handman.

Still nobody liked him.

"He can write anything he hears," they said. "He has talent, but nobody knows what it is." "Watch out for his shorthand," they warned each other, "it's longer than you think." Harry Warren, who worked with him before he left Broadway for Hollywood in 1932, remembered Rose well.

> He was a great feeder. He'd sit with the boys, and he'd say something like, "Come on, gimme another line, nah, nah, you can do better than that, come on, gimme something like . . ." then he'd come up with a phrase or a line, something. Somehow, a song would come out of all this, somehow. And nobody knew what Billy had done, but he'd done something. The difference with him was that was only the beginning. Then he would go out and hustle the publishers. You never knew how much he got, he never told you. He'd go in and make a deal, come out, hand you some money, that was it! You never found out what he sold it for till the statement came. Then you'd find you'd been rigged, he'd taken the biggest bite. And on the one hand you hated him, and on the other hand you knew you wouldn't have gotten any deal at all without him.

Rose would go on to be associated with dozens of hits. He was publishing twenty-five, thirty songs a year, working with so many collaborators that nobody knew any more what he wrote or didn't write. That was the way Rose liked it: he was interested in anything that would make a dollar. He sold a great bill of goods. You did business with him on his terms or not at all. He ran roughshod over everybody, a mixture of larceny and greed that Larry Hart probably found irresistible.

Was Hart the unknown writer from whom Billy Rose got the idea for "Me And My Shadow"? There are plenty of knowing Broadwayites who'll tell you it was. Look at some of the songs Rose put his name on: "You Gotta See Momma Ev'ry Night (Or You Can't See Momma At All)"; "A Cup of Coffee, A Sandwich, And You"; "I Wonder Who's Dancing With You Tonight?"; "Lucky Kentucky"; "If I Had A Girl Like You"; "It's Nobody's Fault But Mine." It's tempting to wonder—because he had such a gift for picking up street phrases and making songs out of them—whether any of them might have been Larry Hart's inventions. Or was Rose's contribution, whatever it was, his own? That was the whole trick: nobody knew, and Billy Rose wasn't about to say.

Back at Brant Lake, Arthur Schwartz realized his ambition to work with Larry Hart on the fortnightly camp shows. One of them was *Dream Boy*, about a fellow who doesn't care for the energetic, athletic activities of summer camp, but would rather stay in his tent and read books all the time—a character not all that far from Hart's own. They decided they needed a theme song for him, and Hart suggested the title "I Love To Lie Awake In Bed." Then, Schwartz remembered:

> I wrote the melody to that title, and it was the first melody that I wrote that I felt was any good at all. I think that I started very slowly, and was very poor in my first efforts, and I grew, as many people grow. But I think Larry was already full-grown. He was not only a brilliant lyricist but a brilliant theatre man. I don't think he advanced any. His gift never improved—it didn't need to. All he needed to do was practice his craft and do more work: he was an absolute genius.

Each show usually had about ten songs. Some of the titles from earlier ones—"Stop! Stop! I Am The Traffic Cop!" "Oh, Mr. Postman, Don't Pass Me By," and "I've Got A Girl In Chestertown"—clearly indicate the tone of the evening. For *Dream Boy*, Schwartz and Hart wrote a waltz about saying goodbye on the "Last Night," and another that utilized the initials of the camp, B.L.C., where everyone was h-a-p-p-y as a k-i-d. The opening chorus, "Down At The Lake," was performed in bathing trunks and towels by the youngest kids, who were called "midgets." Older boys were "seniors."

But the melody Hart kept reprising—he knew a good tune when he heard one—was the first one Arthur Schwartz had written, "I Love To Lie Awake In Bed." In 1929, Schwartz, now a successful theatrical songwriter, was trying to write a rueful sort of melody for the revue *The Little Show* and remembered the tune from *Dream Boy*. His new lyricist, Howard Dietz, almost immediately gave it the title by which it has been known ever since: "I Guess I'll Have To Change My Plan."

So enthusiastic a fan of Rodgers's tunes and Larry Hart's lyrics did Arthur Schwartz become that he actively proselytized in their behalf. He found out that a fellow summer camper named Marshall Rosett had an uncle in the music publishing business. His name was Elliott Shapiro, of the firm of Shapiro, Bernstein & von Tilzer.

> I found out from Marshall that his uncle was going to come up and visit camp. So I told this to Larry very excitedly. I said, "You've taught me a lot of these Dick Rodgers songs you've been working on. Wouldn't you like to have an audition with a real, live music publisher? He's coming up here next Sunday!" So we rehearsed a few of the songs: "Don't Love Me Like Othello," and "Peek In Pekin."
>
> Elliott Shapiro did come up for the weekend. Now Elliott Shapiro was not the man with the cigar that you hear about on Tin Pan Alley, but the grimmest-faced man that you could ever play a song for. And we went to the gymnasium where the shows were put on where there was a piano. Larry and I had rehearsed four or five Rodgers and Hart songs. Elliott Shapiro stood grim-faced at the piano, never looking up. I played and Larry sang, and Elliott Shapiro [not unnaturally thinking Arthur was the composer] said, "It's too collegiate, you fellows will never get anywhere unless you change your style. It's great for amateur shows, but I have to tell you the truth, fellows: change your style or give up."

Dream Boy did not by any means end Arthur Schwartz's association with Larry Hart. Back in Manhattan, he kept on writing his tunes. His father, who knew he had no musical training at all, tried to persuade him to give it up, but Arthur was not so easily discouraged. He kept on seeing Larry Hart, who listened to his tunes and told him which were promising and which were not. He also told Schwartz he couldn't go around trying to get jobs in the musical theatre while he was practicing law; the smart thing to do was wait until he had made enough money to support himself for a year. "You're not ready yet," he said. "You haven't shown enough talent."

Schwartz was called to the New York bar in 1924, becoming a junior partner in the law firm he was working for, and in due course became a successful lawyer. Following Larry's advice, he quit only when he had enough to live on for a year. He saw now that Hart's idea had been completely practical: if he failed in the songwriting business, he could always go back to the law. "In advising me," he said, "Larry was like a big brother who was very careful not to make a mistake."

One night [in 1924], I was at Larry Hart's house trying to play tunes for him, and a friend of his by the name of Eddie Ugast rushed in and said, "Do you fellows want to pick up some quick money?" We said yes. "There is an act called [Joe] Besser and Amy, and they're playing now in vaudeville in Harlem, and they need a comedy song." So Larry and I stayed up most of the night, and we wrote a song, the lyrics of which I don't know. There was a straight man and a Jewish comic. The name of the song was "I Know My Girl By Her Perfume." The straight man sang something like "Estelle has the smell of a rose." Then the Jewish comic had girls with peculiar smells: Becky smelled from herring, somebody else smelled from garlic, and so on.

[While we were writing the song] Eddie Ugast kept telephoning these comedians, whom he knew, saying, "They'll have it, they'll have it! They're coming right over, they're coming right over!" And it must have been three or four o'clock in the morning when we had a verse, and the straight chorus, and the comedy chorus. And we went to see these chaps somewhere in Harlem, because it wasn't far from where Larry lived, and they said, "We'll take it!" So Eddie Ugast, who was the agent, said, "We want a hundred dollars." They said, "Seventy-five," and he said, "Seventy-five is good." Eddie Ugast got twenty-five dollars as the agent, Larry Hart got twenty-five, and I got twenty-five, and that was the first money I ever earned from a song.

8

*The Brief Career
of Herbert Richard Lorenz*

*W*ith the route to Broadway looking more and more like a dead-end street, Hart, Rodgers, and Fields were beginning to lose hope. They had been working intermittently on a show called *Winkle Town*, in which a fellow invents an electronic system that renders electric wires obsolete and tries to sell the idea to the city fathers of the eponymous town. They wrote a dozen songs, put the whole thing together, then concluded that the book (by Fields) wasn't working out as well as they had hoped. Taking their *chutzpah* in both hands, they decided to ask their old pal Oscar Hammerstein II for help.

Hammerstein was now employed full time by his uncle Willie. His most recent musical shows, *Wildflower* and *Mary Jane McKane*, were both highly successful, and he was working on another that would prove even bigger: *Rose-Marie*. But two straight plays in succession, *Gypsy Jim* and *New Toys* (both written with Milton Gropper), failed dismally, fulfilling a dictum of Oscar's grandfather that thereafter Oscar never allowed himself to forget: there is no limit to the number of people who will stay away from a bad show.

Hammerstein read the script of *Winkle Town* and liked it so much he agreed to join the team as a collaborator. As it turned out, he was unable to spend much time on the project, or to lick the problems of the book. Meanwhile, thanks yet again to Helen Ford, Hart and Rodgers got a chance to audition the show for a new producer, Laurence Schwab, whose first production had been *The Gingham Girl*.

Schwab liked the songs but hated the book, whereupon the boys—who had more or less given up on *Winkle Town* anyway—made him a counteroffer. They knew he was about to go into production with a musical he was writing with Frank Mandel for which he had not yet chosen a composer and lyricist. Why not use the songs from *Winkle Town* in his new show?

Schwab liked the idea, but because he didn't consider himself an expert enough judge of music, he asked Rodgers if he would mind playing the songs for a good friend of his, Max Dreyfus, head of the music publishers T. B. Harms. Rodgers agreed, not telling Schwab that he had auditioned years earlier for Max Dreyfus's brother and partner Louis Dreyfus, who had listened politely while Dick played, then brusquely recommended he go back to high school and concentrate on finishing his education.

A date was set, and they went down to the Harms offices, where they were shown in to see Max Dreyfus. The titan of the music business was a small, reserved, soft-spoken, slightly built man of fifty. German-born, Dreyfus had been a cornet player on a Mississippi riverboat, a songplugger for Howley, Haviland and Dresser, a demonstration pianist, an arranger, and a composer. He had become associated with Harms in 1901 when he bought a 25 percent interest in the company, and his first important find was the nineteen-year-old Jerome Kern in 1905.

He listened with his eyes half closed as Rodgers played the twelve songs in the score: "The Three Musketeers," "One A Day," "The Hollyhocks of Hollywood," "Since I Remember You," "Old Enough To Love," "Hermits," "I Want A Man," "I Know You're Too Wonderful For Me," "Baby Wants To Dance," "Comfort Me," "Silver Threads," and a sprightly little schottische called "Manhattan." When he was finished, Dreyfus turned to Schwab.

"There's nothing of value here," he said, as if Rodgers were not even in the room. "I don't hear any music and I think you'd be making a great mistake. You know, we have a young man under contract here who'd be perfect for the job. His name is Vincent Youmans."

The shattered Rodgers slunk away, stunned by Dreyfus's harsh judgment. It would prove no satisfaction at all to him that Schwab didn't hire Vincent Youmans either (George and Ira Gershwin wrote the show, which was called *Sweet Little Devil*). This was worse than just being back at square one. Max Dreyfus, the top man in the music publishing business, had unilaterally decreed that the best work Rodgers and Hart could do was worthless, and that was very hard to take.

Perhaps the answer was to do a straight play instead, something tailor-made for Lew Fields? Larry and Dick hashed over a few ideas with Herb Fields, who again looked backward and remembered a 1904 tearjerker called *The Music Master*, which had run for over five hundred performances at a time when one hundred was considered substantial. Its star, David Warfield, was so successful in the role of the German musician who comes to America in search of his daughter that he played no other until 1907, and then turned down an offer of a million dollars to star in the film.

When the triumvirate got through with it, the story had turned into a satire on Tin Pan Alley called *The Jazz King*. They hoped Lew Fields would play the part of Franz Henkel, the serious composer who is bewildered and hurt when his ambitious Dresden Sonata is revamped as a pop hit; but when his daughter marries a jazz publisher, the father is enabled to return to writing classical music.

To their delight, Lew Fields not only liked the play but decided to put it into production. The authorship of the play—and the two parody Tin Pan Alley songs in it—was attributed to Herbert Richard Lorenz, a pseudonym which seems to have succeeded in fooling hardly anyone. Tryouts were scheduled in Bethlehem, Pennsylvania, on March 24, followed by Harrisburg, Johnstown, Wheeling, Toledo, Detroit, Cleveland, Chicago—and Brooklyn—over the next two and a half months. All being well, the show would open in New York in May.

Fizzing with anticipation, Rodgers and Hart dashed off a Purim entertainment for the Park Avenue Synagogue called *Temple Belles*, a one-act musical comedy directed by Herb Fields which contained four songs, none of them new. There was only one performance, on Thursday, March 20.

Three days later, with Dorothy Fields playing the dual role of Rudolf Rassendyll and King Rudolph the Fifth of Ruritania, Rodgers and Fields presented their adaptation of Anthony Hope's novel *The Prisoner of Zenda* at the Benjamin School. Larry Hart had no hand in the production; all the lyrics were written by Fields. It was yet another example of Herb's creative use of his show business upbringing: a dramatic version of the Hope novel had been successfully produced on Broadway in 1897.

<hr/>

The Lew Fields show, now known as *Henky*, was playing at the La Salle Theatre in Chicago. With less than a month to go, Lew discovered that he did not have enough money to bring the show into New York. The boys certainly couldn't help; during the rewrites, Larry was so strapped he had to hock his diamond stickpin for seven dollars so he could buy cigars. It was decided only one man could produce the thousand dollars Lew Fields said he needed: Larry Hart's O.M. In Rodgers's version, he went to see Max Hart, who called Billy Rose and persuaded Billy to buy into the show.

This begs a lot of questions, the biggest one being that even as persuasive a *hondler* as Max Hart would have found it far from easy to sell such an idea to Billy Rose. And why would Dick, rather than Larry, have been better able to persuade the O.M. to help them?

Billy Rose had moved on from borrowing ideas and writing lyrics. He was running a speakeasy, the Back Stage Club on West 56th Street. He was also investing in plays. One of these was a 1924 production called *The Fatal Wedding*, which, Rose-style, he ended up owning. It was said Billy built the foundations of his investment capital from the proceeds of this production. What simpler than for Larry to ask Billy to help them, and help himself at the same time?

Whichever version is true, Billy decided to invest in the show. Again retitled, it opened on May 13 at the Ritz Theatre in New York as *The Melody Man* with Fred Bickel (later to change his name to Fredric March) in the lead. It featured only two songs, "Moonlight Mama" and "I'd Like To Poison Ivy (Because She Poisons Me)," both deliberately awful, sung by the diminutive specialty act Eva Puck and Sammy White.

The critics split: Woollcott liked the show, and Quinn Martin thought it "tremendously funny." But vitriol-tongued George Jean Nathan put an end to the newborn career of Herbert Richard Lorenz. "The plot," he said, "is not only enough to ruin the play; it is enough—and I feel I may say it without fear of contradiction—to ruin even *Hamlet*."

9

Gilding the Guild

By the beginning of 1925, even Larry Hart's unquenchable optimism was beginning to flag. While he never ceased believing in Dick Rodgers's talent, he had to face the fact that he was thirty years old and getting nowhere. To make matters worse, many of his friends were finding the success which still eluded him, depriving him of the excuse that "they" wouldn't let in new talent.

Howard Dietz, who had not devoted a quarter of the time Larry had to lyric writing, had been invited to collaborate with Jerome Kern on a new—if unsuccessful—musical called *Dear Sir*. Oscar Hammerstein II had an enormous hit as co-writer and lyricist of *Rose-Marie*—so big, in fact, that four touring companies were being formed to take care of the demand for tickets across the country. Even Eddie Mayer—Edwin Justus Mayer in the credits—had triumphed with a stylish comedy about Benvenuto Cellini, *The Firebrand*.

All this, yet the best Rodgers and Hart were being offered was the job of writing a one-act musical called *Terpsichore and Troubadour* for the specialty song-and-dance team of Renée Robert and Jay Velie, who introduced it in vaudeville at the Palace on January 4, 1925, and later took it on the road. Simultaneously, as a favor to his friend Irving Strouse, Larry agreed that he and Dick would write yet another benefit show, this time for the Evelyn Goldsmith Home for Crippled Children. It was presented by Larry and Irving Strouse at the Heckscher Foundation at the corner of Fifth Avenue and 104th Street on February 8, 1925.

Starring (if it can be called that) Mrs. Arthur Bodenstein, Mrs. Leon Osterweil, and Muriel Bamberger, *Bad Habits of 1925*, "The Kind of Show That Men Forget," had seven Rodgers and Hart songs (some dating back two or three years). They also reprised "If I Were King" and "College Baby," a song from *A Danish Yankee* with a lyric by Robert Simon. The most shamelessly showbizzy of them was one called "I'd Like To Take You Home To Meet My Mother," which gave Roselee Steinfeld and Phil Leavitt (and Larry) a chance to skit every musical style in sight—from Victor Herbert and Franz Lehár via Gilbert and Sullivan through German and Italian patter to a Jolson-style finale.

Good fun, but no "mon," as Larry might have said. Dick felt the same way. He was nearly twenty-three and had never held a steady job in his life. Like his partner, he suffered from doubts and insecurity, not to mention a chronic case of insomnia. "If you think about it, it's not too surprising," he said.

We'd spent something like seven years, writing every conceivable kind of show, Varsity Shows, benefits, anything. We'd even had a couple of shots at Broadway. And where had it got us? What made it harder for me, for both of us, was that much as we admired the Gershwins and Vincent Youmans, we couldn't see what their songs had that ours didn't. Yet there they were on Broadway.

⚜

Early in 1924, Larry had come across what he thought was a great idea for a musical. At the corner of 37th Street and Lexington Avenue, outside the house where the events had occurred, he had seen a sculpted frieze below which was a small metal plaque:

MRS. MURRAY RECEIVING THE BRITISH OFFICERS
FROM THE PAINTING BY
JENNIS BROWNSCOMBE

HOWE, WITH CLINTON, TRYON AND A FEW OTHERS, WENT
TO THE HOUSE OF ROBERT MURRAY, ON MURRAY HILL, FOR
REFRESHMENT AND REST. WITH PLEASANT CONVERSATIONS
AND A PROFUSION OF CAKE AND WINE, THE GOOD WHIG LADY
DETAINED THE GALLANT BRITONS ALMOST TWO HOURS: QUITE
LONG ENOUGH FOR THE BULK OF PUTNAM'S DIVISION OF
FOUR THOUSAND MEN TO LEAVE THE CITY AND ESCAPE TO
THE HEIGHTS OF HARLEM BY THE BLOOMINGDALE ROAD,
WITH THE LOSS OF ONLY A FEW SOLDIERS

FLIGHT OF THE AMERICANS ON THE LANDING OF
THE BRITISH, SEPTEMBER 13TH, 1776.

Intrigued, Larry researched the incident to discover that Mary Lindley, the wife of Robert Murray (after whose family Manhattan's Murray Hill was named), had indeed used her feminine wiles to detain the British generals long enough for the American forces under Putnam to escape from lower

Manhattan and join General George Washington's army on Harlem Heights: a turning point in the American Revolution.

He turned the idea over with Herb Fields, who embellished it with a romance between a British captain and an Irish-American girl named Betsy Burke. Herb may have written it that way because spunky Irish heroines were all the rage—his ideas were never startlingly original—or, more likely, because he thought it would be perfect for Helen Ford (Helen's recollection was that Herb wrote it with her in mind). Whatever the reason, they had the bones of a show, which they first called *Sweet Rebel.* According to Rodgers's autobiography, Helen Ford came to his parents' apartment, met Larry and Herb, listened to the songs and the scenes, and fell in love with the show.

Helen Ford's recollection was that Herb stepped up to her in the lobby of the Algonquin Hotel, where she lived, and asked her to read the script. He said a friend of theirs named Cracower, who had gone to camp with them, had promised them twenty five thousand dollars—half the money they needed to put the show on. Since she was out of work anyway—her last show, *No Other Girl,* with a Kalmar-Ruby score, had opened on August 13, 1924, and survived only seven weeks—she agreed to do the show. Something told her it was a winner:

> I knew this would make a star of me, I knew this instinctively. Up to that time I'd been playing these leads in shows where the comedian was probably the most important character, and . . . well, I just wanted a good part. You know what sold it to me? The entrance. We were going through a period when they were sneaking the leading lady on so she'd come on with the chorus and they wouldn't see her. I'd had a couple of shows where that had happened to me, and—it's silly, you know?—but when you're that young you still have your dreams of being the grand lady. So when I read the book, I saw this marvelous entrance, I made the entrance in a barrel. At that time, that was shocking, you know? It was a wonderful entrance, and it was cute and it was funny, with this English redcoat soldier chasing me with one of my slippers in his hand, and obviously I have no clothes. And I did like the music.

Helen quickly set up an appointment with producers Morris Green and A. L. Jones, who had done *No Other Girl.* They liked what they heard—there was no third act yet, and no finale—but having lost money on their last show, they were wary of taking a chance on unknowns like Rodgers, Hart, and Fields, even with Helen Ford. What they did do was to interest the English director John Murray Anderson in doing the show. Anderson, however, was currently immersed in a new Irving Berlin *Music Box Revue,* and that, it appeared, was that. Lew Fields was the next one to turn them down, followed in rapid succession by everyone else in town.

"I spent almost a solid year with them," Helen Ford said, "doing auditions for anybody. At the drop of a hat we did it, man, woman or child, anywhere, all over town, on Seventh Avenue for the cloak-and-suiters, even some men who I think were gangsters. They were putting money into shows in those days, although usually they had a cutie they wanted to put in, too. Would you believe it, we did about fifty auditions?"

One anecdote will serve to illustrate how humbling their experience must have been. In performing the score for one producer, Larry volunteered the remark that the opening number, "Heigh-ho, Lackaday," was a charming pastorale. The producer burst out laughing. No wonder even Dick was "tired of doing amateur shows and more than a little discouraged with the whole field. Don't forget I'd reached the age of twenty-two, and I wasn't earning any money, and I had to make my mind up whether I wanted to go along with another amateur show or . . . well, I had been offered a job in the commercial line, babies' underwear, as a matter of fact."

A friend of Dick's knew a Mr. Marvin, a babies' underwear wholesaler who, although unmarried and fairly young, was looking around for someone he could train to take over his business when he retired. It was a one-man operation; Marvin did all the buying, traveling, and selling. Without telling Larry, without telling anyone, Dick went to see Marvin, who took an immediate shine to him and offered him the job at fifty dollars a week to start. "Something held me back from accepting," Rodgers said, without giving any indication of what the something was. The anticipation of Larry's scorn? The realization that there were few pretty girls in the babies' underwear business? Even so, fifty dollars a week was not to be lightly tossed aside, so Dick asked if he could think it over. Puzzled, Marvin agreed. It was at this point that Dick received the telephone call which was to transform his and Larry's lives.

The caller was Benjamin Kaye, a lawyer and Akron Club member who dabbled in show business; in fact, he had written a song called "Prisms, Plums And Prunes" for *Up Stage and Down*, six years earlier. Kaye told Dick that some of the kids from the Theatre Guild—understudies, bit players, spear carriers all—were putting on a benefit show, and that he'd told them Dick was just the fellow to write the songs for them. "And when Ben Kaye told me about this, it was a little too much for me, because the Theatre Guild was well known, and I couldn't resist it, and I called up the man in the babies' underwear business and told him I wasn't going to be with him, and went into this semi-professional show which turned out to be the *Garrick Gaieties* and put both Larry and me on our feet."

Put like that, it sounds almost predestined. In fact, it was not quite so simple. Edith Meiser, who was a member of the cast and later appeared in several Rodgers and Hart shows, recalled that it came about somewhat differently.

> In those days everybody gave balls for fund-raising things, and everybody went, and everyone subscribed—they were full of fun, everybody dressed to the teeth. All of us who were working for the Guild—because they had many plays, sometimes three plays going concurrently—the small fry, we did takeoffs, not only of the Theatre Guild plays but of other plays, theatrical takeoffs, to amuse the people at these great big balls. And some of our material was written by the very top comedy writers. We were kind of the pets of New York, all we young people, terribly spoiled. We said "Why don't we put on a revue like *The Grand Street Follies?*" [which was] a small satirical revue. "We'll do one for you," meaning the Guild. "Now all we have to do is to find is someone to write the music and lyrics."
>
> Romney Brent and I were appointed a committee. We were supposed to go around and talk to people who could write music and lyrics—as a matter of fact, I was going to write the lyrics. . . . One never knows what one doesn't know when one is young.

Peter Arno, the *New Yorker* cartoonist, who dabbled in songwriting, was among those considered and rejected; it was not until the show was in rehearsal that Benjamin Kaye, "an enchanting old gent," told Edith she ought to talk to a young fellow named Dick Rodgers.

> Romney and I had an appointment to go and see Dick Rodgers. It was in February; Ben Kaye had phoned to set it up. My mother was visiting me, so I borrowed her furs to make an impression. But Romney had a date with this beautiful, beautiful Oriental actress, Anna May Wong. Well, if you think I could get him to break that date to go see Dick Rodgers, well, not at all.
>
> [At that time] I was living in a little walk-up on the East Side, and I went across town to West End Avenue, which was very elegant in those days, and went up to Dr. Rodgers's apartment. I remember it was a day of a February thaw, where everything was dripping; the air was soft but there was still snow on the ground. From the windows of Dick's apartment you could see over to the river which was very hazy. The whole thing was kind of dreamlike.
>
> I walked into this very elegant apartment, and it had an enormous foyer, with a grand piano. Dick was there, and he said, "I'll play you a few of the things I've done for the Varsity Shows up at Columbia." So he did, and he told me, "I did them with Herb Fields and Larry Hart, who is a friend of mine." I said, "Well, you know, I have the ideas for all the songs, so I will naturally be doing some."
>
> We were sitting there round the piano bench together, being very cosy, and he showed me a picture of a show they had done somewhere which was based on the François Villon story, *If I Were King*, in which Dorothy Fields played Huguette. I can remember to this day the picture of her, with the beautiful legs and all that. And he played the music from that, and I wasn't terribly impressed.

The longer Rodgers played, the more Edith began to get the feeling she was wasting her time; none of the songs had that "something" she was looking for. "And then he played 'Manhattan,'" she said. "And I *flipped!* I knew, I knew this was an enormous big hit number. You know, sometimes God sits on your shoulder and you know something that is way over beyond anything you should know." She asked Rodgers if he was interested and, if so, how much of the show he would want to do. All of it, he told her. "And I told him the ideas for the songs, and I said, 'We'll get together and I will write the lyrics'—I'd done a few in my college days."

She went back to the Garrick and jubilantly told the rest of the cast that she'd found the boy to write their show. A few days later, Dick came down to the theatre and played "Manhattan," and everyone felt the way Edith did about it. The first song Dick and Edith started on was "An Old-Fashioned Girl"; after a day or two of kicking ideas around, Rodgers said, "You know, Edith, I usually work with a guy called Larry Hart. Would you let him try a few of the ideas we've worked out?"

"We-ell, all right," Edith said, not a little miffed.

Then he brought in the lyrics!
Well, I did have the sense to know that I was way outclassed, but way out! When those lyrics began to float in, you know, to all these ideas we had, they were so absolutely sensational! I can remember Larry Hart coming into the Garrick Theatre [for the first time], this little bit of an almost-dwarf of a man. I've always said he was the American Toulouse-Lautrec, really. He was that kind of a personality.

But *enchanting!* He had such . . . what is now called charisma. He had such appeal! It was funny really, [because] he was beginning to be bald, even in those days. He had this funny head, this enormous head, and a very heavy beard that had to be shaved twice a day, and big cigars that always stuck out of his mouth, and he would always go round and be washing his hands, this was his great gesture!

Of course, we *adored* him. I mean, Dick we were terribly fond of, but Larry was *adored.* He was a pet; he was something very, very special. That is not to downgrade Dick at all, because we were terribly fond of Dick, too. But Larry was someone you wanted to protect, in a funny way. This funny, little, ugly man who was so dear.

What Edith Meiser didn't know was that when Dick first broached the idea of their doing the show, Larry turned him down flat. He absolutely did not want to do it. He was tired of the amateur circuit, with nothing either on the stage or in the bank to show for it. He didn't give a damn about the prestige of the Theatre Guild—or anyone else's, come to that—and pointed out they would have less than three months to put the whole show together.

On top of that, he told Dick, it was just another revue, just another collection of songs that would be heard for two performances and forgotten. If they were ever going to get anyplace in the musical theatre, it wasn't going to be in revues, but with a book musical, with songs written for specific characters and situations.

They argued. Both of them had terrible tempers. Larry, his sister-in-law said, would jump up and down and scream and tear his hair. Dick wasn't noisy: "the iron fist in the velvet glove," according to Helen Ford. In another context Rodgers said, "Our fights . . . were furious, blasphemous and frequent, but even in our hottest moments, we both knew we were arguing academically and not personally. I think I am quite safe in saying Larry and I never had a single argument with each other."

This time Rodgers won. An appointment was then made for them to audition for Lawrence Langner and Theresa Helburn, directors of the Guild. On an empty stage, Dick played and Larry sang in his bright, piping voice, about going to Yonkers where true love conquers, and to Coney to eat baloney. Next came the song of the brave Musketeers, in the days of good King Louis, whose foes they knocked ker-flooey: songs the boys had performed in front of a dozen or more producers. Unlike Shapiro and Schwab and Dreyfus, however, Langner and Helburn were as enchanted as Edith Meiser had been. They promised to provide the necessary financing—five thousand dollars—and the free use of the Garrick Theatre.

Larry was still reluctant. Only when Dick proposed they incorporate a short, self-contained jazz opera into the revue as a first-act finale would he budge: writing a jazz opera appealed to him. It was a challenge, something new, something different. Typically, once he became involved, he plunged into the show with exuberant energy. Giving the performers the only two songs they had ready—"Manhattan" and "Three Musketeers"—Dick and Larry went to work. In about two weeks they had completed half a dozen new songs; there would eventually be fourteen in all.

The revue, now officially dubbed *The Garrick Gaieties*, with a cast which included Edith Meiser, Philip Loeb, Sterling Holloway, Romney Brent, June Cochrane, Hildegarde Halliday, Lee Strasberg, Betty Starbuck, and Elizabeth (later Libby) Holman, was scheduled for two performances only, a matinee and an evening, on Sunday, May 17, 1925. After a last-minute scare when the police, acting on a complaint by the Sabbath League, tried to close the show down (paid performances were not allowed on Sundays, but Ben Kaye managed to convince them that it really was for charity), the show began. "Soliciting Subscriptions" nailed its colors to the mast for all to see, a bold opener that was but a foretaste of what one critic would call "a witty, boisterous, athletic chow-chow" (if a mixed pickle relish can be said to be witty, boisterous, and athletic).

In the first sketch, Romney Brent, Edith Meiser, and Philip Loeb played Alfred Lunt, Lynn Fontanne, and Dudley Digges in Ben Kaye's lampoon of the current Guild production, Ferenc Molnar's *The Guardsman*. "The Butcher, The Baker, And The Candlestick Maker," a song by Ben Kaye's friend Madame Mana-Zucca (her real name was Agusta Zuccaman—and probably Zuckerman before that—but she had changed it around to make it sound more interesting) came next, followed by Edith Meiser's sketch "The Theatre Guild Enters Heaven," in which Romney Brent as St. Peter rendered judgment on the acceptability of the heroines in recent Guild plays.

Following a scarf dance by Eleanor Shaler, Edith Meiser sang "An Old-Fashioned Girl," the sole effort of the team of Rodgers and Meiser. The first Rodgers and Hart song, "April Fool," performed by Betty Starbuck and Romney Brent, drew a hand big enough to make Rodgers, leading the eleven-piece orchestra, believe he could feel the glow of the audience's enthusiasm.

Next was a spoof of Sidney Howard's play *They Knew What They Wanted*, which had opened at the Garrick six months earlier. It was by Ben Kaye, who called it "They Didn't Know What They Were Getting." By this time, the audience did, and applauded vigorously, as they did for Hildegarde Halliday's impersonation of monologist Ruth Draper. The first-act finale was the piece Larry had fought for, a burlesque on opera in the jazz idiom, set in a department store and called "The Joy Spreader." "The curtain came down," Rodgers said, "on an extremely generous hand."

Act Two opened with "Rancho Mexicana," the production's sole bow to "spectacle," danced by Rose Rolando in a multi-colored set designed by her husband, the distinguished artist Miguel Covarrubias. "Ladies Of The Box Office," which followed, featured Betty Starbuck as Mary Pickford, Libby Holman as a Ziegfeld girl, and June Cochrane as Sadie Thompson, heroine of the hit play *Rain*. Larry had all sorts of fun, writing lyrics that demonstrated brilliantly his flair for putting the social history of the day into his songs.

This was followed by a Morrie Ryskind–Arthur Sullivan sketch about the home life of President and Mrs. Coolidge; Ryskind could still perform it, verbatim, half a century later, and remember the enormous laugh its punch line got. Then Sterling Holloway and June Cochrane sang "Manhattan" in one (that is, in front of an unadorned curtain). There was no question about the audience liking it: the performers had to give several encores, and could have done even more had Larry written extra choruses (a lack he would quickly remedy).

After another sketch Holloway, Brent, and Loeb did the Musketeers number: "Athos, Porthos and Aramis/ We are the kitten's pajamis," followed by Louise Richardson singing the waltz "Do You Love Me, I Wonder?" and Libby Holman doing a torchy version of "Black And White."

The final sketch was a skit on *Fata Morgana*, another arty Guild production which here became "Fate in the Morning." Then musical director Rodgers led the whole cast into the finale, "The Guild Gilded." At its conclusion the audience rose to its feet as one, cheering, stamping, shouting, applauding, whistling. They showed every sign of continuing for an hour, so Rodgers cued the orchestra and cast into a reprise of "Manhattan." And another. And another. By about the tenth reprise, the audience was singing it, too. The stage manager signaled for the house lights. Still the audience lingered, as if reluctant to leave. At last, slowly, they filed out.

Backstage, there was pandemonium, everyone hugging and kissing everyone else. Larry Hart was rushing about, rubbing his hands together and shouting, "This show's gonna run a year! It's gonna run a year!"

His words cast a slight pall on the celebration. The show would not run a year, or even a week. There was going to be only one more performance, that evening; and that would be it. However, when the evening performance was greeted with the same enthusiasm, and the reviews the following day were almost unanimous in their praise, Theresa Helburn agreed to further matinees on the days that *The Guardsman* wasn't playing. There were six more special performances before the Theatre Guild ended the successful run of the Lunt-Fontanne play and turned the theatre over to *The Garrick Gaieties*, commencing June 8.

The principal collaborators now received a small percentage of the gross—in the case of Rodgers and Hart, about fifty dollars each per week. In addition, seven of their songs were bought for publication by Edward B. Marks, "Manhattan," in particular, making a welcome addition to their income. *The Garrick Gaieties* went on to notch up a very creditable 211 performances.

Ironically enough, the "jazz opera" which had been Larry's main reason for agreeing to do the show was dropped for the regular run, along with some other material that included the song "Black And White" and two sketches. In their places went a new sketch by Morrie Ryskind and Phil Loeb called "And Thereby Hangs A Tail," with lyrics by Hart, plus two new songs, "On With The Dance" and "Sentimental Me," and two extra choruses of "Manhattan."

"I can remember, Larry was so very good to us always," Edith Meiser recalled. "If his father was having one of his flush times, Larry would hire a limousine and take us driving, all through the night, to keep cool on those hot, hot summer nights, and wherever we went we sang our lungs out, being kids, we sang like mad. We'd drive out to Long Island and see the dawn, and then come back and go to sleep, and then go to the theatre that night. It was lovely. Lovely, lovely, lovely!"

10

Crest of a Wave

*A*ll at once, Rodgers and Hart were news. Their picture was in the *Brooklyn Daily Times*. "Manhattan" was an enormous hit, the first popular song whose lyrics made headlines nation-wide. "Contagious," said the *Evening Graphic*. The *Morning Telegraph* reported the "unusual lyric" was being reprinted in newspapers all over the country and bringing the city back into the limelight. "The most popular piece in New York," said the *Atlanta Constitution*. "You hear it played wherever you go."

In an interview in the *New York Herald Tribune*, Larry professed surprise that a song with such intricate and elaborate rhymes should be so successful. "The song hit of the show," he pronounced, sticking it to Tin Pan Alley's moon-June-croon-tune brigade with huge relish, "is usually a very simple one with monosyllabic words."

It wasn't only "Manhattan" that became popular; the *Morning Telegraph* also reported that the deliberately unsentimental "Sentimental Me" had the "recording managers of the big phonograph companies breaking their necks to put the record on the market." As if that weren't enough, the producers of the revue *The Greenwich Village Follies of 1924*, which had starred the flamboyant Dolly Sisters, interpolated both songs into the score (mostly written by Owen Murphy and Jay Gorney, but including Cole Porter's song "I'm In Love Again") to beef up the show when it went on tour.

On the crest of a wave, Rodgers and Hart also contributed a new song, "Anytime, Anywhere, Anyhow," to *June Days*, a musicalized version of a frothy 1920 comedy called *The Charm School*. Most of the score was by new-comer J. Fred Coots; quite why Rodgers and Hart wrote an interpolation is unclear, but it must have given Dick special satisfaction to recall that Jerome Kern had written incidental songs for the original play.

As if in response to the success of *The Garrick Gaieties*, things began to happen with *Sweet Rebel*, now retitled *Dear Enemy*. And the one who made them happen was Helen Ford, still carrying the script with her everywhere she went, still trying to find a backer with enough money to put on the show:

> I moved away from the Algonquin. Somewhere in the back of my mind was the notion that, since everybody there was in the same business, trying to do the same thing, I'd never get anyone to put up money for the show there. So I moved to the brand-new Roosevelt Hotel . . . because I thought, I may meet people who have money.

One day she came down in the elevator, and a middle-aged man said, "How do you do, Miss Ford," and asked her what she was doing and whether she had been back to Canada lately. Not sure whether she knew him or whether he was trying to pick her up, Helen asked him if he was from there.

> He said, "No, I'm from New Hampshire." And I told him my husband had gone to school up there, at Dartmouth. "Not Gink Ford!" he said. And it turned out he was Robert Jackson, the brother of [my husband] George's roommate at college. We stood in front of the elevator and talked, and he asked me what I was doing. "I see you have a script under your arm," he said. (Actually, I had been carrying it around with me every day for about a year.) So I said, "Yes, do you know anybody who'd want to put money in a show?" And he said, "Well, I might."
>
> So anyway, that was the beginning. When he met the boys, and heard the music, he was just absolutely in love with the whole thing. But he would drag us, the three boys and myself, to nightclubs night after night—and we couldn't say no. I think that was what taught me I hated nightclubs.

Jackson, a lawyer who had made a million dollars selling his family's grocery store chain, finally agreed to back the entire show. When she told Larry Hart the news he took hold of her hands and danced her around, ring-a-rosy style, chanting, "We've found a butter-and-egg man, we've found a butter-and-egg man!"

As so many of his lyrics testify, Larry had an acute ear for contemporary turns of phrase. This one had come into use early in 1925 when Texas Guinan told her clientele that a big spender at her nightclub who kept insisting on setting them up for the house was "a big butter-and-egg man from the West." The phrase attached itself to any out-of-towner with a big bankroll. George S. Kaufman used it later that year when he wrote his first solo hit, a comedy (some say based on his experiences with Jessel and LeMaire in *Helen of Troy, New York*) about how a hayseed comes to New York and beats the city slickers at their own game.

Now that they had a backer, George Ford returned to New York and agreed to become the producer of *Sweet Rebel*. Because so many people told him that a musical about the American Revolution wouldn't work, he decided to run it through someplace a long way from Broadway. He got in touch with his brother Harry, director of a stock company at the Colonial Theatre in Akron, Ohio, who took the proposition to his boss, a man who clearly knew how to drive a mean bargain; he agreed to let them use the theatre for a week on condition that Helen Ford appear there in *The Gingham Girl*—free! They had little choice but to agree. After Helen completed her engagement in mid-July, they went out to Akron and rehearsed for a week, playing one performance of the show on Monday, July 20.

They certainly picked a week to put on a show: down in Dayton, Tennessee, population fifteen hundred, the legal charade which became universally known as the "Monkey Trial" was being acted out, and all America watched agog. The defendant was John T. Scopes, a freckle-faced, twenty-four-year-old science teacher. The charge was that he had taught contrary to a recently enacted Tennessee statute which decreed it a crime to "teach any theory that denies the story of the Divine creation of man as taught in the Bible."

The trial—Fundamentalism vs. Evolution—turned into a farce. Prosecutor William Jennings Bryan, one-time Secretary of State, three-time Presidential candidate, all-time Defender of the Faith, was ruthlessly revealed to be a pompous mountebank by Scopes's defense attorney, Clarence Darrow, the Great Infidel. The jury brought in its verdict the same Monday *Sweet Rebel*—now retitled *Dear Enemy*—played Akron: guilty as charged. Judge John T. Raulston, "jist a reg'lar, mountaineer jedge," as he called himself, slapped a hundred-dollar fine on Scopes. Darwin was vanquished; Genesis had triumphed. And the rest of America laughed itself silly.

The Akron audience's reaction was sufficiently encouraging to make George, Helen, Larry, and Dick decide to go ahead with their plan to mount an elaborate Broadway production of their show. There was still a great deal of work to do: the second act was weak; there was no finale; new songs were needed; others—"How Can We Help But Miss You?," "Ale, Ale, Ale," "The Pipes Of Pansy," "Dear Me," and "Dearest Enemy"—were discarded.

That last number gave the show its new name when, in September, they did a week of tryouts at Ford's Opera House in Baltimore. The big difference now was that John Murray Anderson, free of his Music Box commitment and encouraged by the fact that they had financing, agreed to direct the show, as well as being in charge of its overall staging (his credit reads "Entire production under the personal direction of John Murray Anderson"). Because of his participation *Dearest Enemy* had a visual style which set it apart right from the start. For instance, Helen Ford said:

For the opening number of the second act, "I'd Like To Hide It," we had four little girls, four mediums, and four showgirls, big girls, you know. I wore a black
dress and a close white wig in front of the other girls, who had on lemon yellow
taffeta dresses, with white wigs that had cherry red ribbon bows on them that went
right down in back. And with that Colonial set, and the staircase that came down
in a circle to the platform on the stage, it looked just so delightful, so pretty.

They opened in Baltimore on Labor Day weekend, September 7. Nearly
fifty years later, what Helen Ford still remembered most about that week was
the heat: when she tried on the costumes, she so wet them with perspiration
that putting them on again was sheer agony. There were other disasters.
Twenty minutes before he was due to lead the orchestra on opening night,
Rodgers was knocked cold by a can of peaches that fell off a shelf in a restaurant. During the first performance, as she started her stage fall in the scene
where Sir John shoots out the light with which she is signaling the American
troops, Helen caught her foot in her negligée and fell down for real; had it
not been for the bun on the wig she had not had time to change, she would
have cracked her skull.

Throughout that frantic week there were changes and rewrites and dele-
tions and interpolations. Orchestra rehearsals were held in a burlesque house
across the street. George Ford recalled that there were much longer lines of
people waiting at the burlesque house than there were for *Dearest Enemy*.
They kept cutting and rewriting, until finally the show had a beginning, a
middle, and a rousing finale—devised by George and Harry—in which General Washington reunites the star-crossed lovers.

Dearest Enemy opened at New York's Knickerbocker Theatre on Friday,
September 18, 1925. As she had known it would, Helen Ford's entrance in
the play's opening scene—apparently clad only in a barrel—proved a sensation. The show opened during one of the most astonishing weeks in the history of the American musical theatre. Two nights before, on September 16,
the out-of-town smash *No, No, Nanette* opened at the Globe. On Monday
the 21st, Rudolf Friml's full-blooded, romantic *The Vagabond King* made
Dennis King a star overnight. The next evening, the curtain went up on
Marilyn Miller at the New Amsterdam in *Sunny*, the first collaborative effort
of Oscar Hammerstein II, Otto Harbach, and Jerome Kern. It might be
more than fairly said that this week witnessed the dawning of the golden age
of the American musical.

No, No, Nanette produced two hits that have become standards: "Tea For
Two" and "I Want To Be Happy." *Sunny*, too, had several lovely, lilting

melodies: "Who?," "D'ye Love Me?," and "Two Little Bluebirds." Rudolf Friml's "Only A Rose" and "Someday" from *The Vagabond King* have become staple amateur operatic society fare. Yet even the best-known of the Rodgers and Hart songs from *Dearest Enemy* is rarely heard today. "Here In My Arms" had a brief vogue in New York and later in London, but the rest of the score has been largely forgotten.

For all that, it contained some delights. The opening song, "Heigh-ho, Lackaday," in which Mrs. Murray's young charges bemoan the absence of their sweethearts, must have given Larry and Dick especial pleasure, for this was the "charming pastorale" which had once been the butt of such derision. From the story of those strange inhabitants of the Bronx known as "The Hermits" to the rousing comedy ensemble about wooden-legged "Sweet Peter" Stuyvesant, Larry's lyrics were subtle, inventive, and sophisticated, and Dick matched him melodically throughout.

Dearest Enemy ran out the season, and if its 286 performances weren't quite up to *Nanette*'s 321, *The Vagabond King*'s 511, or *Sunny*'s 521, its success put Larry Hart and Dick Rodgers very firmly on the map. On November 5, they appeared on the *Morning Telegraph*'s weekly "Up and Down Broadway" radio feature from the eighteenth floor of the Roosevelt Hotel. With Dick at the piano, Larry performed "a medley from the songs the two had written for their three productions," ending with "The Hermits."

Next, Max Dreyfus summoned them to his office, crossly demanded to know why they hadn't brought their earlier songs to him, and invited them to join his stellar group of composers at Harms. Exercising more self-control than many people might have demonstrated in the same circumstances, Rodgers refrained from reminding Dreyfus of his earlier rejections. So T. B. Harms became their publisher, and ASCAP—the then-ultra-exclusive American Society of Composers, Authors, and Publishers—invited them to become members. Dick recalled that he treated himself to a new car, but otherwise his life remained much as it had been. As for Larry, "he was the same, sweet, self-destructive kid I had always known," said Rodgers.

And if you wanted to be self-destructive, there were plenty of people around to help you do it: at the Dizzy Club or the Aquarium, the Hotsy-Totsy, Texas Guinan's, Tony's East 53rd or Tony's West 49th, the Clamhouse, the West 44th Street Club or the Richmond, Frank and Jack's, Felix's, Louis's 21 West 43rd Street, Jack Delaney's, Billy Duffy's, McDermott's, Sam Schwartz's, the Type and Print Club, the Bandbox, and half a hundred more speakeasies, blind pigs, gin joints, and beer flats scattered all over the great

big city. America was afloat on an ocean of bootleg alcohol, a lot of it concocted by money-hungry gangsters who didn't give a damn what was in it or
who died from drinking it.

The social whirl was just as mad: the upper bracket rubbed shoulders with
the underworld, ermine-clad society ladies danced in smoky speaks. Gold
diggers had their sugar daddies; *Follies* stars married gangsters. And, as Edith
Meiser recalled, there were parties, parties, parties,

> dozens of them, huge affairs. The Beaux Arts Ball, of course. Ezra Winters, a
> mural artist with an enormous studio at the top of Grand Central Station, threw
> big, big, big parties. Everybody in New York went to the Palace on Mondays, to
> see the opening. There were always parties, always, always, always. And then there
> were tango Mondays at the Fifth Avenue Theatre, tea dances at the Biltmore,
> dances every Saturday night at the Ritz for show people, down in the old Crystal
> Room with its big staircase so you could make a grand entrance and you and your
> escort paid five dollars which included supper, and everybody in the theatre went
> to them, everybody danced like mad. And of course there were the great
> speakeasies, like Tony's on the West Side, where all the literati went, the Bench-
> leys and Woollcotts and Kaufmans. In back there was a little kitchen, and if Tony
> and his wife were really fond of you they'd give you a good hunk of spaghetti.
> Tony was very good to all of us. He used to keep an eye on how much we were
> drinking, and when he thought we'd had enough, he'd come over and say, "You
> can stay but you've had your quota." Boy, did he ride herd on us—he was the
> demon chaperone!

Rodgers wasn't much of a night owl; Larry Hart didn't come to life until
the lights came on. His favorite hangout was Tony Soma's place on West
51st Street, the same *boîte* where Edith Meiser and the other youngsters went
to eat. "He wasn't a 21-er. He wasn't even a Stork Clubber," his sister-in-law
said. "He went to Tony's, or to little bars. He wasn't the kind of man who
could sit down at a table for four hours. Dinty Moore's, he loved, and went
there a lot. Ralph's [now Barrymore's], Louis Bergen's on West 45th Street,
little places. He didn't want to go places to be seen. He couldn't have cared
less about all that."

It was no secret that Larry's idea of fun was getting tight. According to
expert witnesses, however, he was not a heavy drinker. Being such a little
guy, they say, it took only two or three cocktails to knock him for a loop.
Some indication that Larry was having problems with liquor even then
appears in a letter he wrote to Ira Gershwin shortly after *Tip-Toes* opened at
the Liberty Theatre on December 28, 1925:

> When, the other night at the [Theatre] Guild's menagerie, Joe Meyer told me a
> departing guest was Ira Gershwin, I should have brushed aside your friends,
> grasped you by the hand, and told you how much I liked the lyrics of "Tip Toes"

but, probably because the circus clowns inspired a speedy retreat from a too acute consciousness, I had imbibed more cocktails than is my wont, and so when the coffee-loving Mr. [Joseph] Meyer pointed you out, I could only say "Zat so!"

Your lyrics, however, gave me as much pleasure as Mr. George Gershwin's music and the utterly charming performance of Miss Queenie Smith. I have heard none so good this many a day. I wanted to write to you right after I had seen the show, but—well, I didn't rush up to you at the Guild circus, either.

It is a great pleasure to live at a time when light amusement in this country is at last losing its brutally cretinous aspect. Such delicacies as your jingles prove that songs can be both popular and intelligent. May I take the liberty of saying that your rhymes show a healthy improvement over those in "Lady Be Good."

You have helped to make an evening delightful to me—and I am very grateful. Thank you! And may your success continue!

The Queenie Smith mentioned in Larry's letter was the same actress who had stolen the show from Helen Ford (of Troy, New York), now a star in her own right. *Tip-Toes*, which is also remembered for having introduced Jeanette MacDonald to the Broadway stage, featured largely second-drawer Gershwin songs, the best of which were "That Certain Feeling" and "Looking For A Boy."

Cut to Billy Rose, working at getting out of the smoke-filled world of the Back Stage Club and the bootleg circuit in hot pursuit of something he would never have: class. He was spending fifty thousand dollars remodeling and redecorating the second floor of 683 Fifth Avenue, at the corner of 54th Street, site of the former Criterion Club, which he planned to make a theatre-restaurant-club for the carriage trade. The joint would be strictly legit, he told Larry: no booze (people would bring their own); a five-dollar cover charge (people believed they were getting something good if you overcharged them); no service during the show, which would be a Broadway-type revue starting at 12:40 a.m. And he wanted it to have songs by Rodgers and Hart.

Was Rose calling the debt that went back to when he put up the money for *The Melody Man*? It's hard to imagine Rodgers working for him for any other reason, even for Hart's sake. And even if—as seems likely—Larry was the prime mover in this collaboration, he was far too intelligent not to have known that Billy Rose was using him, just as he used everyone else. But being "exploited" didn't bother Larry—in fact, it amused him.

One time an actor told Larry he desperately needed five hundred dollars to option a play which could make him a star; he just *had* to get that play. Larry lent him the money. Later on, he found out the actor had used it to pay for an abortion for his girlfriend. Larry hooted with laughter: that was the kind of *chutzpah* he admired.

So it was agreed: Rodgers and Hart produced nine songs for the show, which Billy grandly (and as it turned out, accurately) titled *The Fifth Avenue Follies*. Conceived and directed by Seymour Felix—some indication of Billy Rose's ambitions may be had from the fact that Felix, a former child vaude-ville star who had made himself a name as a choreographer, would not work unless he got a five-thousand-dollar advance—the show was a hodgepodge of sketches and songs. Some were topical, like "Lillie, Lawrence and Jack" (the three Britishers were a big hit on Broadway in 1924's *Charlot's Revue)* and "Where's That Little Girl?" (with its nod to Michael Arlen's successful novel *The Green Hat)*. Others were specialty numbers, like "In The Name of Art," "Mike," "Susie," and the much-reprised "Do You Notice Anything?"

Although the reviews were not bad, after the first night in late January 1926, when many public figures turned up, the customers stayed away in droves. Billy Rose cut his losses and sold the club to a bootlegger who bought it for a girlfriend. Later, the police closed it. Rodgers said he and Larry Hart never collected a dime from Rose for their efforts. Maybe Billy figured they owed him: he had that kind of mind.

By the time the club folded, Rodgers and Hart were already involved in other things. The *Herald Tribune* of March 7 reported that Guy Bolton was writing a new show called *Dancing Time* for the English team of Jack Hulbert and Cicely Courtneidge, which would have songs by Rodgers and Hart; whether this was a very early version of *Lido Lady* is not apparent. A few days later, *Variety* indicated the change in their fortunes: Larry and Dick, it sug-gested, were making around five thousand dollars a week, already in the same financial league as George Gershwin and Vincent Youmans. There was a feature about them and their work in the *Herald Tribune*; the *New York Times* called Herb, Dick, and Larry "the Oncoming Triumvirate." Their photograph was in the Atlantic City *Illustrated Boardwalk News*; Larry and Dick were trying out a new show there.

This time Herb Fields was writing with the husband-and-wife team of Eva Puck and Sammy White in mind. The plot was almost mindless: Leonard Silver (White), a Long Island dairyman, hopes to become a great six-day bicycle rider. His girlfriend Mollie (Puck) is his trainer, manager, and pro-moter. A professional manager and his scheming sister try to get Leonard to ride for them, but true love wins the day.

The trio had first taken the show to Lew Fields, whose producing career was in the doldrums. He hadn't cared for *Dearest Enemy*, but he liked this show, which was called *The Girl Friend*, and agreed to produce it. Tryouts were scheduled for March 8 at the Apollo Theatre, and Dick and Larry produced sixteen songs, among them "The Pipes Of Pansy," from the original *Dear Enemy*, and "I'd Like To Take You Home," which had been in *Bad Habits of 1925*.

After five days in Atlantic City, they cut "The Pipes Of Pansy" and "Sleepyhead," another song from the second *Garrick Gaieties*, then brought the show into Lew Fields's Vanderbilt. It was "a tiny little theatre," Edith Meiser recalled, which "only seated seven hundred people, but that of course was not unusual in those days. The top ticket wasn't much more than five dollars."

The Girl Friend opened on March 17, 1926: bright, breezy, and completely of the moment. After the bucolic opening numbers by the ensemble ("Hey, Hey" and "The Simple Life"), Eva Puck and Sammy White introduced "The Girl Friend," with its danceable, singable Charleston-style rhythm. In the next scene, the principals introduced "The Blue Room." Deceptively simple-sounding, it is actually a complex and distinctive melody, complemented perfectly by Larry's triple rhymes—"blue room," "new room," and "for two room"; "ballroom," "small room," and "hall room." Especially notable was the way in which it then so effortlessly avoided the expected repetition in the release, linking "trousseau" to "Robinson Crusoe" and, in the last line—so reminiscent of Schubert's "Moment Musicale"—neatly tying the earlier "little blue chairs" and "worldly cares" with the blue room "far away upstairs."

The rest of the score contained no lasting successes, although one wonders whether Larry knew that George Kaufman and Herman Mankiewicz were writing a play called *The Good Fellow* when he composed the lyric for "Good Fellow Mine." A raucous comedy number Larry wrote for diminutive Eva Puck stopped the show. Called "The Damsel Who Done All The Dirt," it suggested that Washington's real reason for crossing the Delaware was to meet a Jersey flapper, and that when Columbus headed for America it was to flee the King's wrath at his romancing Queen Isabella. Throwing in references to Sophie Tucker and Patrick Henry, Pharaoh and Fanny Brice, Ulysses and Napoleon, the whole thing was full of gloriously outrageous rhymes.

Twelve days later the newspapers noted that "Larry Hart, lyricist of *The Girl Friend*, is writing two new choruses for Eva Puck's comedy song 'The Damsel Who Done All The Dirt.' Larry Hart originally wrote three choruses, but audiences at the Vanderbilt love the song so much that he has been forced to compose additional lines." We even have a—somewhat tongue-in-cheek—word picture of him doing it, as provided by Dick Rodgers to a *Herald Tribune* reporter on March 21, 1926, four days after the opening:

I used to wonder how Larry Hart did write that song ["Manhattan"]. Now I know. He wrote it on a dirty envelope four minutes and twelve seconds before the show opened at the Guild Theater last June. That, I think, is the truth. Larry says he really worked on the song, polished it as it grew from a tadpole to a frog. I don't believe it. I saw him write a sparkling stanza to "The Girl Friend" in a hot, smelly rehearsal hall, with chorus girls pounding out jazz time, and principals shouting out their lines. In half an hour he fashioned something with so many healthy chuckles in it that I just couldn't believe he had written it in one evening.

Whether Larry ever actually wrote those extra choruses is another matter: only three seem to have survived.

In spite of quite favorable reviews, however, the public did not seem to like *The Girl Friend* at all. After a poor first two weeks, Lew Fields began to wonder aloud if they should cut their losses and close. Larry, Dick, and Herb offered to waive their royalties if Fields would give the show a chance. He agreed, and after a couple of weeks more, word of mouth and the success of the title song and "The Blue Room" brought the audiences around. *The Girl Friend* ended up with a nine-month run of 301 performances, closing on December 4 before beginning a successful eight-city tour. The boys from Morningside Heights had another hit.

11

More Gaieties

On December 28, 1925, a bright London revue called *By the Way* starring an English couple, Jack Hulbert and Cicely Court-neidge, opened at the New York Gaiety. The show was the lantern-jawed Hulbert's first venture into producing, and while he was in New York, he and his partner Paul Murray scouted Broadway looking for shows they could take across to London.

There was plenty to look at: not only *No, No, Nanette, Sunny,* and *The Vagabond King* but also the Shuberts' lavish production of *Princess Flavia,* a musical version of *The Prisoner of Zenda* scored by Sigmund Romberg which starred Harry Welchman and Evelyn Herbert. Larry, Dick, and Herb must have gone to see it with more than casual curiosity.

At the Selwyn, Bea Lillie, Gertrude Lawrence, and Jack Buchanan ("Lillie, Lawrence and Jack") were appearing in *Charlot's Revue of 1926,* which featured the hit interpolation "A Cup Of Coffee, A Sandwich, And You," by Joe Meyer, Al Dubin, and Billy Rose. *The Greenwich Village Follies* was at the 46th Street Theatre, and two Gershwin shows, *Tip Toes* and *Song of the Flame,* had opened within two nights of each other, the former on the same evening as *By the Way.*

At the Lyric, Irving Berlin and George S. Kaufman (aided, without credit, by Morrie Ryskind) had a big hit with *The Cocoanuts,* starring the Marx Brothers, a show famous for being the one that didn't have "Always" in the score. There are two stories about how that song was born. One has it that Berlin's musical secretary, Arthur Johnston, had a girlfriend called Mona who heard them talking about how much money a hit song could make. She asked

Berlin if he could write a song about her. And he said, sure, one of these days. "Why not right now?" chirped Mona, whereupon Irving hummed, and Arthur jotted down, a tune called "I'll Be Loving You, Mona" on a paper napkin.

The other story, which sounds much more authentic, suggests Berlin sat up all night and wrote the song while in Atlantic City working on the show. In typical fashion, he couldn't wait for a reaction and shook George Kaufman awake to hear it. Kaufman didn't get up right away—he confessed that five in the morning was not one of his best times—and suggested, not altogether flippantly, that always was a long time for a romance, and that maybe "I'll be loving you Thursday" would be nearer to real life.

This was not at all what Irving wanted to hear, and thereupon he decided—to Kaufman's intense displeasure—not to use the song in the score. The acerbic Kaufman remarked that the songs Irving did write for the show were so forgettable they didn't even reprise them in the second act: the actors couldn't remember them that long.

Perhaps the real reason Berlin's songs for *The Cocoanuts* weren't up to his usual high standard was that he was in love, and his particular path to it was rockier than most. The story of his courtship of Ellin Mackay, daughter of one of the wealthiest men in America, Clarence Mackay (whose father had left him a fortune estimated between thirty and sixty million dollars), reads like something that might have been concocted by Damon Runyon on speed.

"The day you marry my daughter, I'll disinherit her!" Mackay is supposed to have told Irving. "The day I marry your daughter, I'll settle two million on her!" Irving is said to have retorted.

And so on. So much so on, that there were skits about them in Broadway shows. Four days after the New Year arrived in New York with fog and rain and the temperature in the forties, Irving Berlin and Ellin Mackay took the subway—the first time she had ever seen such a conveyance, much less ridden it—to Municipal Hall where, amid uproar, they were married. They fled to Atlantic City and literally hid; there, they heard the news that Clarence Mackay had cut Ellin out of his will—a ten-million-dollar insult. Countering it, Irving dedicated "Always" to Ellin and assigned its copyright to her, suggesting that this was the true reason he had withdrawn it from the Marx Brothers show—he smelled a big hit, and he was right. "Always" went on to earn three hundred thousand dollars in royalties in its first year, and seventy years later it's still going strong.

On Friday evening, January 6, Alexander Woollcott—whose adoring biography of Berlin had been published the preceding year—gave the couple supper at his West 47th Street apartment, from which they went directly to the liner *Leviathan* at Pier 46. The next day, at 1:00 p.m., they sailed off on a two-and-a-half month honeymoon tour of Europe, while behind them Tin Pan Alley ground its gears and came up with Al Dubin and Jimmy

McHugh's "When The Kid Who Came From The East Side Found A Sweet Society Rose."

⌦━◈━⌫

On Broadway, meanwhile, it was business as usual. That spring, Hulbert and Murray could have seen such dramatic hits as *Cradle Snatchers*, which had premiered September 7 and lifted young Humphrey Bogart to stardom; George Jessel in *The Jazz Singer* (September 14); the Pulitzer Prize play *Craig's Wife* (October 12) by George Kelly; Victor Moore and Otto Kruger as a pair of lovable crooks in *Easy Come, Easy Go* (October 26) by Owen Davis; Eugene O'Neill's own production of *The Great God Brown* (January 23); tennis champion Bill Tilden making his stage debut in *Don Q., Jr.* (January 27); a stage version of the preceding year's F. Scott Fitzgerald novel *The Great Gatsby* (February 2); and the definitive 1920s camp melodrama *The Shanghai Gesture* (February 1), in which Florence Reed gave her most famous performance, as Mother Goddam.

If they had been *very* quick indeed, Hulbert and Murray might also have caught a new play by Larry's friend Henry Myers. *Me*, which opened at the Princess on November 23, 1925, lasted only thirty-two performances.

While the particularly American appeal of *Dearest Enemy* seemed to them unlikely to survive the Atlantic crossing, Hulbert and Murray liked some of the Rodgers and Hart songs from it so much—particularly "Here In My Arms"—that they had already invited Larry and Dick to do the score for a libretto in which Hulbert was planning to star. This was the *Dancing Time* project written by Guy Bolton. Clearly that didn't work out, because the show the boys eventually wrote was *Lido Lady* with a book by Ronald Jeans based on an unproduced book by Bolton, Harry Ruby, and Bert Kalmar.

The story, which also went by the name *The Love Champion* for a while, was no strain on the intellect. It concerned the tribulations of a duffer at sports, played by Hulbert, who wants to marry the lady of the title, a tennis champion. The problem is that her father is determined she will marry a man who is as brilliant an athlete as she. Love, of course, conquers all.

⌦━◈━⌫

Although it meant working for the first time away from Herbert Fields, Rodgers and Hart accepted the commission and made arrangements to go to London as soon as they were finished with their next assignment, a new edition of *The Garrick Gaieties*. Larry and Dick had been reluctant to do a follow-up, because the element of spontaneity and youthful insouciance the first show had traded on would be missing. But Terry Helburn and Lawrence Langner of the Theatre Guild won them over by pointing out that revues, from the *Ziegfeld Follies* on down, were doing big business, and by promising them star billing. Larry and Dick argued no more: star billing at the Theatre Guild was very attractive. Working with many of the same talented young people—Edith Meiser, Romney Brent, Philip Loeb, Hildegarde Halliday, Betty Starbuck, Sterling Holloway—they set out once more to kid the theatrical profession in general and the Theatre Guild in particular.

"Most of the *Gaieties* has been written in rehearsal," said Hart. "We had only two numbers ready for last year's *Gaieties* when they started rehearsing. We got them off on those two numbers and then kept about a day ahead of them all through the five weeks [of] rehearsal."

> Our musical comedy burlesque ["Rose of Arizona"] in the current *Gaieties* was done in half an hour. . . . We began with the idea of a burlesque of the "Vagabond Song" from *The Vagabond King*. We liked "To hell with Mexico" for a refrain and started off with that. Then of course the scene had to be on the American border, and from there we were off with the flag flying. The sentimental refrain, the flower number—after the Winter Gardens pattern—and that's all there was to it. I scribbled off the lyrics while Herb wrote the lines, and Dick didn't even sit down at the piano—he just wrote off the notes.

As well as "Rose of Arizona," which contained six songs, Rodgers and Hart produced eleven songs for what would become *The Garrick Gaieties of 1926*, including a recycled version of "Lillie, Lawrence and Jack," and "Sleepyhead," which had been dropped from *The Girl Friend* in Atlantic City. More notable, pun intended, was "Keys To Heaven"—a "hot" number whose opening bars might easily be mistaken for an early version of George and Ira Gershwin's 1930 show-stopper "I Got Rhythm."

They opened at the Garrick on May 10, 1926. As in the first *Gaieties*, in which the program had boasted that the show had neither principals nor principles, the audience again knew exactly where it stood, despite the cast's musical assertion "We can't be as good as last year."

The show began with "Six Little Plays," in which Theatre Guild "mortician" Philip Loeb bemoaned the sad fate of the year's productions: *Arms and The Man* (represented by Betty Starbuck); *The Glass Slipper* (Blanche Fleming); *Merchants of Glory* (Jack Edwards); *Androcles and the Lion* (Romney Brent); *Goat Song* (William M. Griffith); and *The Chief Thing* (Edith Meiser). After a Ben Kaye sketch, Sterling Holloway and Bobbie Perkins (June Cochrane was now in *The Girl Friend*) strolled on to sing, not this time about the delights of Manhattan, but the pleasure of getting away from it.

"Mountain Greenery," with its bright, insistent triplets, is quintessential Rodgers and Hart; throughout, Larry's multiple rhymes marry perfectly with every nuance of Rodgers' lilting melody. Yet rarely does anyone remark on the fresh uniqueness of the title or the rhymes Larry contrived to follow it: "greenery" and "scenery," "keener re(ception)" and "beanery," "cleaner re(treat)" and "machinery." Later he would add others even more inventive, such as "life its tone" / "Heifetz tone," and "map her own" / "chaperone."

The first-act finale, "Rose of Arizona," was a full-blooded spoof of every kind of musical in sight, from Ziegfeld's showgirls to risqué vaudeville comedians, from *Sally* to *The Vagabond King*, and especially of *Rio Rita*, a show that would not even open until the following season. Nothing was sacred: in "It May Rain (When The Sun Stops Shining)" Rodgers actually went so far as to parody his idol Jerome Kern's song "Till The Clouds Roll By," and Hart obliged with boneheadedly optimistic lyrics of the same "some sunny day love will find a way" variety.

Then, best of all, came the rousing finale of "Rose," so perfectly, so mercilessly guying the stentorian—and equally empty-headed—male choruses of Shubert productions like *The Vagabond King* and *The Student Prince in Heidelberg*. No one who had ever heard Rudolf Friml's "To Hell With Burgundy!" could suppress a giggle when the butch male chorus exhorted Shriners and Elks and Pythian Knights and Babbitts of low degree to fight for law and order, to follow 'cross the border, and "To Hell With Mexico!"

Even so, some of Larry and Dick's misgivings were realized; the reviews were less than enthusiastic. In spite of them, the show ran for almost six months, but it taught Rodgers and Hart the valuable lesson of never trying to beat themselves at their own game.

Pausing only for Dick to record some songs from *The Girl Friend* for Ampico piano rolls, and to close *Dearest Enemy* at the Knickerbocker Theatre and send it off on a twelve-city tour, Rodgers and Hart embarked on the flagship of the Lloyd-Sabando Italian Line, *Conte Biancamano*, for Naples. Larry delighted in rolling the ship's name around his tongue as if he were speaking Italian. He had a mixed crossing: when he wasn't hanging over the rail, Rodgers recorded, he was hanging out at the bar.

They went from Naples to Venice, where they met Noël Coward, who introduced them to Linda and Cole Porter. Although he had not yet made his major breakthrough on the Broadway stage, Porter charmingly performed some of his songs for Larry and Dick after dinner at his home, the celebrated Palazzo Rezzonica, where Robert Browning had died.

Larry Hart was a terrible tourist. He no sooner arrived than he wanted to leave. Mountains to him were just a lot of rocks, and he was more aware of the fleas than the beauties of the Costiera Amalfitana. Broadway was and always would be his beat: he was lost without the noise, the bars, the Runyonesque mixture of sharpies and deadbeats that provided his daily fix. He and Dick had a very dull time in Paris, and they headed for England with something like relief. At least the natives were friendly and spoke more or less the same language.

Their fame had, in a small way, preceded them: one of the numbers from *Fifth Avenue Follies*, "Maybe It's Me" (with a new lyric by Donovan Parsons), became the first of their songs to be sung in a British show. As "I'm Crazy 'Bout The Charleston," it was performed at the Palace, Manchester, on March 17, 1926, in *Charles B. Cochran's Revue of 1926*, starring Hermione Baddeley and Elizabeth Hines (who had sung "Anytime, Anywhere, Anyhow" in *June Days* and would soon be involved in a bitter lawsuit with Florenz Ziegfeld). Soon after the show got to the London Pavilion in April, the song was dropped to make room for a spot by Will Rogers, making his first appearance in the British capital.

London turned out to be as much of a disappointment to Rodgers and Hart as Paris had been. Part of the reason was that the libretto of the show they had come to write, *Lido Lady*, was already finished, so there were none of the give-and-take sessions Larry and Dick were accustomed to having with Herb Fields. They were further dismayed to discover that the leading lady, Phyllis Dare, was considerably longer in the tooth than they had been led to believe. And worst of all, their producers, Jack Hulbert and Paul Murray, virtually ignored them. Some indication of how this impasse came about may be gleaned from Hulbert's autobiography, in which he refers condescendingly to Larry and Dick as "my two college boys."

All this is not to say they didn't take the job seriously. "It occurred to us," Larry wrote, "that we might not be able to reach the British audience with no more than a rudimentary knowledge of what that audience cared to see. We had about five weeks in which to work. Part of that time was devoted to finding out where we were, familiarizing ourselves with places, names, colloquialisms, the popular news topics of the day."

The results of this research are there to be seen in "A Cup Of Tea" (which adds further, even more outrageous rhymes for Philadelphia: "tea does as well f'ya" and "bootleggers yell f'ya"), "A Tiny Flat Near Soho," and some of the other songs, including "Morning Is Midnight." "I also dared attempt to rhyme the typical London 'What ho!' in this fashion: 'I have a motto/ From Aesop—What ho,'" said Larry. "And finally this bit which will probably be intelligible only to someone who lives within ten miles of Soho. 'You can say what/ You may, what/ Care I? Eh, what?'"

What threw them both completely, he went on, was the English approach to rehearsals, which were "quite the most leisurely things in the world. They are halted each afternoon for tea, and the vibrating energy which distinguishes American preparations for the fateful night is missing. At first we missed the scurrying to and fro, the uncertainty about details, the maddening stretching of working hours into early morning day after day."

The reason for this calm approach, they learned, was that instead of the two- or three-week booking American shows exploited for road tryouts and minor changes, *Lido Lady* had deliberately been given an extended tryout— one week each in Bradford, Southsea, Leeds, and Newcastle, and two weeks in Liverpool and Manchester. At the end of that period, as Larry observed, a producer could "know with fair definiteness whether he is to have a London success or not."

They saw some shows: *By The Way*, Archie de Bear's *R.S.V.P.*, and *The Co-Optimists*—the fifth annual production of what was basically an end-of-pier revue (they even called it "A Pierrotic Entertainment") featuring composer-lyricist Melville Gideon, Davy Burnaby, Gilbert Childs, and Stanley Holloway. But English musical comedies were too light, Larry said. "Their music is too feathery. The English composer strives to imitate American jazz, and because his feet do not touch American soil, he falls just short. Whether we live in the North or the South, the American Negro's music has influenced us. Lacking that influence, the English musical writer can only echo an echo."

Rodgers and Hart polished off their contributions to *Lido Lady* as quickly as possible. Apart from "Here In My Arms" from *Dearest Enemy*, "I Want A Man" (resurrected from *Winkle Town*), and "What's The Use?" (another recycling from the Billy Rose *Follies*), they wrote fifteen new songs, handed over the score, and told Hulbert that if it was all the same to him, they'd be just as happy to pack and go home.

Hulbert "thanked them heartily for their excellent contributions to *Lido Lady*, expressing the genuine pleasure I had had in working with two such amiable chaps, and wished them the best of luck when they got back to America." A so-so actor whose dancing would never give Astaire any sleepless nights, Hulbert made a specialty of playing "goofs." Maybe it wasn't an act.

The "college boys" sailed from Southampton September 6, 1926, on the *Majestic*, without even waiting for the show's opening in London on December 1. Perhaps it was just as well: by the time he brought the show in, Hulbert had cut "Morning Is Midnight," "I Want A Man," "Ever-Ready Freddy," and some other songs, and interpolated Con Conrad's "But Not Today" and the Henderson-Brown-DeSylva hit "It All Depends On You." It certainly did no harm: the show ran until the following July, chalking up a very respectable 259 performances.

The White Star liner *Majestic*, formerly the Hamburg-Amerika liner *Bismarck* handed over as part of Germany's war reparations, was the world's largest ship. White Star publicity boasted that she had the "dignity and proportions of a noble mansion," swimming pools where green-lighted water washed beneath marble pillars, re-creating "the luxury of Imperial Rome," "a grand piano of exceptional size and beauty" in the salon, and "crusaders in mail flank[ing] the stone fireplace" of the gentlemen's smoking room.

Unimpressed, Larry spent most of the Atlantic crossing in the bar. He had even less opportunity than usual to spend time with his partner. Dick had discovered that among the passengers embarking at Cherbourg were his friend Ben Feiner, Ben's parents, and most important of all, Ben's sister, the lovely, dark-eyed, willowy Dorothy. "By the time we reached New York," Rodgers said, conveniently forgetting his relationship with Helen Ford, "we had managed to get to know each other well enough to know that we wanted to know each other better."

As the *Majestic* steamed past the Statue of Liberty and up the Hudson toward the West Side docks on Monday, September 13, a tugboat appeared. The sound of singing and band music brought everyone over to the side of the ship, to see a huge banner that read GARRICK GAIETIES. Aboard the tug was the entire cast, singing the songs from the show.

12

The Great Ziegfeld

The summer of 1926 was a hectic one. In July, a three-week subway strike brought the worst traffic jams ever seen in New York. Manhattan socialites were in a whirl over the visit of Queen Marie of Rumania and her daughter, Princess Ileana. When Trudy Ederle, daughter of a New York delicatessen owner, became the first woman to swim the English Channel on August 6, the city turned out to give her the wildest ticker-tape welcome ever staged for a woman.

On August 23, New York was treated to an even gaudier display of mass hysteria when more than thirty thousand people descended on Campbell's Funeral Church on Broadway for a glimpse of the body of the Great Lover, Rudolph Valentino, who had died at the height of his popularity. Silent-movie star Pola Negri, wearing a three-thousand-dollar set of weeds, had a well-orchestrated breakdown beside the coffin. As usual, Tin Pan Alley jumped shamelessly upon the bandwagon: before Valentino was even in his grave, there was a song on sale called "There's A New Star In Heaven Tonight."

Prohibition was in its seventh year. Someone estimated the annual profits of bootlegging at $3.5 billion. Among notable theatrical debuts were those of Barbara Stanwyck, Muni Wisenfreund (later Paul Muni), Jack Benny, Ray Bolger, Jack Haley, Jessie Royce Landis, and Claude Rains. The Charleston was all the rage: Gimbel's did land-office business when they announced "Charleston flare dresses" on sale at $1.58 each. The tune everybody was dancing to was "The Black Bottom," a DeSylva, Brown, and Henderson song from *George White's Scandals of 1926*. Contrary to popular belief, the song had nothing to do with anatomy, but referred to the color of the mud at the "Black Bottom of the Swanee River"; "when it began to shake and shiver" it would "start to get the darkies struttin' around."

In September, Gene Tunney beat the hitherto-unbeatable Jack Dempsey, replacing the "Manassa Mauler" as heavyweight champion of the world. A hurricane killed 372 people in Florida and the other Gulf states. George Abbott, collaborating with Philip Dunning, had his first big success as author-director with *Broadway*, starring Lee Tracy (understudied by the young James Cagney, who had been noticed in the preceding season's *Outside Looking In*). Fanny Brice appeared in David Belasco's production of Willard Mack's *Fanny*, her only "straight" Broadway play. It was a disaster and lasted only sixty-three performances. "Oy, it was a terrible play," said Fanny, and she was right. In October, Ernest Hemingway's *The Sun Also Rises* became a best-seller; Al Jolson appeared at the Colony Theatre in a Vitaphone one-reeler with synchronized sound singing "April Showers," "Rockabye," and "When The Red, Red Robin Comes Bob, Bob, Bobbin' Along"; and "Peaches" Browning's divorce scorched the town's tabloids.

Back on Broadway, Larry Hart was delighted to discover that Arthur Schwartz had finally got some of his songs into a show, among them the cheekily titled "Baltimore, Md., You're The Only Doctor For Me" (lyric by Eli Dawson), which Arthur said was so bad even *he* wouldn't play it at parties. Another of Larry's pals, Phil Charig, contributed a notable song, "Sunny Disposish," lyric by Ira Gershwin, to the revue *Americana* at the tiny Belmont Theatre. The show also featured the diminutive Helen Morgan singing "Nobody Wants Me" (by Morrie Ryskind and Henry Souvaine) while sitting on top of an upright piano. It's said that when Jerome Kern was casting about for a Julie for *Show Boat* he recalled Miss Morgan performing this number, and perhaps he did, although Helen had been doing the top-of-the-piano routine since November 1924, when she had appeared in Billy Rose's Back Stage Club, which she also helped finance. It was with the money he made there that Billy opened the short-lived Fifth Avenue Club.

Especially notable was the Broadway debut of Larry Hart's onetime collaborator W. Franke Harling. Striving for recognition in both the popular and serious worlds of music (later, in Hollywood, he would write "Beyond The Blue Horizon," "Sing, You Sinners," and the Oscar-winning incidental music for the John Wayne movie *Stagecoach*), Harling had written the score for *Deep River*, one of the most fascinating failures of that, or indeed any recent, season.

With a libretto and lyrics by Laurence Stallings, a critic-turned-playwright who had co-authored the 1924 success *What Price Glory?*, the dark, brooding story was set in New Orleans in 1835. More opera than musical, and making no concessions to the high kicks and jazz-baby songs of most Broadway offerings, *Deep River* lasted only four weeks. As critic Burns Mantle put it, "The opera-going public would not come down to it, nor could the theatre public rise to it."

Oscar Hammerstein II had been having more of his mixed luck: the public didn't care for the new Friml-Harbach-Hammerstein show at all. *Wild Rose* was no *Rose-Marie*; it folded after sixty-one performances. Much better liked was a new George and Ira Gershwin offering, with a book by P. G. Wodehouse and Guy Bolton cutely named *Oh, Kay!* In addition to launching Gertrude Lawrence as a Broadway star who would continue to shine for thirty years, *Oh, Kay!* presented the musical theatre with a still-unresolved conundrum.

When Ira Gershwin fell ill during the writing of the show, it was the untried Howard Dietz, and not (as might have been expected) "Plum" Wodehouse, whom George Gershwin invited to lend a hand with the lyrics. Wodehouse tactfully omitted any mention of this in his own memoirs. Gershwin's biographers have also shied away from what was undoubtedly one of the wonder man's less glorious moments. It seems he picked Dietz because Howard couldn't demand the money or the credits that George could scarcely avoid giving Wodehouse. Dietz himself said his contributions to the show included the verse of "Clap Yo' Hands" and two songs, "Heaven On Earth" and "That Certain Something You've Got" (which Ira rewrote as "Oh, Kay, You're Okay With Me"). Most important, it was Howard Dietz who came up with—at least—the title of the show's most enduring song, "Someone To Watch Over Me." He never received any credit for it. As Dietz wryly remarked, "I'd have worked for the Gershwins for nothing. And I did."

Oscar Hammerstein II bounced back on November 30, collaborating with co-producer Frank Mandel and Otto Harbach. They took as the basis of their libretto the true story of a Berber chieftain, Abd-el-Krim, who had surrendered in the summer of 1926 and been sent into exile after five years of leading the Riffs in revolt against the Spanish and French. Featuring Vivienne Segal as Margo Bonvalet and Robert Halliday as Pierre Birabeau/The Red Shadow, and endowed with one of the best scores Sigmund Romberg ever wrote, *The Desert Song* turned out to be a smash hit.

By the time it opened at the Casino, Rodgers and Hart were enmeshed in not one but two musicals, destined to open within twenty-four hours of each other. How this came about requires some explanation.

The *New York Times* had announced Rodgers and Hart's involvement in *Peggy* on August 29, prior to their return; they wasted no time getting back together with the show's librettist, Herb Fields. In their absence, Fields had written a successful show with Vincent Youmans called *Hit the Deck*, which featured "Sometimes I'm Happy" and "Hallelujah!" Now he had dug out of his father's archive the script of a 1910 production called *Tillie's Nightmare*.

Like all of Pop Fields's shows, it had been more a superior vaudeville-with-a-story than a musical. It had starred Marie Dressler as Tillie Blobbs, a drudge in her mother's boardinghouse, who dreams of a trip on a yacht, a ride in an airplane, marriage to a millionaire. Then she wakes up and philosophically goes back to being a drudge.

In Herb Fields's new version, Tillie Blobbs became Peggy Barnes, and the boardinghouse was owned by her aunt, Mrs. Barnes. Peggy's dreams took her to a Fifth Avenue department store, on a yachting trip, and to a Havana racetrack. But now, where the earlier show had been rumbustious, *Peggy* gently mocked the whole Cinderella genre. The plot itself was one long dream, which incorporated all the latest Freudian thinking: animals talked, people had pink hair, small objects grew immense, and so on. More Lewis Carroll, as one critic later remarked, than Earl.

Rodgers and Hart, who had already made use of Freud's theories in "You Can't Fool Your Dreams," set to work on the score, which was to contain eighteen songs. The show was a "family" affair. Edith Meiser had finally obtained her release from the Theatre Guild—she had to turn down the part of Mrs. Murray in *Dearest Enemy* because they wouldn't release her for it, nor for *The Girl Friend*, the title song of which, Larry told her, had been written with her in mind. ("Of course," Edith smiled, "he probably told all the girls that.") Another *Garrick Gaieties* graduate, Betty Starbuck, was in the cast, as was Lulu McConnell, the lovable comic of *Poor Little Ritz Girl*.

Seymour Felix, who had staged the *Fifth Avenue Follies*, was handling the choreography; he is said to have taken only a week to stage the sensational dances featured in the production. Roy Webb, who had taught Rodgers the basics of notation and conducting during his amateur show days, was director of music. Arthur Schwartz, Doctor of Jurisprudence, brought in by Hart so he could get the experience of seeing a show being put together, was rehearsal pianist. It looked like they were going to have a happy show, except for one thing: they didn't have a leading lady.

Their original choice was Helen Ford, but Helen was still on the road with *Dearest Enemy*. They tried to get Ona Munson, who had taken over the lead in *No, No, Nanette* after Louise Groody left the show, but Munson had been offered a Hollywood contract (she would later play Belle Watling in *Gone With The Wind*). Audition after unsuccessful audition was held; they were getting desperate, not only because they had to bring the show into New York in a month, but because while they were still writing it they were summoned by Florenz Ziegfeld and invited to write the score for his new show, *Betsy*. And Ziegfeld wanted the score ready in three weeks, so he could open in New York the day after Lew Fields opened *Peggy*.

It is difficult to understand why they did it. Ziegfeld was known to be a tyrannical monster, and even for Rodgers and Hart two full theatrical scores inside three weeks was ambitious. Rodgers implied Ziegfeld flattered him into doing the show by telling him he wouldn't think of doing *Betsy* with anyone but Rodgers and Hart. "I was only twenty-four, and it was the Great Florenz Ziegfeld speaking. If he felt that way about us, how could I possibly refuse?" He went on to record that Larry "grumbled a bit"—a masterpiece of understatement, one suspects—but that even he was impressed by the Ziegfeld name. Since Larry's attitude toward the Great Ziegfeld was demonstrably short of awe, perhaps there was some other reason. An incident from the life of Sigmund Romberg may help to illuminate what actually happened.

The following season, Ziegfeld produced a show called *Rosalie*, which took as its basic premise the visit to America of a queen (not too much unlike Queen Marie of Rumania) whose daughter falls in love with a West Point cadet (Princess Ileana had visited the Academy). Several attempts to secure composers for the show failed, so Ziegfeld approached Romberg, who turned him down because he was just about to go into rehearsals with *New Moon*. Ziegfeld was adamant, so Romberg suggested a collaboration with George Gershwin. Gershwin was summoned to the Ziegfeld office, but when he arrived, he told Ziegfeld he was too busy, too: he was working on *Funny Face*. "Gentlemen," Ziegfeld said, "it is hopeless to argue with me. Both of you had better agree now. It will save a world of argument."

Faced with this ultimatum, Romberg and Gershwin capitulated, whereupon Ziegfeld told them, more or less as an afterthought, that he had to have the music in three weeks, and not a day later. As cynical a manipulator as anyone along the Great White Way, Ziegfeld doubtless had reason to be confident of their capitulation: he'd probably already pulled the same trick on Rodgers and Hart.

Whatever the reason, Larry and Dick agreed; and on October 15, the *New York Sun* announced that Ziegfeld had contracted Rodgers and Hart to write not only a play in which Belle Baker would appear, but also another musical. A week later the *Morning Telegraph* revealed that this would be Ziegfeld's *Follies*, which would open New Year's Eve at his new theatre. Not until it was far too late did Larry and Dick learn that Ziegfeld felt he had been manipulated into producing the Belle Baker show and really didn't have his heart in it at all.

It had all begun when, coming back from Europe, Ziegfeld and his backer Leonard Replogle had seen vaudeville comedienne Belle Baker perform at a ship's concert. Belle was a tiny woman with a great big voice, something in the Fanny Brice style (one of her big hits had been Irving Berlin's "Cohen Owes Me Twenty-Seven Dollars") who could also handle big ballads. Ziegfeld was convinced she had what it took to become a top Broadway star.

Casting around for something to put her into, he learned that his top writer, the flamboyant, hard-drinking William Anthony McGuire, was doing a rewrite on a libretto hacked together by lyricist Irving ("Swanee," "Tea For Two") Caesar and gag-writer David Freedman, who wrote revue sketches for the likes of Fanny Brice and Bert Lahr. McGuire—whom Ziegfeld continued to indulge in spite of McGuire's debts and drunken irresponsibilities—persuaded Ziegfeld that this would make a good vehicle for Belle Baker, and against his better judgment Ziegfeld agreed to produce it.

The story—first titled *Buy Buy Betty* and later *Betsy Kitzel*—was another of those simple-minded affairs Larry Hart loathed: Mama Kitzel wouldn't allow any of her other children (three sons and another daughter) to get married until her daughter Betsy, who had no boyfriend, found a husband. Any thoughts Larry and Dick might have had of collaborating with the writers were soon dispelled. They had maybe two meetings with Belle Baker, and—apart from a sumptuous dinner at the Ziegfeld château in Hastings-on-Hudson (where they discovered the real reason they had been invited was to perform the score for nine-year-old Florenz Patricia, Ziegfeld's daughter)—Ziegfeld more or less left them to their own devices.

He was far more interested in the realization of a personal dream: his own theatre, a place where he would be in complete artistic control, away from his costly squabbles with Abe Erlanger. His friend William Randolph Hearst, who owned a piece of land at 54th Street and Sixth Avenue, decided to defy the off-Broadway hoodoo—nothing that far from the Great White Way had ever succeeded—and build a theatre there in partnership with Arthur Brisbane.

Hearst elected to have Ziegfeld supervise its construction, and Ziegfeld decided to make it the loveliest theatre New York had ever seen—an egg-shaped auditorium designed by Joseph Urban, decorated with panels representing legendary figures of the Middle Ages, set against black walls with chrome yellow borders. The seats would be upholstered in gold. There would be a terrace on which patrons could stroll on balmy evenings. The foyer on the second floor would be a symphony in white and gold, and when it was completed, the theatre would carry the name of its creator.

Meanwhile, the other Rodgers and Hart show, *Peggy*, now in rehearsal, was still without a leading lady. In sheer desperation, Herb Fields decided to give Helen Ford one last phone call. To his infinite relief she told him *Dearest Enemy* was closing in Columbus on November 27, and she would after all be available to play the lead. Herb dashed off to Ohio with a script and a contract.

When troubles come, the saying goes, they come not single spies, but in battalions. Dick and Larry loved *Peggy* but—to use Dick's topical phrase—thought *Betsy* "so much applesauce." The Ziegfeld show was an enormous thing with dozens of scenes, a large cast, and phalanxes of showgirls. It should have been sent out on tour for months to be whipped into shape before it came to New York. Instead, Ziegfeld allotted it only a five-day try-out in Washington's National Theatre.

Long before it got there, it had become abundantly clear how much trouble the show was in. Caesar and Freedman were fighting with each other and everyone else. Caesar wanted to interpolate some of his songs into the score; Rodgers and Hart fought tooth and nail to prevent it. At their insistence one Caesar number, "Don't Believe," was thrown out, but Caesar managed to keep in "Tales Of Hoffman" and also contributed lyrics—he claims it was at their invitation, but that stretches credibility—to a comic march Rodgers wrote for one of the characters.

As if all this were not enough, Ziegfeld became progressively more tyrannical and unreasonable. In spite of their having attended three different rehearsals on Thanksgiving Day, he excoriated Dick and Larry for not being around enough. The Great Glorifier said some inglorious things, his general manager added a few more. Dick and Larry threatened to walk away from the show, and were about to do so when sanity—of a sort—was restored by the intercession of Max Dreyfus.

Ziegfeld's problem was that he had no time to devote to the problem of *Betsy*; he was totally preoccupied with his theatre. He cannot have become any less choleric when he read in the *Morning Telegraph* of November 5 that Rodgers and Hart were to write the score of an Arthur Hammerstein production, *Polly with a Past*, with a book based on the 1917 play by Guy Bolton and George Middleton. They didn't (it eventually appeared in 1929 as *Polly* with a score by Phil Charig, Herbert Stothart, and Irving Caesar); but even though the news item was no more true than the one that said they were going to write the *Follies* score—Irving Berlin got that little plum—it certainly didn't improve relations between them and Ziegfeld.

On December 9, 1926, just five days after *The Girl Friend* closed at the Vanderbilt, the cornerstone of the new Ziegfeld Theatre was laid before an enormous mid-afternoon crowd. Among the fifteen hundred celebrities who braved the icy wind were Marilyn Miller, Gene Buck, Elizabeth Hines, and Will Rogers, who also acted as master of ceremonies. The proceedings began with the national anthem, played by Vincent Lopez and his band. Ziegfeld's wife Billie Burke and their daughter Patricia placed an iron box containing a collection of theatrical mementoes into the cornerstone, which she then cemented into place. Following the ceremony, all the guests proceeded to the nearby Warwick Hotel for a typically lavish Ziegfeld reception.

Rodgers and Hart now found themselves shuttling between the two shows. During *Betsy*'s tryout week at the National Theatre in Washington, seven of the songs Larry and Dick had written were excised. They couldn't wait to be done with it, and both of them were glad to escape after the opening. The book was lousy, and the show, although handsomely mounted, was a mess.

On December 27, the curtain at the Vanderbilt went up (one of four openings that night) on the Helen Ford show, now retitled *Peggy-Ann*. Of the eighteen songs originally written for it, four had been dropped in Philadelphia (Larry and Dick had even tried one of them, "Come And Tell Me," in the Ziegfeld show, but it didn't work there, either). Also making its penultimate appearance before they gave up on it for good was "The Pipes Of Pansy."

There was no opening chorus, no razzmatazz, nor indeed any songs until the story was well under way. The first was an ensemble number followed by a wistful "charm" song, "A Tree In The Park." When the dancing arrived, it was a surprise; Seymour Felix had taken it much closer to ballet than the usual Broadway routines. As Peggy-Ann set off from the country to her idea of New York, singing "Howdy To Broadway," the chorus made the trip with her in pantomime, changing from country to city clothes en route.

In the next scene, Peggy-Ann found herself on Fifth Avenue, where Helen Ford sang a delightfully "naughty" song full of advice on how to avoid the perils of the big city, "A Little Birdie Told Me So." Many years later, at her home in Pasadena, she recalled:

> We had a wonderful run with the show, ten months, something like that [the show ran 333 performances, from December to the following October], and we were going to Boston. Dick came to me and said, "We're going to have to do something about that song," because, you know, Boston was a puritanical city in those days, and nothing risqué was permitted. Well, Larry dug in his heels and said no, he was damned if he was going to rewrite the song for Boston. And I supported him, because I'd been singing it like that for the best part of a year; you know, you get so you play a song a certain way, it was all worked out. But Dick said we had no choice. Larry was pretty upset, but he finally agreed to do another lyric. I remember, it was closing night [October 29, 1927], and I was packing. And Larry Hart burst into my dressing-room, waving a roll of toilet paper and shouting, "Have I got a lyric for you!" He'd gone into the men's room and written the lyrics on the toilet roll. I was shocked, but he just laughed—he loved to shock people, anybody. He was like a naughty little boy.

Leading into the first-act finale was one of *Peggy-Ann*'s most charming songs, the winsome "Where's That Rainbow?" with its unforgettable middle eight adapted from the verse of *Lido Lady*'s abandoned "Camera Shoot," and its mocking allusions to all the "wish" songs of the time—their own "Blue Room," "I'm Always Chasing Rainbows," Lou Hirsch's "Love Nest," and

Kern's "Look For The Silver Lining"; even the line "Everything's gonna be all right" was a Harry Akst song title.

The second act opened with Peggy-Ann dreaming she owns a yacht. When the crew mutinies (having discovered that Peggy-Ann and Guy are not married!) a wedding is scheduled. Peggy-Ann's mother officiates, using a phone book; the bride turns up in her underwear. After a shipwreck, the survivors' lifeboat is towed to Cuba by a huge, talking silver fish.

This provided an excuse for a big, colorful production number ("Havana"), after which Peggy-Ann and Guy sang "Maybe It's Me," yet another number resurrected from *Fifth Avenue Follies*, followed by Lulu McConnell's specialty number, "Give This Little Girl A Hand" (Larry again shamelessly appropriating one of Texas Guinan's catchphrases). The finale was as unusual as everything else: a slow comedy dance, performed on a darkened stage (because Peggy-Ann wakes up in the dark).

The critics were unanimous in their praise. "From the beginning," said the *New York Times*, "Fields, Rodgers and Hart have brought freshness and ideas to the musical-comedy field, and in the new piece they travel a little further along the road." The powerful and influential Alexander Woollcott gave them a rave, as did Walter Winchell. In the *New York Daily Mirror* Robert Coleman placed them "in the foremost ranks of our youthful and talented show builders" and praised their originality, humor, cleverness, and unbounded enthusiasm.

If the boys were euphoric about such reviews, their euphoria quickly disappeared. That night, *Betsy* opened at Ziegfeld's New Amsterdam Theatre. The show was a beautifully mounted mess, top-heavy with ensemble numbers in the Ziegfeld fashion and heavy-handed "Jewish" routines. Al Shean, formerly part of the vaudeville comedy duo Gallagher and Shean, played a social lion called Stonewall Moskowitz, who has money in the bankowitz so's as good as any mankowitz (Mankiewicz, get it?). That may have been one of the best jokes in the script.

The audience had a long wait for the first "charm" song, when Belle Baker (Betsy) and Allen Kearns, as boyfriend Archie the pigeon-fancier, sat on a fire escape to sing the cute duet "In Our Parlor On The Third Floor Back." Then Belle Baker—drab in the clothes of a cleaning woman in a vaudeville house—sang "This Funny World," a song whose downbeat sentiments offer a glimpse of the other side of Larry Hart's mercurial character. It didn't go down well.

"Follow On," an ensemble by the "Daughters of the Belles of New York," was followed by the first of two grandiose solos by Borrah Minnevitch and his Harmonica Symphony Orchestra (later to become known as the Harmonica Rascals), a knockabout specialty act. This in turn was followed by an ensemble of "National Dances" which looked lovely but hardly moved the so-called plot forward. After these, Belle Baker sang her second solo, "Push Around." More ho-hum. The uninspired ensemble which followed, "Bugle Blow," purported to show how jazz had turned a nation of immigrants into "good Americans." It led directly into the first-act finale; the audience's response was less than rapturous.

Act Two, set on Coney Island, opened with Evelyn Law and ensemble doing the bright "Cradle Of The Deep." Next came "If I Were You," with its cleverly juxtaposed sentiments ("I'd tell me that I really loved me"). It was at this point that Dick and Larry received the shock of their lives. Belle Baker stepped up front and belted out a song they'd never heard before!

It turned out Rodgers and Hart weren't the only ones who'd smelled turkey. The night before the show opened, Belle was in despair. This was her Broadway debut, and her two solo numbers gave her no chance to show off that great big voice of hers. She called her old friend Irving Berlin, who came over to see her. Belle told him her troubles and begged him to come up with something.

Berlin told her he had an idea for a song, but he couldn't get a middle eight for it. He played it for her on the piano: two notes, then four, then five, then three. He kept banging away one-fingered on the black notes (Berlin could only play in one key) halfway through the night till he found what he wanted, and then put words to the whole tune. Next morning, he and Belle played it for Ziegfeld, who turned it over to an arranger.

It was this song that Belle Baker now stepped up and sang. It was called "Blue Skies," and it stopped the show cold. The audience wouldn't let her off. She did twenty-seven encores. On the twenty-eighth, she blew the lines. Ziegfeld had them put a spotlight on Irving Berlin, who stood up and fed her the words. It was a great moment—provided you weren't Rodgers and Hart. Nothing they had written came anywhere close to the verve of the Berlin song, and nobody knew it better than they.

When the show was over, Dick slunk off in one direction, and Larry, who always brought his mother to every opening, slunk off in another. On the sheet music of "Blue Skies," published shortly after the opening, there is a drawing of Belle Baker tying strings around three embarrassed-looking men. Two of them look surprisingly like the composer and lyricist. It isn't difficult to imagine the reason for their embarrassment.

The critics tore *Betsy* apart, and the show died on its feet with only thirty-nine performances to its credit, a catastrophic flop. Three days before it closed on January 29, 1927, Larry and Dick sailed for England. In reply to the obvious question, Rodgers always said no, there was no connection between the two events.

13

"One Dam Thing After Another"

O n October 10, 1926, a new Charles Dillingham production, the Jerome Kern–Otto Harbach musical *Criss Cross*, opened at the Globe Theatre. It was one of Kern's least successful scores, but that hardly mattered, since audiences were there primarily to see the antics of comedian Fred Stone. The show is important in musical theatre history for a meeting which took place at intermission, the circumstances of which are related by Edna Ferber (who misremembered the premiere at which it happened):

One night I went to a first night with Alexander Woollcott, then dramatic critic on the New York Times. We saw Fred Stone in Stepping Stones at the Globe. After the first act we drifted out to the lobby and my courtly cavalier bounded off to talk with someone else, leaving me to my own devices, of which I had none. A pixie-looking little man with the most winning smile in the world and partially eclipsed by large black spectacles now fought his way through the lobby throng toward Woollcott. He said (I was later told):

"Look, Aleck, I hear you are a friend of Edna Ferber. I wonder if you'll kind of fix it for me to meet her. I want to talk to her about letting me make a musical from her Show Boat. Can you arrange an introduction or a meeting or something?"

Mr. Woollcott, with a dreadful relish for the dramatic plum which had thus fallen into his lap (if any) said, musingly, "M-m-m, well, I think I can arrange it if I play my cards right." "Thanks," said Kern. "Thanks awfully, Aleck, I'll be—"

Woollcott now raised his voice to a bellow. "Ferber! Hi, Ferber! Come over here a minute!" Then, "This is Jerome Kern. Edna Ferber."

Only later did Edna Ferber discover that Kern and Hammerstein had begun working on a musical version of her novel before she even met him. Oscar's uncle Arthur was very keen to produce, but on December 11,

Ziegfeld released the news that he had signed Kern and Hammerstein to write the musical, advancing the composer fifteen hundred dollars and Hammerstein a thousand. His office also announced that Elizabeth Hines would play the part of Magnolia (unwisely, as it turned out: she did not, and sued Ziegfeld for breach of contract). On December 13, suggesting that even at this early stage Oscar had decided to greatly augment the part, Paul Robeson was signed to play Joe.

Amid all the excitement about *Show Boat*—*Variety* reported the as yet nonexistent production already in rehearsal in January—and the imminent opening of the season's most eagerly awaited new musical, *Rio Rita*, in the fabulous new Ziegfeld Theatre, the career of Rodgers and Hart struck an unaccustomed low. Apart from the interpolation of "Rose Of Arizona" in producer Louis Macloon's *Hollywood Music Box Revue* starring Fanny Brice, for the first time in their ten-year partnership they did not have a new show in preparation. What to do?

Word from London was that *Lido Lady* was a smash hit at the Gaiety. It might be instructive, Dick suggested, to find out what it had that was pulling in the crowds. Larry needed no persuading to embark on a larky first-class-all-the-way vacation that would give him a chance to see all the new plays in London. They sailed from New York on January 26, 1927.

The book of *Lido Lady* was still tosh, the jokes were as corny as they remembered, and Phyllis Dare was still singing "Here in my arms, it's a-dora-bool." Yet the show was as big a hit as *Sunny*. It had broken all records for the theatre and was playing to full houses every night. If Hulbert and Murray felt any gratitude toward the writers for this unexpected success, however, they managed to keep it successfully concealed, leaving Larry and Dick severely to their own devices. Apart from a few meetings—first with Lee Ephraim and Jack Buchanan, who were bringing *Peggy-Ann* over in the summer, and then with Herbert Clayton and Jack Waller, who wanted to produce a London version of *The Girl Friend* using some of their songs and the title, but not the book—the boys had little to do.

Through Larry's friend lyricist Desmond Carter, they met Firth Shephard and Greatrex Newman, who were preparing a revue featuring the combined talents of Phyllis Monkman, who'd captivated audiences in *The Co-Optimists*, the ubiquitous Leslie Henson, and Laddy Cliff, with Cyril Ritchard and his wife Madge Elliott as juveniles. Larry and Dick happily agreed to the interpolation of a couple of their songs—"Sing" (for Henson, Cliff, and Ritchard) and "If I Were You" (for Monkman and Henson). *Lady Luck*—perhaps the only musical comedy to have featured three pianos (and no orchestra) as accompaniment—was due to open at the Carlton on April 27. It would run for nine months, giving the Rodgers and Hart songs a better airing than they'd ever had in *Betsy*.

The rest of their time in London was spent going to parties—brunch at the Guinnesses, cocktails with the Prince of Wales—and trying to persuade Dorothy Dickson that she'd be ideal for *Peggy-Ann* and vice versa. Dickson, whose huge success in 1921's Kern show *Sally* had marked the beginning of a three-decade reign over the London musical stage, wasn't terribly impressed by the invitation. "In fact," she said, "I was downright condescending."

It was at this point that C. B. Cochran called Rodgers and Hart and asked them to come and see him. Portly, red-faced "C.B."—also known as "Cocky" for reasons not solely to do with brevity—lived like Larry Hart in a permanent aura of cigar smoke; he was the nearest thing London had to a Florenz Ziegfeld. At his antiquated offices in Old Bond Street, he invited Larry and Dick to provide the score for a new revue he was staging at the London Pavilion in May. They said yes on the spot.

Since Ronald Jeans hadn't even begun writing the book, Larry and Dick decided to go back to Paris and then on to the south of France until after Easter. In Paris, they ran into a couple of girls from New York named Rita Hayden (sometimes spelled Heiden) and Ruth Warner, and squired them around the capital. It was during this stay that the most famous—or the most apocryphal—near-accident in songwriting history occurred.

Larry chartered a cab to take them all down to Versailles. Having spent most of the day seeing the palace, they realized they were going to be late for the opera that night. The rush-hour traffic on the way back was very heavy, so Larry urged the taxi driver to hurry. Suddenly a truck came straight at them. A crash seemed inevitable, but it missed them by centimeters. It was at this point that Rita uttered her immortal words of relief: "Oh," she gasped, "my heart stood still!" Seeking to pass the incident off, Larry—who hadn't seen how close a call it had been—remarked that there was a great song title there. Dick—showing impeccable sangfroid—wrote the words in his notebook.

They got back to London in mid-April and, at the suggestion of Max Dreyfus's brother Louis, head of the Harms music publishing company in England—the same Louis Dreyfus who had once told the fledgling composer to go back to school—rented a service apartment at 29 St. James's Street with separate quarters for both of them on the same floor, Dick's conveniently equipped with a piano.

The Cochran revue was to be called *One Dam Thing After Another*. It would star American pianist Edythe Baker, a former vaudeville dancer whose trademark was a white grand piano; Mimi Crawford, a dainty blonde beauty; the comedians Douglas Byng and Morris Harvey; elastic-limbed Max Wall; dancer Sonnie Hale, husband of musical comedy star Evelyn Laye; and Jessie Matthews, a discovery of Cochran's.

Daughter of a costermonger, one of a family of eleven children, born over a butcher shop in Berwick Street—then and today the open-air market of London's Soho—Jessie was to become one of Britain's best-loved stars; she spent the rest of her life covering up her origins. Getting a part in Cochran's *Music Box Revue* had been her first big break.

She was "an interesting looking child with big eyes," Cochran wrote later, "a funny little nose, clothes which seemed a bit too large for her, and a huge umbrella. It may have been an ordinary sized umbrella but it seemed to dwarf her. 'You're engaged, my dear,' I said when she had finished her song and dance."

Gamine and graceful, a cute singer and a splendid dancer, Jessie Matthews went on to make her mark on the London stage in *André Charlot's Revue of 1926*, later accompanying the show to New York. The revue ran only 138 performances in America, and, after a brief appearance in the *Earl Carroll Vanities* in January 1927, she returned to England. Cochran remembered her and put her under contract. She was just twenty years old.

As they began work, Rodgers and Hart decided to dust off a few of their earlier songs, reasoning that although some of them had already been aired in the States, they would be fresh to British ears: the Atlantic was much wider in those days. "Shuffle" was used as a production number for Edythe Baker and a dancing ensemble. "Paris Is Really Divine," sung by Sonnie Hale, and "Gigolo," sung by Jessie Matthews, neatly fitted into a sketch about Paris. "Idles Of The King" was also tried but was dropped before the opening, as was a non-vocal piece called "Danse Grotesque à la Nègre."

The new songs were an ensemble number, "The Election"; "Make Hey! Hey! While The Moon Shines," performed by Max Wall; another jazzy number for Edythe Baker and ensemble, "I Need Some Cooling Off"; a duet, "My Lucky Star" (a reworking of "When I Go On The Stage" from *You'll Never Know*), sung by Hale and Mimi Crawford; and the title song, "One Dam Thing After Another."

Interestingly, two of the "Dam Things" that followed each other were Cole Porter's "Play Me A Tune" and Ray Henderson's "The Birth of the Blues," which Edythe Baker performed with great élan. The Porter song had originally been in a revue, *Hitchy-Koo of 1922*, which folded out of town; "Birth of the Blues" was the big hit number from the preceding year's *George White's Scandals* on Broadway. Since Larry and Dick's hostility toward producers who interpolated songs into their scores has already been noted, it is safe to assume the reason the two numbers were allowed in was because they were part of Baker's repertoire.

The big hit of the show, of course, was the Parisian taxicab number. During the writing of the score, Dick was leafing through his notebook and came across the phrase he had written down in Paris. He went to the piano and, in that uncannily fluent way of his, wrote a melody to the title, thinking to fit it to the first four notes. Later, when Larry got up, he came into Dick's apartment, and Dick said, "I've got a tune for that title of yours."

"What title is that?" Larry asked.

"'My Heart Stood Still,'" said Rodgers.

"I love it," Larry said, "but I never heard it before in my life." He had completely forgotten the Paris episode.

Rehearsals began, as all Cochran rehearsals began, at the Poland Rooms in Poland Street. The producer tended to put his shows together by guess and by God, trying things out with various artistes until he felt they would work. Apart from the specialty acts like Edythe Baker's piano playing, Rich Hayes the Eccentric Juggler, or ventriloquist Borem and "Cherry," no decisions were made in advance about who would sing which song, who would be better or funnier in that sketch or this routine. He tried everyone in turn, Jessie Matthews said, in an effort to get the best from them.

> One day, Cochran introduced me to the composer of the songs, Richard Rodgers, and Larry Hart, the lyric writer, a tiny little man who rubbed his hands together constantly and always smoked a big cigar. He said they had a new song, and asked Dick to play it. Then he said, "Jessie, you try it." So I sang the song, and he said he liked it. "But we've got to have a verse," he said. "Where's the verse?" And Larry Hart, who was sitting there with his hat on the back of his head and the inevitable cigar sticking out of his mouth, rushed over, and said "Verse, verse, you wanna verse?" He pulled an envelope out of his pocket and scribbled on it.
>
> "How's this?" he said, putting it into my hands. "How d'ya like this, babe? Think you can read my writing?"
>
> We never altered one word of it.

Actually, they did: in the first published version of the song, the first four lines differ from the well-known verse they finally settled on. With minor emendations the version Larry wrote that day became verse two. The refrain, once set down, was never changed, and Larry was always proud of it. Walking down Broadway with Arthur Schwartz one day he told him why.

"Everybody knows me for triple rhyming," he said. "Well, you just take a look at 'My Heart Stood Still.' I showed 'em. I could have written, 'I took one look at you, I threw a book at you,' but I didn't. It needed a simple lyric. So that was what I wrote."

The show opened at the London Pavilion on Thursday, May 19. It was a gala royal premiere, attended by the Prince of Wales, who was friendly with Edythe Baker. Rodgers, who had also made the Prince's acquaintance at parties thrown by Edythe's beau (and later, husband) Gérard d'Erlanger, was elated; Cochran less so. He predicted that the audience would spend more time watching the royal party than the musical, and he was absolutely right. The final curtain fell to no more than polite applause.

Confessing himself unable to decide whether to close immediately—he was up against some pretty stiff opposition which included the Gershwins' *Lady, Be Good!*, starring the Astaires, at the Empire, *The Vagabond King* at the Winter Garden, and *The Desert Song*, starring Harry Welchman and Edith Day, at the Theatre Royal, Drury Lane—and substitute a movie at the London Pavilion, Cochran was sufficiently encouraged by the reviews to keep the show on for a while.

While Cochran vacillated, Charles "Lucky Lindy" Lindbergh made the first solo flight across the Atlantic, landing in Paris thirty-three and a half hours after leaving New York to a rapturous reception from a crowd of a hundred thousand people. *One Dam Thing After Another* tottered along. Then about a month later, the Prince of Wales, visiting the Royal Western Yacht Club in Plymouth, asked society bandleader Teddy Brown to play "My Heart Stood Still." Neither Brown nor any of his sidemen knew the song, so the Prince hummed the melody until they could busk it. The story found its way into the London papers, together with some of the words and music. Sales of the song—and of tickets at the London Pavilion—soared. The show eventually ran for 237 performances, and in the process made Jessie Matthews a star.

Long before the Prince "dictated" that deceptively simple melody for Teddy Brown, Rodgers and Hart had taken the *Berengaria* back to New York. They arrived Friday, June 3, and were immediately buttonholed for an interview. Photographs of them, taken on board, show Larry cradling a wire-haired terrier named John—perhaps a birthday present to himself (he was thirty on May 2). He took advantage of the opportunity to plug the lyric for "Paris Is Really Divine" and to sit still for a cartoon by Larry Sobel which appeared in the *Morning Telegraph* eight days later.

Immediately after their return to New York, Larry of necessity immersed himself in family matters. The problem, as always, was his sixty-one year old father. A complete reversal in all his business fortunes had put Max into bankruptcy; for all the remaining years of Larry's life he would be dunned and hounded by his father's creditors, real or imagined. Now the O.M. was suffering from heart trouble and could neither negotiate the three flights of stairs in the house at 119th Street nor go to the Cayuga Club for his nightly game of pinochle. With the utmost reluctance, Larry made arrangements to sell the now-famous house and move his family and entourage to a penthouse on 101st Street and Central Park West.

The most notable member of this entourage was the Harts' housekeeper Mary Campbell, a buxom, black Jamaican who had joined the family the same year Larry met Dick, and who ran things *her* way. Irving Eisman said she cooked Jewish food better than anyone, but you had to be careful not to get on her bad side. After a decade of the Hart family's eccentricities, Big Mary—encouraged by Max—had developed a sharp tongue and an outspoken nature, and she rarely hesitated to express her opinion, invited or otherwise.

It's well known that Mary was the prototype for the maid Rheba in Moss Hart and George S. Kaufman's 1935 play *You Can't Take It with You*: she was a character in every sense of the word. A small flavor of her vernacular—she was almost certainly Larry's source—is captured in parts of the lead-in to an unpublished Rodgers and Hart song called "Good Provider." It has to have been from Mary's lips that Larry stole gems like "Frankfurters? Un-hunh. They make me noxious," and "Take yo' feet offen my sofa! Who you think you are—Gloria Swansong?"

Her rise-and-shine routine with Larry and his younger brother was, "Get up, Mr. Lorry, Mr. Teddy, you bums, it's time for dinner already!" Other gems included "You all is laborin' under a misdemeanor" and "You all said a heapin' teaspoonful." Someone asked her how come she wasn't married. "I don't have no man because the kind I want, I don't git," she replied, "and the kind that want me, I don't want." It's said that during a 1935 visit to the Hart establishment Josephine Baker, the St. Louis–born black singer who'd made it big as a chanteuse in Paris, asked for "*café au lait, s'il vous plaît.*" To which Big Mary snapped: "Speak with the mouth you was born with!"

In spite of all her husband's outbursts, Frieda and Max lived an ordered life: every day, high holy days and holidays included, Max brought his wife breakfast in bed. The family ate chicken every Sunday, served on the dot at one, with Uncle Willie and Aunt Rachen and whoever else was visiting as guests; Monday was sweetbreads, Tuesday pot roast, and so on. Someone came in twice a week to wind the grandfather clock; every summer, the rugs were taken up, dust sheets were draped over the furniture, and Mr. Seligman the drapery man came in to take down the drapes and clean them. Perhaps it was Frieda's way of imposing some kind of order on the anarchic world of her husband and her sons.

For Larry, life began at dinnertime, his takeoff point for the night's revelries. Sunday was special: sturgeon from Barney Greengrass; delicatessen from the Tip Toe Inn; plenty of gossip, jokes, booze, and cigars bought by the hundred from Dunhill's. His largesse was almost thoughtless. The story goes that on one occasion he invited playwright Lawrence Riley—author of the 1934 smash hit *Personal Appearance*, later filmed by Mae West as *Go West, Young Man*—to dinner.

"About seven?" Riley asked.

"Hell," Larry replied, "bring as many as you like."

He was never happier than when throwing big, everyone-invited, midnight-till-exhaustion parties in the portion of the penthouse he had isolated from his parents' quarters by installing massive, almost soundproof doors. Larry would invite theatrical friends, producers, directors, people associated with whatever show he was working on; and always, these days, Doc Bender.

According to Larry's sister-in-law, one of the few people who seem to have had a soft spot for Bender, Doc was an amiable, humorous, and rather likable sort who left a flourishing dental practice to become an agent, and proved to have an eye and ear for talent. His enthusiasm for his clients was contagious, she said, but as he started to become successful he became arrogant and vituperative, and as time passed, he was intensely disliked. Except for his clients—and Larry—no one on Broadway had any use for him.

People tolerated Bender only because of Larry, who professed to be amused by Doc's blatant flouting of the conventions and seemed indifferent to being tarred with the same brush as his gay dentist friend. Through Larry, Doc gained entrée to people and places that would otherwise have been denied him. When Larry threw one of his parties, Doc would always arrive accompanied by good-looking young actors anxious to meet the right people, adding fuel to the rumor that he was a pimp. If such rumors bothered Bender—or Larry—neither of them gave much sign of it.

The problem was work. After a night on the town with Bender—and Larry spent most of his nights on the town with Bender—the last thing in the world he wanted to do was work. And that, of course, became a problem for Dick Rodgers and Herb Fields. Very early on in their partnership, they had learned never to arrange working meetings at either of their homes: Larry would simply not turn up, or if he did, it would be several hours past the appointed time. They took instead to going up to West 119th Street, telephoning from the 86th Street subway that they were on their way so that Big Mary could get "Mr. Lorry" out of bed.

Fat chance. Nobody could get Mr. Lorry out of bed until Mr. Lorry was good and ready. After about the third attempt, Mary would explode. "Mr. Lorry, it's Mr. Dick again, and it's goddamn time to get the hell up!"

Then Larry, black with the night's beard, would stumble into the day in Peter Pan–collar pajamas and black robe, letting Dick and Herb cool their heels while he ate breakfast, read the papers and his mail, lit the first of the twenty cigars he would smoke that day, and tried to get them talking about whatever was top of the news. Whenever either of them would try to hurry him along, Larry would grin and say, "Okay, okay, take it easy, I'll be with you in a minute, just want to read this."

Two, three, sometimes four hours later, he might be ready to begin work. That didn't mean the session would last very long. Frequently, Larry would discover that he didn't have a cigar, or matches; that he needed to go to the bathroom or make a phone call or place a bet or get a shave or mail a letter. If they let him go, he would simply vanish, and they would have to find him, bring him back, and start over. Sometimes, Rodgers recalled, he would disappear while they were walking in the street, and "you'd suddenly discover you were talking to yourself."

Success hadn't changed Larry Hart a bit: he was still the same lovable, infuriating bundle of contradictions he'd always been. From his point of view, the trouble with Dick was he always wanted to work, and never wanted to have fun; Dick, of course, saw it exactly the other way around. "When the immovable object of his unwillingness to change came up against the irresistible force of my own drive for perfection," Dick said, "the noise could be heard all over the city." Larry had become the errant husband, he the nagging wife. It was not a role he relished.

———※———

The story goes that one day Larry breezed in especially late for one of their meetings; this time, both Dick and Herb let him see how much they resented the cavalier way he treated their partnership. Larry got a fit of the sulks and trashed every idea they came up with.

"God damn it, Larry!" the exasperated Rodgers snapped at last, "there has to be something you'd like to do!"

"All right, you want to know what I'd like to do?" Hart yelled back. "I'll tell you what I'd like to do! I'd like to do *A Connecticut Yankee*!"

14

A Great Big Beautiful Hit

*D*ick Rodgers and Herb Fields were not enthusiastic. The idea of making a musical based on Mark Twain's whimsical novel had come to them six years earlier at a movie theatre on 64th Street. After seeing the silent version of *A Connecticut Yankee in King Arthur's Court*, starring Harry Myers (best remembered as Charlie Chaplin's drunken benefactor in *City Lights*), they had been fired by its possibilities. Before they could go further, however, they needed the permission of the lawyer who handled Twain's estate, a flinty-eyed old gent with the mellifluous name of Charles Tressler Lark.

To their surprise, Lark had given them an option. To their astonishment, he wanted no payment. Herb worked diligently on the idea, but he just couldn't make it jell, so he laid it aside. By 1923, needing ideas for the annual show at the Institute of Musical Art, they had decided to use some of the material in what finally became *A Danish Yankee in King Tut's Court*. If it hadn't worked in 1921, why should it work now?

Having been maneuvered into taking a position, Larry now typically proceeded to defend it to the death. Here was something that could break the mold, an opportunity for experimentation, innovation, and originality, something with a cohesive story line for once instead of another revue with songs hung on it like Christmas tree decorations, or worse yet, another Cinderella or Romeo-and-Juliet or boy-meets-girl/boy-loses-girl/boy-gets-girl plot.

His vehemence won them over; they went back to the original story, and agreed that by updating the modern part of Twain's novel to the present day, and using the dream sequences in Camelot to introduce anachronistic humor, they had the basis for what could be a great show. There was only one problem: the five-year option had expired.

Full of excitement, the boys trooped uptown to renew it, but this time

Charles Tressler Lark—no fool he—was not quite as generously disposed as he had been five years earlier. Perfectly cognizant who Rodgers, Hart, and Fields were, he stipulated a hefty advance and a generous royalty to the Twain estate should the show be produced. That agreed, Herb rushed a copy of the novel across to his father, who was about to leave for London to stage the English production of *Peggy-Ann*, starring Dorothy Dickson (who'd overcome her condescension) and ace comediennes Maisie Gay and Elsie Randolph. Fields cabled back from mid-Atlantic that he didn't see anything in the book that would make a show.

Undaunted, the boys decided to go ahead anyway, convinced that even if Pop Fields didn't want to do it, they'd have no trouble finding a producer who would. They set to work; and when Fields returned from England after the July 27 opening of *Peggy-Ann*, now settling in for a six-month run, they presented him with a completed script.

═══☀═══

A small historical aside: while Lew Fields was in London he almost certainly looked at the Herbert Clayton–Jack Waller production of *The Girl Friend*, which opened September 8 at the Palace—after all, he had produced the New York version. He would have found that, apart from four Rodgers and Hart songs—"The Girl Friend" and "The Blue Room" had been kept, and "Mountain Greenery" and "I'd Like To Take You Home" added from elsewhere—there was absolutely no resemblance to the original show.

The score now included "I'm In Love" and "Step On The Blues" by Con Conrad, lyrics by Otto Harbach and Gus Kahn; the book was of the "Couldn't you squeeze me in a little back room somewhere?"/ "I'm afraid I'm not that sort of boy" variety, adapted by R. P. Weston and Bert Lee from Philip Bartholomae and Harbach's *Kitty's Kisses*, which had opened at the Playhouse on Broadway a couple of months after *The Girl Friend*.*

═══☀═══

Pop Fields loved the script of *A Connecticut Yankee*, and again partnered with Lyle Andrews to bring the show to the Vanderbilt in November. To direct he signed Alexander Leftwich, fresh from a failed play called *Set a Thief*

---

*For some strange reason, no English recordings seem to have been made of any of the songs from this production of *The Girl Friend*, which featured Roy Royston, George Gee, Emma Haig, Louise Brown, and Sara Allgood, nor does any mention of it appear in any of the standard reference books.

which had starred Rudolph Valentino's widow, Natacha Rambova, and also Margaret Wycherly of happy memory. Leftwich and Fields were old friends; Leftwich had co-directed *The Melody Man* in 1924.

For the part of the Yankee, Fields decided on William Gaxton, a well-known vaudeville performer who had yet to make a mark on Broadway. His love interest would be Constance Carpenter, an English dancer Dick and Herb had admired in the Gershwin show *Oh, Kay!* Nana Bryant, who had played the lead in Eddie Mayer's *The Firebrand*, was tapped for the role of the wicked Morgan Le Fay. The choreography was handed to a relative Broadway neophyte with the unlikely name of Busby Berkeley.

"Buzz"—whose attitude to the female sex was at least as ambivalent as Larry Hart's—had scored his first success as an actor in the 1923 musical *Irene*, in which he played the part of an effeminate dress designer. Although he had never had a day of formal choreographic training in his life, he was already getting a reputation as an innovative dance director, in spite of detractors who scoffed at his "drill routines." Rehearsals—which Buzz loved to break off so he could tell smutty jokes to the girls—were scheduled for late June, and Dick and Larry got down to some real work on the score.

The story opened with a prologue, in which Martin (the Yankee) is visiting his former fiancée, Alice Carter, on the eve of his marriage to Fay Morgan. Discovering them together, Fay knocks Martin out with a champagne bottle; while he is unconscious, Martin dreams he is back in Camelot in the days of King Arthur. He falls in love with the Demoiselle Alisande le Carteloise (a.k.a. Sandy), "a damosel who's as dumbas'ell—but sweet." Dubbed "Sir Boss," he is put in charge of industrializing the country.

Working on a percentage basis, Sir Boss creates a one-man revolution by introducing telephones, advertising, efficiency experts, and radio to the astonished Knights of the Round Table. King Arthur talks suspiciously like President Coolidge, and Merlin speaks a hybrid language somewhere between Malory and Damon Runyon ("Yon damsel is a lovely broad"). Hoardings proclaim "I Would Fain Walk a Furlong For a Camel" (guying the real-life cigarette ad-line "I'd Walk a Mile for a Camel") and the merits of a musical called *Ye Hibernian Rose of Abie* (the record-breaking five year run of *Abie's Irish Rose* had ended that summer). The King's evil sister, Queen Morgan Le Fay (a non-singing role in this original production), has Alisande kidnapped, but Sir Boss rescues her just before awakening. Back in the present day, he realizes it is Alice, not Fay, whom he really loves.

While they were still in rehearsal with *Yankee*, Larry and Dick received an approach from the producer Charles Dillingham (whose name Larry rhymed, appositely, with "willing ham") to write the score for a new show starring the English comedienne Beatrice Lillie. The book, he told them, would be by Guy Bolton, Harry Ruby, and Bert Kalmar, who had teamed successfully the preceding year in *The Ramblers*, which produced one of Kalmar and Ruby's best-known songs, "All Alone Monday." A more recent show for Dillingham with Jerome Kern and Otto Harbach, *Lucky*, had been anything but; so why Rodgers and Hart agreed to do something they knew would be superficial remains a puzzle. More than likely the highly work-oriented Rodgers simply couldn't pass up any opportunity to write songs, especially for an important producer like Dillingham, who was Ziegfeld's only rival in fame and prestige.

Right away they encountered a problem: Bea Lillie wanted to introduce "My Heart Stood Still" on Broadway. While she was unquestionably a great clown, Bea didn't have the voice for a big song, so Rodgers and Hart lied to Dillingham and told him it was already slated for *A Connecticut Yankee*. Now all they had to do was get it back from Cochran, since it was still being performed in his revue.

Then a bombshell fell: on August 10, *Variety* reported that Flo Ziegfeld, who was preparing a new edition of the *Follies* to open August 16, and perhaps felt (with some justification, as it turned out) that Irving Berlin's songs were not up to his usual standard, had cabled Cochran in London and offered him ten thousand dollars for the American rights to the song.

Benign and gentlemanly though he might appear, Cochran was a tough negotiator. Rodgers and Hart had a five-thousand-dollar buy-back option on the song; here was Ziegfeld offering twice that. Cochran now made Larry and Dick an offer which they were in no position to refuse; on September 16, without mentioning a price (although it is not difficult to guess what it must have been), the *New York Times* reported that they had bought back their song. *Variety* printed the authorized version on October 19:

> Despite the story cabled to *Variety* from London several weeks ago that Charles Cochran was offered—and refused—$10,000 for the American rights to "My Heart Stood Still". . . and denied by Florenz Ziegfeld who is alleged to have made the tender, the English producer says it is true he refused the money. Cochran has assigned the American rights to Hart and Rodgers, receiving as a consideration a reduction of royalties on the entire revue, which already totals $5000 with indications of another $5000 before the show finishes in London and a substantial total of deductions on provincial tours.

On September 30, Pop Fields took the show to Stamford, Connecticut, for a couple of days' tryout. The critic of the local paper, the *Advocate*, clearly caught up in the spirit of the thing, wrote, "Mark Twain hath been done in ye song and dance, and ye may not believe it, but in King Arthur's court, they now doeth the Charleston and other weird hoofing, and lo! the lingo of ye modern times is spoken."

From there they moved to Philadelphia's Walnut Street Theatre for final tryouts. While they were doing so an event of considerable significance took place in New York: on October 6, Warner Bros. premiered the first "talking picture." Starring Al Jolson, *The Jazz Singer* caused a sensation (doubtless adding to Larry and Dick's chagrin, one of the songs Jolson featured was "Blue Skies"), although no one realized yet just how much of a revolution had begun.

A Connecticut Yankee opened at the Vanderbilt on West 48th Street, replacing *Peggy-Ann*, on Thursday, November 3. A reworked version of "A Ladies' Home Companion" and "My Heart Stood Still" (which replaced "You're What I Need") were the only retreads in the score; both were in the prologue. To Rodgers, who was conducting the orchestra ("mostly with the spine and hips," as an onlooker commented), the response to their big song was decidedly muted, and nothing like he had anticipated. Had they squandered ten thousand dollars on a dud?

The next number was one Lew Fields had wanted to drop from the score during the tryouts in Philadelphia, because he felt the audiences were cold to Larry Hart's nimble juxtapositioning of Broadway slang and archaic English. A major disagreement ensued. After all, they'd dropped a lot of good songs already: indeed, one of them, a breathtaking waltz called "Nothing's Wrong," was dropped so far that it has never been recovered. Larry and Dick furiously insisted the song Fields wanted cut should stay in. A compromise was reached: if the New York audience's reaction was as cool, out it would go.

"Thou Swell" not only stopped the show: it made it. Every phrase of the song, with its multiple internal rhymes and its dexterous marriage of medieval archaism and modern slang, was daringly superior, both lyrically and musically, to even the best Broadway songs of the time. But more was to come: in the second act, Larry proceeded to top himself with a merry little roundelay called "On A Desert Island With Thee," famous for another beautifully wrought bridge which coupled "forget them for the nonce" with *"Honi soit qui mal y pense."*

A Connecticut Yankee had a "big" score, and as a result there were quite a few cuts. With its in-jokes on other Broadway hits of the moment, "Evelyn, What Do You Say?" was one; "Britain's Own Ambassadors" was another. Perhaps they were no great loss, although it is difficult to understand what was wrong with "You're What I Need," which was dropped in Stamford, or why no place could be found for a wonderfully tongue-in-cheek minor-key waltz entitled "I Blush," or the wistful "Someone Should Tell Them."

Was it difficult to cut such good songs from a score, songs which in later years would prove a treasure trove for musical theatre archaeologists and not a few record producers? "I'm not married to my tunes," Rodgers said. "I never was. And Larry wasn't married to his words. There were plenty more where those came from. We were interested in making the show work. If a song didn't fit, or if the audience reaction told us it was wrong, out it came."

Except, perhaps, in the case of "Thou Swell"?

"Ah," he grinned. "That was different. I was stupid. I liked it."

A Connecticut Yankee immediately became a great big beautiful hit. The critical reaction was almost unanimous, and was perhaps best expressed by Frank Vreeland of the *New York Telegram*, who urged his readers:

> GO THOU SLUGGARD
> AND ENJOY
> A CONNECTICUT YANKEE
> AND TELL
> YE COCKE-EYED WORLDE
> THOU HAST HAD
> YE HELUVA TIME

15

A Willing Ham for Dillingham

Connecticut Yankee was a resounding hit: only one other Rodgers and Hart show would run longer in Larry's lifetime. Bookings were over capacity, around twenty-five thousand dollars a week; on January 11, 1928, the show broke the weekly house record with receipts of thirty-two thousand. In addition to its 418-performance year at the Vanderbilt, it ran another year and a quarter on the road, touring forty-nine cities in all. For the first time Rodgers and Hart had a big, successful show, and the celebrity and fortune that went with it. A year earlier, during tryouts of *The Girl Friend, Variety* had suggested they were earning around five thousand dollars a week. Now they were really in the money.

Success was like a fine wine. They did interviews. They posed—as a trio, Herb and Larry perched on Dick's knees, and as members of the Lew Fields "family"—for the famous theatrical photographer Florence Vandamm. Dick bought himself a La Salle coupe. On January 28, he cut a piano-roll recording of "My Heart Stood Still" for Ampico. He was making plans for another trip to Europe; he was also paying serious court to slender, lovely Dorothy Feiner.

Larry appeared as the subject of one of the *New York Sun*'s "Times Square Tintypes" which "revealed" that he hated first nights, the radio, vaudeville, society, plays with a message, and home cooking. He was fond of mountains, good cigars, Bea Lillie, tropical scenery, Shelley, and chop suey. "Is very nervous at his own opening nights. Paces the back aisle continuously," the writer added. "Asks standees, 'How do you like it?' and won't take no for an answer. During intermission he shaves himself to look neat for the second act. Stays up all night waiting for the reviews."

As always, Broadway continued to entertain and astonish: on November 11, a play so bad it was terrific debuted at the 48th Street Theatre. In *The Squall*, a gypsy waif named Nubi invades the home of a sex-starved Spanish family and proceeds to seduce every male in sight. The dialogue was as purple as the plot, and when the waif said, "Me Nubi. Nubi good girl. Nubi stay," critic Robert Benchley made for the exit, audibly announcing to the audience as he went: "Me Benchley. Benchley bad boy. Benchley go." The show's press agent, Richard Maney, promptly labeled it "The Play That Made a Street-Walker of Robert Benchley" and, with that and one or two other hypes, managed to milk a 262-performance run out of it.

There was a scintillating sequence of musicals. On November 22, the recently formed producing team of Alex Aarons and Vinton Freedley opened their new theatre on West 52nd Street, the Alvin—its name a combination of the first syllables of their Christian names—with a new Gershwin show, *Funny Face*, starring Fred and Adele Astaire. The score contained a set of especially memorable Gershwin tunes, including " 'S Wonderful," "He Loves and She Loves," "My One and Only," and "High Hat," which gave Fred Astaire the white-tie-and-tails image with which he would thereafter forever be associated.

On the last day of the month, another new musical theatre, built by Arthur Hammerstein as what turned out to be a somewhat short-lived memorial for his father, the first Oscar, premiered an Emmerich Kalmán–Herbert Stothart–Robert Stolz musical with a book by Otto Harbach and Oscar Hammerstein II called *Golden Dawn*, a Viennese operetta-ish story set in Africa, with white actors in blackface portraying blacks. Although the show managed 184 performances, it was a disappointment, as was Rudolf Friml's *The White Eagle*, which opened on the night of December 26. Based on the highly successful 1905 melodrama which had become a famous early movie starring Dustin Farnum and William S. Hart, it was an unqualified flop.

That same night, eleven plays—an all-time record, just as the total number of productions for 1927, 268, was the highest ever—opened on Broadway. Of them all, only one, *Excess Baggage*, stayed the course. But the following evening, Tuesday, December 27, witnessed the opening of a show destined to become perhaps the jewel in the crown of the musical theatre: Jerome Kern and Oscar Hammerstein II's *Show Boat*. From its first performance at the National Theatre in Philadelphia—when in spite of having run four and a half hours, it won instant recognition and sellout status—the big question was: Would it conquer New York?

The final dress rehearsal was typically chaotic. "Costumes were wrong or missing," Edna Ferber remembered,

> lights refused to work; Norma Terris and Howard Marsh as the love-stricken
> Magnolia and Ravenal struggled with the duet, Only Make Believe, while a vast,
> vital figure hurled itself, shirt-sleeved, down the theater aisle.
> It was Joseph Urban, the scenic designer. "Wo ist mein himmelblau!" he bawled

to the electricians. "Gott verdammt! Wo ist mein himmelblau!" The famous
Urban heaven-blue having been found, the duet emerged more clearly.

"For sheer endurance and bravery," Ferber added, "I've never seen anything in the theater like that Show Boat company."

After the opening night performance, with the play running an hour and a half
overtime, they had stayed for notes. They knew that there would have to be
terrific line cuts. They took their cuts, rehearsed them and actually put them into
the matinee performance. The matinee ran until nearly six. There was another
rehearsal after the matinee. More cuts. These went into the evening performance.
They now were down almost to proper curtain time. By midnight of the second
day they were standing on their ankles as they took their curtain calls, but they
were smiling.

But would they conquer New York? No one needed that assurance more
than Ziegfeld. On opening night he sat on the stairs to the balcony, weeping
because there was so little applause for the big numbers from the over-
whelmed audience. "They don't like it," he sobbed at the end of the first act.
"Goddamn it, I knew they wouldn't!"

The second act was no better, and as Ferber said, the play was too long.
The audience left the theatre in an almost funereal silence. Ziegfeld went
down to Dinty Moore's on 46th Street for corned beef and cabbage; he sat
stoically amid the polished brass, white paint, and evergreen shrubs waiting
for the reviews. They were the finest for any show for years. He didn't—
couldn't—believe them. Only the next day, when he arrived at the New
Amsterdam and saw thousands of people lined up waiting to buy tickets, did
he finally realize he had a hit.

Starring Helen Morgan, Norma Terris, and Howard Marsh, mounted
with lavishness and taste, and blessed with one of the finest scores ever written for a musical, *Show Boat* remains Ziegfeld's crowning achievement as a
producer. Many of its songs have become standards, among them "Make
Believe," "Bill," "Can't Help Lovin' Dat Man," "You Are Love," "Why Do I
Love You?," and the sweeping, sonorous, unforgettable "Ol' Man River."
Kern and Hammerstein would never have a bigger success.

All through this period, Rodgers and Hart were hard at work on the Dillingham show, *She's My Baby*, which had begun rehearsals November 7 and had
tryouts of a week each in Washington, Baltimore, and Newark. As well as
Bea Lillie, the producer had lined up the debonair Clifton Webb, who'd

made a mark as far back as 1916 in Cole Porter's Broadway bow, *See America First*, then graduated to featured roles in *As You Were* with Irene Bordoni and *Sunny* with Marilyn Miller. More recently he had won plaudits as an adagio dancer partnering Mary Hay at the Palace and at Ciro's nightclub. The ingenue and juvenile were Irene Dunne and smiling, red-haired Jack Whiting, the poor man's Fred Astaire.

To their dismay, Larry and Dick discovered that Dillingham had lost interest in the show. "I remember we had a runthrough in Washington, with a dress rehearsal the following night," Rodgers recalled. "At the end of the first act, I saw him calmly strolling out of the theatre. 'Mr. Dillingham,' I called after him. 'What about the second act?' Without breaking stride he said, 'Oh, I saw it last night. I don't have to see it again.' I can't recall . . . ever hearing another word from him about the show or the score."

As if that weren't bad enough, the Kalmar and Ruby book was another trite little Broadway squib. Bob needs money to put on a show. His wealthy uncle, Mr. Hemingway (yes, it was a joke), won't give him a loan until he's convinced Bob has a wife and child. He persuades Tilly the maid to pose as his wife and passes off the janitor's infant as his baby. Not the kind of libretto likely to have brought out either Larry's or Dick's best.

As a result, neither of them seems to have tried too hard: of the eighteen songs used in the show, no fewer than eleven were from other productions, notably *Lido Lady*. They even tried "The Pipes Of Pansy" again; again, and for absolutely the last time, it was discarded. The new songs were only so-so, adding nothing noteworthy to the Rodgers and Hart catalog.

It was suggested around this time that Larry Hart was experimenting with colloquialisms in the lyrics for this show; if he was, there's little overt evidence of it. "Wasn't It Great?," with its rhyming roster of the current Broadway critics (which fell flat on its face opening night), and "When I Go On The Stage," which had fun with some of the reigning names of Broadway— Irene Bordoni would look like a big boloney, Charles King and Louise Groody like Punch and Judy, and so on—were "typical" Larry Hart pairings; all too typical, perhaps.

With the exception of "You're What I Need," salvaged from *Yankee* and sung by the so-called Nightingale Quartette, and Bea Lillie's shamelessly mugged show-stopper, "A Baby's Best Friend Is His Mother," none of the songs did a thing for the show. *She's My Baby* came to the Globe on Tuesday, January 3, 1928. The critics were by and large not thrilled. Although Burns Mantle thought it "hilarious," and Benchley loved Lillie—as who did not?— Woollcott dubbed it "dreary rubbish," and George Jean Nathan thought Larry's lyrics "as self consciously recherché as so many Greenwich Village lampshades." "In for a while, but never a smash" was *Variety*'s accurate prediction.

Relying almost totally on Bea Lillie, the show was no match for the Ziegfeld contribution which came in the following week. Taking as their inspiration the 1926 visit of Queen Marie and Princess Ileana, Guy Bolton and William Anthony McGuire (book), "Plum" Wodehouse and Ira Gershwin (lyrics), George Gershwin and Sigmund Romberg (whom Ziegfeld, remember, had given just three weeks to produce the music) had—somehow—collaborated successfully to produce a new show for Marilyn Miller in which the queen of musical comedy herself played a princess: *Rosalie*.

The plot was the usual nonsense: Rosalie falls in love with a West Point lieutenant. They cannot marry unless her father the King abdicates. Our hero proves his love by flying solo, like Lindbergh, all the way to his true love's country, Romanza. Let the *New York World* explain why the public ate it up like honey:

> Violins tremble; trumpets blare. Fifty beautiful girls in simple peasant costumes of satin and chiffon rush pellmell on to the stage, all squealing simple peasant out-cries of "Here she comes!" Fifty hussars in a fatigue uniform of ivory white and tomato bisque march on in a column of fours and kneel to express an emotion too strong for words. The lights swing to the gateway at the back and settle there. The house holds its breath. And on walks Marilyn Miller.

Without producing any songs as memorable ("The Man I Love" was dropped during rehearsals), *Rosalie* went on to rack up 335 performances, a longer run than *Funny Face*. *She's My Baby*, first musical of the year to open, was also the first to close, managing only a miserable seventy-one performances.

Rodgers wasn't around to see it happen. To spare them what had turned into a particularly cold and miserable New York winter, he had decided to treat his parents to a vacation in the south of France, and on January 10 he sailed for Southampton on the *Columbus*. Larry Hart, accompanied by Bea Lillie and Joan Clement from the cast of *She's My Baby*, came down to see him off. Their adieu was dutifully photographed. Bea and Dick shake hands, while Clement—whose sole Broadway appearance this seems to have been—has the adoring look one often sees on the faces of young women in close proximity to Rodgers. And Larry? Larry looks, as he so often looks, as if he wishes he were somewhere else.

16

Oscar Hammerstein Was Right

*D*uring the next two years, six more Rodgers and Hart musicals were produced on Broadway, none of them remotely in the class of *A Connecticut Yankee*. The first was a retreat to familiar territory: Herb Fields had a new idea, which was again basically an old one with a—more or less—new twist. A year earlier, while Larry and Dick had been in Europe, he had teamed up with Vincent Youmans on a successful (indeed, Youmans's last successful) musical, *Hit the Deck*. The show was produced by Lew Fields, who cast Youmans's original Nanette, Louise Groody, in the part of Loulou, owner of a Newport coffeehouse who falls in love with Bilge (Charles King), a sailor who patronizes her establishment.

The show had produced two durable hits, "Hallelujah" and "Sometimes I'm Happy," although neither was written for the show. Youmans had written "Hallelujah" years earlier, when he was in the Navy himself; and "Sometimes I'm Happy," rescued from Alex Aaron's *A Night Out*, a rewrite of a British musical which had died a-borning, was a reworking by Irving Caesar of the tune Oscar Hammerstein had set as "Come On And Pet Me" in 1923's *Mary Jane McKane*.

Herb Fields's "new" idea was to use the marines instead of the navy, this time making the hero, Chick, a tough guy rather than the shy sailor of *Hit the Deck*. The setting would now be Honolulu, and the plot would have something to do with Chick's courtship of the English Lady Delphine, who is also being courted by a rich German, Ludwig von Richter. Buck private Chick promotes himself to captain to impress Delphine, is exposed and disgraced, but redeems himself in a yacht disaster.

Pop Fields liked it enough to agree to produce, and again brought in Alexander Leftwich to direct and Roy Webb as musical director. The hero's role went again to Charles King. Buzz Berkeley not only handled the choreography but tried out for, and got, the featured part of Chick's sergeant,

Douglas Atwell. The part of Delphine went to Flora LeBreton, a winsome English blonde, and the second female lead to another English import, Joyce Barbour.

One of the show's better ideas (no prizes for guessing whose) was to assemble, instead of the usual line of winsome chorus boys, the burliest and most butch group of singers and dancers ever seen onstage. Just reading the lyric of "A Kiss For Cinderella," one of the ensemble numbers, gives a clear picture of what raucous fun they had onstage. The trouble was they acted like genuine marines in real life, too, and on the final Saturday night of the Wilmington tryouts, April 28, they embarked on a memorable pub crawl which ended with an uproarious bread-roll fight in the lobby that nearly got them thrown out of their hotel.

Rodgers and Hart produced an almost completely fresh score for the show. One of the songs, "I Love You More Than Yesterday," was inspired by a line from Goethe. The number they put their money on was the lilting "Do I Hear You Saying 'I Love You'?" It was sung in the first scene, reprised in the third, played during intermission, and formed part of the finale. As it turned out, the public preferred "You Took Advantage Of Me."

Present Arms opened at Lew Fields's Mansfield Theatre on April 26, 1928, and went on for a creditable but not brilliant 155 performances. Of course, it had some pretty stiff competition. Marilyn Miller was still packing them in at Ziegfeld's New Amsterdam in *Rosalie*. Down the street at the Lyric, Dennis King and Vivienne Segal were starring in Rudolf Friml's *The Three Muske-teers*, with lyrics by P. G. Wodehouse and Clifford Grey and a book by the prolific William Anthony McGuire. At the Liberty, Herb Fields's little sister, Dick's old flame Dorothy Fields, had begun a brilliant career as a lyricist by collaborating with Jimmy ("When My Sugar Walks Down The Street") McHugh on the songs for a smash-hit revue called *Blackbirds of 1928*. The score also included the immortal "I Can't Give You Anything But Love, Baby," a line which according to legend Dorothy Fields heard a boy say to a girl looking in the window of Tiffany's. And at the New Amsterdam, grossing a tremendous fifty thousand dollars a week, was the wonderful, lilting *Show Boat*, whose songs were being played and sung everywhere.

These were the halcyon days of the Broadway theatre. Over 250 productions opened each year, and the "road" was also in good shape. Eugene O'Neill's Pulitzer Prize–winning play *Strange Interlude*, starring Lynn Fontanne and featuring Larry's old friend Tom Powers, was the talk of the town: it began at 5:15 p.m., broke for dinner, and resumed at 8:30. Mae West's *Diamond Lil*—"pure trash, or rather impure trash," as Charles Brackett described it—was at the Royale. The daring *him* by e. e. cummings—daring because of its theme, homosexuality—lasted long enough for Lionel Stander to give twenty one performances as the First Fairy.

The new medium of "talking pictures" was whisking away talented players to Paramount's production-line studio out at Astoria, Long Island, as fast as they appeared. Claudette Colbert, Archie Leach (soon to be known as Cary Grant), Spencer Tracy, Chester Morris, Barbara Stanwyck, Miriam Hopkins, and Lee Tracy were among those who forsook the footlights for the camera during the year, notwithstanding the pronouncement in *Billboard* by Joe Schenck, head of United Artists, that talkies were no more than "a passing novelty" which was "not meeting with public favor."

But the hottest new medium was radio, growing rapidly with audiences of two and three millions—car radios were all the rage—and making heavy inroads into the sale of phonograph records and sheet music. On March 19, a new comedy show debuted, airing five times a week; it was called "Amos 'n' Andy." On May 27, accompanied by Edwin Frank Goldman's Orchestra, Rodgers played, and Olive Klein and Lewis James sang, some of his songs on "Family Party," a program sponsored by General Motors and broadcast from station WEAF New York, part of the NBC radio "Red Network" of thirty one stations formed in 1926 by David Sarnoff. Later in the year William S. Paley, son of a cigar manufacturer, would purchase sixteen radio stations and merge them into the Columbia Broadcasting System. Dinosaur-like, vaudeville continued to function, without realizing it was already dead.

During the weeks preceding the opening of *Present Arms*, Larry Hart was again beset by worries about his O.M. Although, as in the early years, the Harts lived in considerable style—there were still parties and a constant stream of visitors—Max was no longer the outrageous old devil of yore. Suffering from hardening of the arteries, cardiac problems, and general debility, he had become thin, gray, and frail. To make matters worse he was hounded incessantly by creditors and lawsuits from his speckled past; for the rest of his life, Larry would have to fight off their attentions. Even when he made his last will in 1943, Larry had to insert into it spendthrift clauses that would make it impossible for these claimants to get their hands on his money.

A couple of weeks after the opening of *Present Arms*, English dancer-actor Jack Buchanan, who was getting into producing with the Gillespie brothers and Louis Dreyfus of T. B. Harms, cabled Lew Fields querying whether rights were available for an English production and asking him to send copies of the songs over. Seeing an opportunity to publicize the show, Dick and Larry went one better: posing Flora LeBreton, Charles King, and Joyce

Barbour on a grand piano, Dick played while Larry and the stars sang the tunes over the transatlantic telephone, an event the *Evening Post* dutifully photographed.

Almost simultaneously, Larry could be found in the same newspaper defending himself from the accusation embodied in Howard Dietz's snide observation: "Larry Hart can rhyme anything. And does." Talking to a reporter on May 19, he protested that he had "never tried to be clever in the sense that polysyllabic and intricate rhyme schemes were a goal." He and Dick had tried to adjust their technique to the requirements of the situations in the play and the characters who sang the songs, he said.

> Of course as we grow older we learn. And we have tried to be simpler in what we hope to make song hits. The music in our biggest sellers is becoming more melodic and more definite in rhythm, the lyrics less complex . . . Mr. Rodgers and myself are striving for the public consumption, and this means not for the intelligentsia but for men, women, girls, and boys from every walk of life . . . We feel we are getting away from the sophisticated style that we adopted in the first Garrick Gaieties, but believe us, writing music and lyrics for a commercial box office is no easy task. If you think so, just try it on your own piano.

If those were truly his sentiments in May, it is difficult to understand how, while he was expressing them, Larry came to believe there was a musical in a novel he came across by Charles Petit called *The Son of the Grand Eunuch*. In it Li-Pi Tchou, the eponymous son, and his wife Chee-Chee flee from Peking because the young man, about to succeed his father, doesn't care to meet the usual conditions of employment. On their journey they are set upon by Tartars, monks, and brigands. Chee-Chee is carried away, but Li-Pi Tchou rescues her. Then the Grand Eunuch catches up with his son and orders him to prepare for high office. Chee-Chee and her friend Li-Li Wee arrange for a friend to first kidnap and then substitute for the surgeon. "Patient" and "doctor" play dominoes during the "operation."

Bearing in mind that several of his friends are on record saying Larry often told them he was himself a eunuch, it takes no stretch of imagination to visualize him capering about, giggling and chortling over the verbal puns and double entendres he would be able to work into the show. He always took the greatest delight in getting his sexual innuendoes past Mrs. Grundy; the comical possibilities inherent in the plight of a young man in love, about to be deprived of his masculinity, were irresistible. He had to tell someone, but who?

Dick Rodgers had taken his mother and brother to Colorado for a trip after the death of his much-loved maternal grandfather, Jacob Levy, so Larry excitedly called Herb Fields, then in Hollywood finalizing the sale of *Present Arms* to the movies. Herb loved the idea; on their way back east he and his sister Dorothy stopped off at the Broadmoor Hotel in Colorado Springs, bringing Dick a copy of the book. He read it, then told Larry and Herb they were both crazy.

They were taken aback. But Pop Fields was ready to produce it, they told him. It would be another show by the "family": Alex Leftwich directing, Roy Webb to do the music. They had Betty Starbuck in mind for the part of Li-Li Wee, and Helen Ford would be a natural for the part of Chee-Chee.

Still Rodgers held out: "You just can't talk about castration all evening," he said. "It's not only embarrassing—it's downright dull!" At this juncture Larry pointedly observed that Dick was always the one complaining about the appalling monotony of subject matter in musicals. Finally, not wishing to be the lone holdout, Dick relented. The triumvirate took themselves off to Valley View Farm in upstate New York to adapt the story and score it, and by early July they were ready to go into rehearsal. Herb cabled Helen Ford, who was vacationing in Paris.

I had foolishly signed a contract with old man Fields, he was producing again. I read the novel, and I thought they were absolutely crazy. I cabled Herbie and told him there was no part in that for me. But when I got home, about a month later, they were down at the gangplank to see me, and old man Fields had arranged it with some politician so I didn't have to go through Customs. In spite of what I'd told Herbie, they were all ready to start on the show. The costumes were being made, the chorus were in rehearsal; it was a foregone conclusion that I'd do it. Then they wouldn't let me see a script, I couldn't get hold of a script! It wasn't until the Saturday before we went into rehearsal that they gave me one.

When George Ford read the script—he vetted everything that was sent to Helen—he said, "My God, this is awful!" He took it over to his brother Harry and asked him to look at it. Harry agreed; it was as bad as George thought. "Helen can't play in that thing," he said. "Even if it was good, I don't think she ought to do it." So on the Monday, Helen came back to the theatre and told the boys she was not going to do the show.

Their faces! Shock! Disbelief! They said, "Helen, you're crazy!" They spent the whole morning arguing with me, and finally somebody said, "Let's go to lunch." So I left the theatre and walked up 47th Street, and they caught up with me, one on each side of me, and Larry running along in front of me, rubbing his hands, you know. I told them I wanted to go to lunch by myself, if at all. In fact, I wanted to go home. But they steered me into a restaurant on Broadway, and they all started talking at me: "Helen, we've already sunk so many thousand dollars into it, you've got to do it," and I said, "But I told you I didn't want to do it." And they said, Dick, I think it was, said, "Just open with us, just until we can get somebody else." So I finally—stupidly—agreed to do that because they were booked into [the

Forrest Theatre in] Philadelphia in three weeks' time. And they never tried to get anybody else. And I could hardly rehearse it, I was so unhappy in it.

⌦＝▩＝⌫

As *Chee-Chee* took shape, Larry and Dick tried much harder than ever before to integrate the songs into the story. In fact, there were only six fully realized numbers in the show: "I Must Love You," "Moon Of My Delight," "Singing A Love Song," "Better Be Good To Me," "Dear, Oh Dear," and "The Tartar Song." The rest of the music, sometimes no more than four bars, was used only to advance the plot. The production itself was lavishly produced and mounted, with settings and costumes that were as colorful and exotic as the locale.

On August 26, Larry, always the advance man, took a reporter from the *Philadelphia Public Ledger* to one side and told him musical comedies were changing style.

Forty years ago when Vienna was Vienna, Johann Strauss played the tunes in the Prater that were fiddled all over the world. Ten years later Arthur Sullivan's more intellectually nuanced tunes ruled the musical waves for Britannia. Popular music moves in cycles. . . . The American production has reached a stereotyped form. What is known as the American musical consists most often of the Cinderella legend thinly diluted, with hot chorus dance specialties and low comedy scenes sanctified by age.

It is time for renaissance. The persistently rhythmic unmelodious foxtrot is becoming a bit tiresome to ears that still love music. The wishy-washy legend of the saccharine and poor little heroine who wins a fairy prince has lost its meaning by constant repetition. The Brothers Gershwin with George Kaufman attempted a brave revolt with "Strike Up The Band." They were a little ahead of their time but they will try again.

Herbert Field[s], Richard Rodgers, and myself were luckier with "Peggy Ann" when we satirized the Cinderella legend out of countenance. . . . We went a lot further with "Connecticut Yankee" where the conventionalized love story was left out altogether and now in "Chee Chee" . . . we are doing away with the ordinary idea of the musical comedy dance routines and chorus number stencil. Here we dare to write musical dialog not as opera bouffe with recitative, but with little songs, some of them not a minute long.

This new musical technique of Richard Rodgers insures the continuity of our story. The libretto by Herbert Fields is far from commonplace with a heroine who dares to be a sophisticated and even a naughty little baggage. In decor too we shall attempt no garish overloading of the stage, but we shall decorate our stage with pictures that are refreshing in simple and pristine beauty.

Chee-Chee opened at the Mansfield on September 25, 1928. Despite its simple and pristine beauty, it turned out to be the greatest failure any of them had ever experienced. Although the critical reaction was in the main understanding, even supportive, the public's view coincided with that of critic St. John Ervine of London's *Observer*, guest critic for the *New York World*, who headlined his review NASTY! NASTY! and went on to say, "I did not believe any act could possibly be duller than the first—until I saw the second."

For the first time Rodgers and Hart experienced the bitter truth of impresario Oscar Hammerstein's dictum about bad plays. The public stayed away in droves, and *Chee-Chee* died after thirty-one performances. With it died the "family." Although they would inevitably continue to meet in the small world of the theatre, Rodgers and Hart, Lew Fields, Helen Ford, Herb Fields, Alexander Leftwich, and Roy Webb never worked together as a team again.

17

Makers of Melody

The last few months of 1928 were a particularly bad time for Larry Hart. The break-up of the old gang and the failure of *Chee-Chee* coincided with the final illness of Max Hart. O.M. had been suffering from a failing heart for more than ten years, and arteriosclerosis for the last five; he had been pretty much an invalid since the preceding spring, attended regularly by Dr. Albert S. Hyman of 1235 Park Avenue. He died at 7:19 a.m. on the morning of October 9, following a chronic myocardial failure. Frieda, Larry and Teddy were at his bedside.

Max's death "did not perceptibly affect Larry, except possibly to make him a bit more philosophical," said Henry Myers, who met him in New York shortly after O.M. died. Larry described Max's demise with respectful reverence. "He really admired him," Myers said, "for his ruthless ingenuity."

> He was a wonderful character, Larry said. You know how fat he was? Well, he wasted away until he was so skinny you'd never have known him. He couldn't lie down—had to sleep sitting up. He kept asking, over and over, "Why must I suffer like this?" But at the end—his very last words, with his last breath—he said: "I haven't missed a thing!"—and died.

Max was buried two days later next to his mother's and sisters' graves in a family plot at Mount Zion Cemetery in what was then rural Queens. There was no newspaper obituary: Larry Hart hated them with a venom that suggests they made him only too aware of his own mortality. Just two weeks later, he was given a further reminder of it when the the family returned to Mount Zion, this time to bury Max's older brother Harry.

The show perforce went on. Hardly more than a month after Max was buried, and hard on the heels of the election of Herbert Hoover—"a chicken in every pot, a car in every garage"—as President, the *New York Times* announced:

> *Loving Ann*, a musical by Owen Davis, Richard Rodgers, and Lorenz Hart, will soon be produced by Aarons and Friedley. Mr. Davis will provide the libretto. It will be his first work for the musical comedy theater, although he is the author of two straight plays, *The Nervous Wreck* and *Easy Come, Easy Go*. Another work, *Ladyfingers*, is impending. *Loving Ann* will open in Philadelphia on February 4 and is expected to follow *Treasure Girl* at the Alvin.

George Gershwin, the composer of *Treasure Girl*, had been working on something a little more sophisticated during its genesis, an orchestral piece which he completed ten days after the Gertrude Lawrence vehicle opened at the Alvin. *An American in Paris*, a "tone poem for orchestra," received its first public performance in Carnegie Hall on the evening of Thursday, December 13, 1928, preceded by Franck's *Symphony in D Minor* and Guillaume Lekeu's *Adagio for Strings* and followed by the Magic Fire Scene from Wagner's *Die Walküre*. In the celebrity audience that night Larry Hart found himself sitting next to Florenz Ziegfeld, who told him he had just signed George and Ira Gershwin to write a new show.

After the concert there was a star-studded party at the Jules Glaenzer apartment on Fifth Avenue, where Gershwin was presented with a silver humidor by financier Otto Kahn, who made a lengthy speech. Among other things, he declared that although George expressed the genius of young America, Gershwin—like America—had not yet experienced "the ordeal of deep anguish, besetting care and heart-searching tribulations" necessary for true genius. Ziegfeld turned to Larry and gave him a sly wink. "Kahn needn't worry, he'll get plenty working for me," he promised. "He'll suffer."

A few days after Gershwin's Carnegie Hall concert, busily cementing his status in the upper-class social circles to which he so ardently aspired, Dick Rodgers decided to throw a party to see the old year out. The four-hundred-name guest list for his supper dance at the Park Lane was a New York Who's Who that included Dick's playboy producer friend Alex Aarons and his wife Ella, Aarons's partner Vinton Freedley, the William Randolph Hearsts, the Gerald Warburgs, *Vogue* publisher Condé Nast, Cartier's Jules Glaenzer and his wife Kendall, writer and publisher Bennett "Beans" Cerf, newspaperman Herbert Bayard Swope, cartoonist Peter Arno, and Paramount's Long Island studio boss Walter Wanger.

The party was a huge success. Dick's father had fun greeting famous guests like Mayor Jimmy Walker with a vague "I'm sorry, I didn't catch the name." Gershwin played; Bea Lillie, Gertie Lawrence, and Noël Coward sang. Dick and Larry previewed the songs from their new show and broke the news that Walter Wanger had signed them to make a movie short for Paramount. Inez Courtney, one of the cast of their new show, would appear in it with them.

The New Year of 1929 started with a sensation: Jerome Kern, who had produced nothing for the Broadway musical stage since *Show Boat*, put his collection of first editions—probably one of the most delightful and valuable private libraries in the world—up for auction. The sale took place at the Anderson Galleries on Park Avenue at 59th Street commencing Monday evening, January 7, and continuing for three further days. There were 1,488 items in all: Kern and auctioneer Mitchell Kennerley anticipated a total of perhaps a million dollars. In the event, the sale realized nearly double that amount.

Lady Fingers, an Eddie Buzzell adaptation of Owen Davis's *Easy Come, Easy Go*, opened at the Vanderbilt on January 31, 1929. Buzzell, one of the many Eddie Cantor clones, played a sociable bank robber who gets nice, decent, rich John Price Jones into all sorts of trouble when out of gratitude he involves him in his latest heist. Although it was no masterpiece, the show managed a respectable 132 performances. The score was nominally by Joe Meyer and Edward Eliscu, but for reasons no longer on the record, producers Lew Fields and Lyle Andrews felt it needed beefing up and asked Larry and Dick if they could use a couple of their earlier songs, "Sing" and "I Love You More Than Yesterday."

"Sure, go ahead," Larry told them. He would have let them have the songs for nothing, Milton Pascal said, had not Dick insisted on an advance. As it turned out, the Goethe-inspired ballad was dropped during the play's run. When the show was filmed at Paramount in 1930 as *Only Saps Work*, featuring Leon Errol and Stuart Erwin, the songs were not used.

Eleven years Larry's senior, Owen Davis, whom Rodgers remembered as "a man of great humor and kindness," was another odd duck. One of the most prolific playwrights in the annals of American theatre, he had begun his theatrical life as stage manager for impresario A. M. Palmer; to justify his salary he sometimes played four or five of the walk-on parts. Later he became office boy, sketch-writer, and sometime director for J. J. Shubert at the Baker Theater, Rochester, where Jake produced Davis's first play, *Under Two Flags*. It was a great hit in Rochester, and the Shuberts planned to bring it to New York only to find Davis had already made a deal with David Belasco. An adaptation of Ouida's purple-prose novel about the Foreign Legion, it later became a successful Ronald Colman–Claudette Colbert movie.

Davis went on to write a series of hits and misses that included *Lazybones*, which starred George Abbott (who incidentally described it as "claptrap," a view clearly shared by audiences), *Icebound*, which won a Pulitzer Prize in 1923, and a 1926 adaptation of Scott Fitzgerald's *The Great Gatsby* directed by George Cukor (and filmed the same year starring Warner Baxter). Davis's most recent plays, however, had been considerably less successful: *Sandalwood* thirty-nine performances; *Gentle Grafters* thirteen; *Carry On* eight; *Tonight at Twelve*—which featured Davis's son, Owen Junior—fifty-six.

Then all at once his luck changed. In December 1928, after reconciling their differences (yet again), Eddie Cantor and Florenz Ziegfeld teamed up as star and producer of *Whoopee*, based on Davis's *The Nervous Wreck*. The score, one of only two written for Broadway by Walter Donaldson, gave Cantor one of his biggest hits, "Makin' Whoopee." The standout ballad "Love Me Or Leave Me" was sung by Ruth Etting, a lanky blonde songstress whose husband, gangster Moe "the Gimp" Snyder, had the disconcerting habit of sneaking up on people backstage and poking a gun into their ribs, saying "Stick 'em up!" A real joker, the Gimp.

His collaboration with Rodgers and Hart was Davis's first original work for the musical theatre. *Spring Is Here* was based on his play *Loving Ann*, in turn based on his earlier farce *Shotgun Wedding*, which had never made it to Broadway. Competent but unoriginal (even the final title may owe something to a tune that appeared in a Kern-Kalmar-Ruby score for *Lucky* in March 1927) it featured Plot 347: Betty (Lillian Taiz) thinks she loves Stacey (John Hundley), but when their elopement is foiled by her father she realizes that it was Terry (Glenn Hunter) she was really meant for. This is bad news for her sister Mary Jane (Inez Courtney), who also loves Terry; but it all works out happily in the end.

Twelve songs were written, two of which were cut before the show reached New York. Another which was nearly jettisoned was a soaring melody Rodgers wrote—most untypically; he was as dispassionate about composing as everything else, and rarely "inspired"—while still exhilarated by his first flight in a seaplane from the Hamptons to Port Washington on Long Island. Larry composed one of his most affectingly ardent lyrics, and when Dick played the tune for Alex Aarons, the producer fell in love with it: no matter what, he told Dick, "With A Song In My Heart" had to be in *Spring Is Here*. Next day Aarons had Dick play the song for Vinton Freedley, who hated it so much that Aarons's insistence it be kept in the show almost caused a permanent split between the two partners.

During rehearsals Larry and Dick ran into a problem: Glenn Hunter, who had made his name in the stage and screen versions of the Kaufman-Connelly success *Merton of the Movies*, couldn't carry a tune. The result of this anomaly was that the show's best numbers—and they included "With A Song In My Heart"—had to be given to Taiz and her unsuccessful lover, Hundley.

To accommodate Hunter's lack of voice, yet another song, "Yours Sincerely," was constructed with long verses so that he could "talk" it as a letter he was writing about how much he loves Taiz; she in turn would "reply" that she is in love with Terry and intends to use Hunter's charming endearments to tell him so. The show's title song, by the way, is not the well-known ballad recorded by Frank Sinatra and many others. It's a flighty, upbeat tune with an undistinguished lyric and remarkable only for one thing: it is one of the few instances of Rodgers and Hart writing two completely unconnected songs with the same title.

Spring Is Here was scheduled into the Shubert Theater, Philadelphia, for two weeks (from Monday, February 25, to Friday, March 8), with a weekend for final adjustments by director Alexander Leftwich prior to the New York opening on Monday, March 11. The show opened to few raves. Indeed, one critic complained that most of the songs were of the "you need romance, something in pants" variety. This is more than a little unfair, because in fact those words are taken from the verse of one of the show's prettier songs, "Why Can't I?"—but since when did a critic have to get it right? Once again, Rodgers and Hart's readiness to try doing something different probably lost them ticket sales to tired businessmen. The show managed only a disappointing 104 performances.

On April 30, Arthur Schwartz had his first real success at the Music Box, with Howard Dietz as lyricist. Others who contributed included George S. Kaufman, comedian Fred Allen, Henry Myers (still hoping for that big break), and Herman "As Time Goes By" Hupfeld. The revue was called *The Little Show*. Among the Schwartz-Dietz numbers was the first love song ever written to a department store, the ineffable "Hammacher Schlemmer, I Love You." Also in the show was the recycled version of "I Love To Lie Awake In Bed," the tune Schwartz had written with Larry Hart years earlier at Brant Lake. As "I Guess I'll Have To Change My Plan," it failed at first to please the crowd, who preferred "Can't We Be Friends?," written by George Gershwin's soul-friend Kay Swift, and Dietz's and Ralph Rainger's "Moanin' Low." This, the climactic number of the evening, was sung by Libby Holman to a drunken Clifton Webb in a squalid Harlem apartment set; thus awakened, he proceeds to perform a particularly torrid dance with Holman, and then, to prove the truth of the line about his being as mean as can be, strangles her!

Other notable events of that springtime included the St. Valentine's Day Massacre in Chicago, the inauguration of President Herbert Hoover, and the publication of Ernest Hemingway's *A Farewell to Arms* and Thomas

Wolfe's *Look Homeward, Angel*. Humphrey Bogart and his first wife, Mary Phillips, appeared in *Skyrocket*; Archie Leach and Jeanette MacDonald played leading parts in *Boom Boom*; Edward G. Robinson was in *Kibitzer*; and Bette Davis made her off-Broadway debut in the short-lived Provincetown Playhouse production of *The Earth Between*.

Broadway was reeling under the onslaught of the talkies. Of eighteen theatres on the Great White Way between 42nd and 53rd Streets, all were showing movies except the Palace. MGM's *Broadway Melody*, starring Charles King, Bessie Love, and Anita Page, had taken the town by storm. Technicolor was just around the corner. *Billboard* reported that Fox had forty-eight dialogue features on its production slate; Paramount's Jesse Lasky announced a program of some sixty-five features and eighty one- or two-reel shorts. Universal's Carl Laemmle promised fifty-five pictures, Nicholas Schenck of MGM sixty-three, RKO Radio Pictures another thirty.

Rodgers and Hart, meantime, had themselves become (very minor) movie stars in a twenty-minute two-reeler written and directed by the former theatre critic of the *Globe*, S. Jay Kaufman. *Makers of Melody* appears to have been a pilot for a series of shorts about songwriters. The basic idea was simple: Larry and Dick, as themselves, would answer the perennially unanswerable question "Where do you get your ideas from?" by demonstrating how they had come to write "Here In My Arms," "Manhattan," "The Girl Friend," and "The Blue Room." Featured as singers were Inez Courtney from *Spring Is Here*, Robert Cloy, Allan Gould, Ruth Tester, and Kathryn Reece (later in *Animal Crackers* with the Marx Brothers).

Principal photography was effected early in April at the Paramount–Famous Players–Lasky studios, supervised by Walter Wanger. It was situated not far from the eastern end of the Williamsburg Bridge on 35th Avenue between 34th and 36th Streets in Astoria, Queens; talkies were turned out there on a production-line basis. Feature films using established Broadway stars and comparative newcomers like Edward G. Robinson and Claudette Colbert were made on weekdays; tests were done at night; and short subjects—such as the Rodgers-Hart "starrer"—were shot over the weekend.

Opening on a "backstage" scene—presumably during *The Girl Friend*, since that's what the orchestra is playing—Dick and Larry are found working on something amid the clatter. The stage manager brings in a Miss Merrill, a feature writer for the "United Syndicate" who wants to do a story. Fine, says Dick, as long as you don't ask us which we write first, the words or the music. She prevails on them to tell her how they got started, and we dissolve to our heros acting out a scene from their early days of rejection. Seeking ideas, Dick scans the newspaper.

"Murder, suicide, robbery, blackmail," he intones funereally.

"Which one are you going to do?" asks Larry gloomily.

"The way I feel right now, I'd like to do 'em all," groans Dick.

"What a beautiful, tough hard-hearted town it is," Larry sympathizes.

"What? Oh, Manhattan."

"Manhattan!" Larry says, as inspiration strikes. "Manhattan! We'll have Manhattan, the Bronx and Staten."

Wipe.

Ruth Tester and Allan Gould perform three choruses of the song. Anyone familiar with only modern recordings would find them arcane: Miss Tester sings it in tick-tock rhythm, every syllable doubtless supervised by the ever-watchful Rodgers. A recording issued some years ago retains some of the period charm of the original; it was also responsible for propagating the legend that it was Larry, not Allan Gould, who sang—perhaps chanted would be a better word—the male part. Alas, no.

Next, Miss Merrill asks the boys where "Here In My Arms" came from.

"One summer afternoon . . . " says Hart.

"Boy, it was hot," adds Rodgers.

Dissolve.

They are at the piano. Dick plays a line that sounds not dissimilar to "Here In My Arms." "Here in my arms I think it's adorable," Larry declaims, blithely ignoring the fact that the words don't fit.

"No, no, Larry, too long," Dick says impatiently.

Miffed, Larry says, "Oh, all right then, here in my arms it's adorable."

"That's better," says Dick. "Now let me try that." He plays another line of melody.

"Wait a minute, Dick," Larry interrupts. "The beginning of that is sweet but the ending is sour."

Dick tries again; this time it's all sweet, and Larry recites the lyric as Dick plays. Next comes a performance of "The Girl Friend" by the talking Gould and a sometimes off-key Inez Courtney, followed by a leaden close-harmony version from an uncredited male trio.

In a third scene the boys explain how "The Blue Room" was written in Washington: something to do with Dick getting an invitation to go to the White House, and Larry forgetting his dress pants. Segue to Kathryn Reece and Robert Cloy duetting the song, followed by another appearance of the singing eunuchs, and it isn't difficult to see why this pilot project remained just that. Reviewing *Makers of Melody* for *Variety* following its May 31 release—and naming as its principals Richard *Rogers* and *Lawrence* Hart— "Bige" buried it as "a novel try that missed."

18

Wall Street Lays an Egg

*A*round the same time they were filming *Makers of Melody*, Rodgers and Hart sat still for a newspaper profile in the *New York World*. Instead of the usual "and-then-we-wrote" stuff, each was asked to give his impressions of the other. According to Dick, Larry was erratic and temperamental; but

a steam riveter could be going full blast outside the window and not disturb him when he is writing lyrics. It practically is impossible to start him working, but it is a feat of genius to make him stop once he has begun. He never has been known to show up for morning rehearsals, but the moment we start through the stage door for lunch there is Larry rushing in. I'm beginning to think he waits around the corner until he sees us coming through the door. His taste in ties is sartorial blasphemy. He is careless and rumpled in dress and when I've endured it as long as possible I grab his hand, rush to the children's department—as he is only five feet tall—and have him completely outfitted.

Besides a mutual admiration for our own work, we share another ardent weakness—an affection for chorus girls. Larry likes them tall, preferably over six feet, and mean-looking. His avocation is giving impromptu parties. When he is lonesome he suddenly descends on a musical show, any one will do, abducts half the chorus, bundles them into taxis and holds unholy whoopee at his penthouse apartment until dawn. The easiest way to arouse his venom and gain his everlasting resentment is to leave one of his parties sober. This he takes as a personal insult. His most pernicious habit is a proclivity to disappear off the face of the earth while one strolls with him. Not a vestige of the man can be seen until, having abandoned him for lost, he bobs up serenely, without explanation, some blocks from the scene of his evaporation.

[Once] we were walking along the Ringstrasse in Vienna when Larry went into his disappearing act. After gazing about for several minutes I finally discerned two enormous Teutonic demi-mondaines standing in front of the Bristol Hotel. Hidden between their voluminous skirts, practically lost to view, was Larry, wildly gesticulating and emitting floods of excited Viennese.

His conception of the hereafter is a huge Turkish bath. The only time he appears at work on schedule is after he has spent the night at his favorite one, where they put him out at nine o'clock in the morning. For years he has cherished a secret passion to become an actor. It is his boast that if he ever sang a song, no orchestra in the world could drown his voice. He constantly infuriates me by insisting that the orchestra is playing too loudly—the one subject on which we quarrel intermittently and never agree because I know the orchestra is not playing too loudly. His sophistication is much more sincere than his sentimentality and he is generous to a fault. The characteristic and subtle sense of humor that he exhibits in his lyrics colors his entire life.

And Dick? He was the tops, said Larry:

Of all the people engaged in any kind of artistic work, Dick Rodgers is the most methodical and the least temperamental, but he becomes infuriated if a cat walks across a carpet while he is composing. He has regular working hours and never varies his schedule—from eleven in the morning until six at night. He is practically the Beau Brummel of Broadway. His ties are an everlasting delight and a perfect enigma. He adores chorus girls. Blonde. And very innocent looking. Brains not essential, but they must be innocent looking. His reticence on matters romantic is a continual provocation and impregnable. He loathes but two things—dialect comedians and my cigars.

On April 21, soon after their Paramount short was completed, Rodgers sailed for Europe aboard the *Homeric* for a short vacation. Some indication of the changes that had taken place in his lifestyle may be had from the fact that the main purpose of his trip was to attend a dinner party being given in his honor by Jules and Kendall Glaenzer at the famous Champs Elysées restaurant Laurent; *le tout Paris* was invited, as well as visiting firemen like Noël Coward and Elsa Maxwell.

After a few days in the south of France, Rodgers returned to Paris to throw a thank-you bash for the Glaenzers at the Ritz, inviting very much the same crowd, as well as opera singer Grace Moore and Paramount moguls Walter Wanger and Jesse Lasky. Dick was moving with the moneyed crowd these days: on June 26, *Variety* would announce he and Larry had signed to do an original musical for Metro; they were to be paid five thousand dollars each for a minimum five weeks. Although this romance does not seem to have been consummated, it might explain one of the mysteries of their partnership: why they began work on, but never completed, a musical adaptation of Ferenc Molnar's 1926 hit *The Play's the Thing*.

Adapted for the stage by none other than P. G. Wodehouse—from Molnar's *Játék a Kastélyban* (literally, "A Play in the Castle"), first produced in Budapest in 1925—it had starred Holbrook Blinn as the crafty librettist Sandor Turai. In order to save his young composer Manski's marriage, and thus their forthcoming operetta, Turai orders opera star Ilona Szabo (Catherine Dale Owen) and her former lover, the actor Almady (Reginald Owen), to declare that the love scene between them, overheard the preceding night by the composer, was only a play rehearsal. And so on.

Molnar was enraptured by Wodehouse's adaptation, a hit when it opened at the Henry Miller Theatre on November 3, 1926. "I met him once at the Casino in Cannes," Wodehouse said years later, "and he stopped the play at the table for about five minutes while he delivered a long speech in praise of me—in French, unfortunately, so I couldn't understand it."

Bearing in mind that at the same time Wodehouse created this unexpected success he was working on *Oh, Kay!* with Guy Bolton, who in turn was also working with Rodgers and Hart on *Lido Lady*, a tenuous connection appears. What remains unclear is (a) whether they had the idea of adapting the play as a movie, (b) whether they had Molnar's agreement (unlikely: he was renowned for having turned down the proposition of an opera based on his play *Liliom* with music by Giacomo Puccini on the grounds that he wanted it remembered as a Molnar play, not a Puccini opera), (c) whether they had any clear commission for it at all, and (d) exactly when it was they worked on it.

The only clues are about twenty pages of dialogue in Larry's writing, some typed pages, and three songs (all undated)—the Act One opening, a song that might have been called "Music Is Emotion," and another, "Italy"— which were donated by Rodgers to the Music Division of the Library of Congress in the mid-1960s. Even in these rough draft pages there are some lovely felicities. Anticipating Alan Jay Lerner by more than half a century, Larry offers not only the rhyming of Budapest and "rude a pest" but goes him one better with "so crude a pest."

What is really interesting about these pages, of course, is that not only do they illustrate Larry actually at work (looking at them it is easy to imagine him scribbling furiously, crossing out lines, substituting others), but they are quite probably his first attempts at what he and Dick would later call "rhythmic dialogue." When they were actually written is impossible to say.

Undaunted by the relatively short run of *Spring Is Here* (or perhaps encouraged by their share of the income from the movie sale to First National), Aarons and Freedley produced another Owen Davis show, *Me For You*, in September. The cast included Victor Moore as the principal comedian, Jack Whiting as the lead, and, in support, a newcomer named Ray Bolger who had been noticed in Gus Edwardes's *Ritz-Carlton Nights* at the Palace a couple of years earlier and was attracting attention as a dancing comedian. Once again the reliable if unimaginative Leftwich was tapped to direct.

The show went into the Detroit Shubert on Sunday, September 15, for two weeks, and closed in a shambles. Writer Jack McGowan—an ex-actor and singer who had appeared in vaudeville and musicals—and veteran play doctor Paul Gerard Smith were brought in during a second series of tryouts scheduled from Friday, October 25, for nine days at the Shubert in Philadelphia. They tossed out the original story in which the heroine threw over a district attorney to marry a crook (Victor Moore cast against type) and concocted something to do with a yacht being used for rum-running unbeknownst to its owner, Mrs. Trumbell, who is saved by the heroism of Coast Guard officer Jack Mason (Whiting). Janet Velie, sister of the song-and-dance man for whom Dick and Larry had once written "Terpsichore and Troubadour," played Martha Trumbell; also aboard was another graduate of the Rodgers and Hart academy, Betty Starbuck. As *Heads Up!* they took it into the Shubert at Philadelphia for a week's tryouts beginning October 25.

Four days later, to paraphrase *Variety's* immortal headline, Wall Street laid an egg. Since 1924, with only one or two hiccups, America had fallen for the legend of the Big Bull Market in which everyone just kept on getting richer and richer. It was easy: you borrowed money to buy securities; you put up the securities as collateral. If the stocks went up, you collected; if down, you used them as security to raise more money—margin—to cover your investment. And by 1929 so many people were buying on margin that there was six billion dollars outstanding in broker's loans.

On September 3, an all-time high was reached: steel was 261 compared to 138 a year earlier. General Electric went to 396 (128 the preceding year). AT&T was at 304 (179), Westinghouse 289 (91). Two days later, the first warning note was sounded when steel and other key issues fell off. By October 4, margins were being called in on all sides. There was just one small problem: the people who had the margins only had one way of raising the money to pay them off, by selling stocks. On Monday, October 21, an avalanche of selling began. On October 23, more selling drove values down by five billion dollars. The next day, Black Thursday, action by the money men partially stopped the rot; but the following Monday morning the panic began in earnest, and the following day the bottom fell out of the market. The toll was stupendous—over thirty billion dollars in open market values wiped out. Just around the corner was something that would bury the American Dream for a decade: the Depression.

Although this was hardly the best time to be writing a happy-go-lucky show for thousands of tired businessmen who might very well now lack the five dollars needed to buy a ticket, there wasn't much the creators of *Heads Up!* could do except soldier on. The show had been a major investment in time and creative effort for Larry and Dick, who had written at least twenty songs for it. Few of them have survived: "My Man Is On The Make" enjoyed a brief vogue when boop-boop-a-doop girl Helen Kane (inspiration for the cartoon character Betty Boop) sang it in the movie version of the show a couple of years later; and Lee Wiley rescued "As Though You Were There" for a 1940 album; but the only ones with any staying power were "A Ship Without A Sail" and "Why Do You Suppose?"

Milton Pascal, who lived with the Harts for about six months after he got out of college, recalled that one day Larry called him to read the lyric of "Ship" over to him and ask him what he thought of it. Milton said he thought it was great. A few hours later, Larry called him in utter panic. "How did it go?" he screamed. "What did it say, for Chrissake?" He had lost the scrap of paper he wrote the lyric on, and he couldn't remember a word of it. Fortunately, Pascal could.

Later in the year, irrepressible as ever, Larry favored the *New York World* with the following "true" account of how "Why Do You Suppose?" was written, headlined A LESSON IN SONGWRITING:

Editor's Note: Herein the lyricist of the songwriting team of Rodgers and Hart exposes the way in which the most catchy song of their present show came into being. And if you think you can do better, just try it and see what happens.

Mr. Richard Rodgers, Esq., archly surveyed me and said something which I interpreted as a rebuke. Sure of my ground, I turned quickly and gave him a piece of the lower depths of my mind. And all over a lyric. A lyric means the words of a song that you never can and never need to hear.

This happened in Philadelphia, a place in which they try out our shows. The particular show was *Heads Up!* It is now playing at the Alvin Theatre. (This means that this article is designed to help the show at the box office.)

But to return to the lyric question. It seems Mr. Aarons and Mr. Freedley, our producers, wanted a new theme song. We already had a pretty good theme song, but nobody stormed the song seller in the lobby in hopes of buying copies. And it so happened that we found out the truth on the last Friday of our tryout in Philadelphia. So Mr. Aarons and Mr. Freedley gave us a few hours in which to grind out a hit.

Dick Rodgers thought of using a melody which we had written for an operetta called *She's My Baby* but which we had discarded before the New York opening. So we went to our hotel and began to grope for a lyric.

Larry then cited in full a song called "They Sing! They Dance! They Speak!" which had been dropped before the tryout.

I showed this lyric to Mr. Rodgers, and that was why he made the remark which I was careful not to quote in the first sentence of this monograph.

"It doesn't fit the music," said he, "and furthermore it has the quality of a certain insect."

So we started to think. "Why do you suppose we need another theme song?" I queried.

"That's it!" said Mr. Rodgers.

"What's it?" I queried, for once in my life perplexed.

"The title," he said again. "'Why do you suppose?'"

"Why do you suppose robins have red breasts?" Again I queried.

"That's great!" he answered.

And so the immortal poem was written. We finished it in time to catch the curtain coming down at the theatre. It was Friday, remember, and tomorrow was our last chance, our final out-of-town performance. . . . Well, this was better, so we kept the actors, Jack Whiting and Barbara Newberry, after school that night, and Dick taught them the song.

I forgot to add that while I was working on the lyric, Dick was finishing the musical transcript, called "the piano part" in our native Tin Pan Alley vernacular. Before the lyric was completed, he dispatched one of the musicians to take the piano part to New York on the midnight train.

Mr. Russell Bennett made the orchestrations in New York without knowing what the lyric was all about. Yet by a strange coincidence, on the musical line which corresponds to the words "cats meow" in the lyric, there is a distinct and authentic caterwauling in the instrumentation. Such is the telepathy of great souls.

Our musician came back to Philadelphia bearing the orchestration on the Saturday morning train. We called an orchestra rehearsal at noon, and that afternoon the song was sung to the great joy and edification of the five Quakers in the audience.

Thus "Why Do You Suppose?" was written, rehearsed, orchestrated and performed within sixteen hours in spite of the obstacles of distance and nervous tension.

In spite of the fact that it was a complete and utter fabrication—"Why Do You Suppose?" was nothing more than a new lyric set to the melody of "How Was I To Know?," which had been dropped from *She's My Baby* prior to the New York opening—and in spite of the testimony of Milton Pascal, who watched Larry write an extra chorus of it on an envelope during rehearsals because they were having trouble with the travelers and Jack Whiting couldn't make his costume change, Larry and Dick obviously loved this story. They "sold" it again to the *New York Times* where it appeared, posing as a news item, on December 29, and yet again to the *Herald Tribune* the following January 5. Aarons and Freedley must have had a good press agent.

Heads Up! opened at the Alvin on November 11 to generally friendly reviews; as unoriginal as apple pie, it ran a modest 144 performances, which was not at all bad considering that ticket sales had come to a virtual halt in the wake of the stock market crash. Jerome Kern and Oscar Hammerstein's *Sweet Adeline*, starring Helen Morgan—introducing "Why Was I Born?"— had opened to standing-room-only audiences but suffered a similar fate and was gone by April. *Fifty Million Frenchmen*, which came into the Lyric a fortnight after *Heads Up!* and featured a book by Herb Fields and a Cole Porter score that included "You Do Something To Me," managed what was in the circumstances a miraculous 254 performances. Despite the presence of another Cole Porter hit, "What Is This Thing Called Love?" (sung by Frances Shelley and danced by Tilly Losch), the English revue *Wake Up and Dream*, starring Jessie Matthews and Jack Buchanan, could only attract the public for eighteen weeks.

In the face of all these disasters, Larry did what he always did: he threw a party, a huge gathering at his apartment to celebrate Dick's December 7 engagement to Dorothy Feiner, news of which appeared in the *New York Times*, the *Morning Telegraph*, and the *Sun*. Producers having become somewhat thin on the ground on Broadway, Larry and Dick signed to do a new show for Ziegfeld. On December 17, Larry told the papers its star would be Broadway's beloved clown Ed Wynn (no news at all: Ziegfeld had released that information in November, together with the news that the score would be by Kern and Hammerstein). Larry added that he was planning a trip around the world; if he was, he never made it.

Rodgers and Hart bowed out of 1929 with a photograph and a feature in the *New York Times* of December 29 which showed them celebrating ten years of collaboration and ran—as such pieces usually did—through all their successes and their latest plans. Whoever their press agent was—and all the evidence suggests it was Larry—he wasn't bad either.

19

Ten Cents a Dance

*T*imes were hard: but for Larry Hart and Dick Rodgers the sun was shining brightly. That same November, Charles B. Cochran, over from London with Noël Coward and Evelyn Laye in *Bitter Sweet*, invited them to do the music for a show which would star Jessie Matthews, now a major name in London. If Larry and Dick had any hard feelings over their previous experience with Cochran, they gave no sign of it. Here was a chance for them to do something they had always wanted: music, lyrics, and book.

Seizing on a topic of the day, they came up with an idea about a woman whose beauty has been preserved by the science of modern cosmetics; Jessie would play herself and her own grandmother in the show, which would have the title *Ever Green*. They sent the outline to Cochran, who loved it but told them that he couldn't get to it until after he produced another Coward play, *Private Lives*.

Just prior to the New York opening of *Heads Up!*, Larry talked to a reporter from the *Boston Herald* about their changing approach to songwriting. He had already declared—and demonstrated—his intention of getting away from the polysyllabic rhyming for which he was renowned. In addition, he said, "Dick and I have tried to get away from the old canned formulas. I don't mean that we want to do the cynical and superficial thing, but we want to do away with all dead feelings and made-to-order phrases. We have tried to get back to real life, to express fundamental emotions in our own way. I think we got something of that in 'My Heart Stood Still.'"

It was laudable; but it would turn out to be entirely the wrong approach with Ziegfeld. The Great Glorifier only wanted one thing for his new show: hits. And right now he needed them more than at any other time in his career. For Ziegfeld's finances were at such a low ebb he could not—and not, as was more often the case, *would* not—pay George and Ira Gershwin the royalties he owed them on *Show Girl*, the Ruby Keeler musical, which closed when Ruby quit because of illness after three weeks. With his usual disregard for the facts, Ziegfeld blamed the Gershwins for the show's failure; not that it mattered. The following month the stock market crash wiped him out.

The circumstances of his fall were of the kind that make the Gods laugh. Counter-suing a signmaker who had sued him for non-payment, Ziegfeld—in typically tyrannical fashion—decreed his entire staff must agree to testify that the electric sign made by the complainant, priced at sixteen hundred dollars, was worthless. Every one of them was in court that black October Thursday. The switchboard operator was ill; so no one answered when Ziegfeld's brokers frantically tried to reach him by telephone to beg him to cut his losses. By the time he came out of court, two million dollars' worth of stocks, for which he had put up about fifty thousand dollars on margin, were worthless. He was ruined. Charles Dillingham, also wiped out, remarked philosophically, "Well, at least that finally makes us even."

Ziegfeld was down but not out. There's a story to the effect that when he heard his great friend and backer Jim Donohue had thrown himself out of a window after losing everything in the Crash, he wired Donohue's widow YOUR LATE HUSBAND PROMISED ME $20,000 JUST BEFORE HE FELL. Two days later the money arrived. Perhaps it was thus he managed to get *Simple Simon* into rehearsal immediately after the Christmas holidays.

With a book by its star Ed Wynn and Guy Bolton, it told the story of a Coney Island newsvendor who escapes his drab life by dreaming fairy tales about Cinderella and her adventures in the kingdoms of Dullna (ruled by Blackbeard the Pirate) and Gayleria (ruled by Old King Cole). The whole show was tailored almost exclusively to Wynn's zany talents. Imagine him: Simple Simon on his way to rescue Cinderella, riding a bicycle with a piano built on to it, lolloping off into a Joseph Urban–designed forest, brandishing a sword as large as himself, while declaiming his famous catch phrase "I love the woodth, I love the woodth!" Got that? Now write a song to fit the scene.

On January 12, just before they took the show to Boston for its shake-down, Dick and Dorothy Rodgers threw a posh pre-nuptial reception (they planned a very small, family wedding) in the ballroom of the Park Lane Hotel; from 4:00 till 7:00 p.m., Emil Coleman's orchestra provided music for dancing. The wedding proper was scheduled for March 5; every newspaper in town recorded Rodgers's trip to City Hall for the license on February 24. First, however, Rodgers and Hart had a show to write.

They composed two dozen songs for *Simple Simon*, some of them so-so, one or two retreads like "I Can Do Wonders With You" from *Me for You*, many of them a lot better than they needed to be: "Don't Tell Your Folks"; the plaintive "He Was Too Good To Me"; "I Want That Man"; and a rewrite of two songs from *Chee-Chee*, "I Must Love You," now called "Send for Me," and "Singing A Love Song," reworked affectingly as "I Still Believe In You."

Unbeknownst to Ziegfeld, perhaps even to Dick, Larry worked a private joke into the lyric to one of them. A year or so earlier, Dorothy Parker—like Larry a patron of Tony Soma's 52nd Street speakeasy—had decried the growing trend toward songs written to match and promote, no matter how clumsily or unaptly, the title of a movie. Referring to *The Woman Disputed*, a steamy Norma Talmadge starrer which had just opened, Parker said disgustedly, "I suppose the next thing you know, there'll be a song called 'Woman Disputed, I Love You!'" Which is no doubt why those very words appear in the verse of "Sweetenheart."

Simple Simon moved into the Colonial Theatre in Boston for tryouts on January 27. Worried and ailing, Ziegfeld was as impossible to work with as always. Famous—or infamous—for hating to open, much less reply to, letters, he would scream abuse at his doggedly loyal assistant, Matilda "Goldie" Golden ("the last virgin on Broadway"), if she attempted to take away the mail piling up on his desk. Let there be so much as a tea stain on a chorine's costume, and he would come down from the balcony screeching with anger. He spent hours writing penciled letters of complaint or instruction to the cast, crew, and composers on that famous red- or blue-bordered stationery; no one could read them.

Some stories about Ziegfeld suggest the producer's intransigence bothered Larry less than it did his partner. "Flo's idea of a non sequitur is anything that doesn't relate to him," he told friends—and anyone else who'd listen. At the opera, Ziegfeld had turned to him and said, "That's the greatest voice I ever heard, isn't it?" At a rehearsal of *Sally*, when Larry remarked on the splendor of the costumes, Ziegfeld smiled. "You think that's something?" he said. "Wait till you see the rags!"

Larry's favorite Ziegfeld story concerned a rehearsal in which there was an orgy scene. To the tune of something exotic from the orchestra, scantily clad Ziegfeld "harem girls" were pirouetting sinuously around a solitary man, their sheikh. Then down the aisle raged Ziegfeld. "What's that sonofabitch doing on the stage?" he yelled. "Get him out of there so those girls can have their goddamned orgy by themselves!"

Simon was too long; Ziegfeld arbitrarily cut song after song, abruptly demanded rewrites or fresh material, trashed dance routines, remained inflexible when it came to staging, was high-handed about what went in and what didn't. Although it would have made a wonderful production number— and just a few months later did—he tossed out "Dancing On The Ceiling." Other casualties included "Say When—Stand Up—Drink Down," "He Was Too Good To Me," "Prayers of Tears and Laughter," "I Can Do Wonders

With You," and "Sing Glory Hallelujah," a number that thumbed its nose at petty censorship—for which Boston was, of course, notorious. Perhaps for that very reason, Hart persuaded Isaac Goldberg of the *Boston Evening Transcript* to include the lyric in an article that appeared February 1, 1930.

Larry also offered Goldberg—known to be writing a biography of George Gershwin—an interestingly elitist view of the current crop of movie musicals: they would "prove to be a boon for the finer things of the playhouse," he said. "You see, the cheaper stuff finds its natural audience now in the picture houses. Well, the rest follows logically: the finer things will be left for the finer audiences."

If Ziegfeld believed Rodgers and Hart songs were finer things for finer audiences, he did a great job of concealing it. The story goes that one day he rounded on Larry and Dick. "Everything you fellows write is clever!" he complained. "Everything is fancy. Why can't you just write me a nice, simple hit?" He didn't add "like Irving Berlin," but he might as well have. Stung by the implication, Larry and Dick determined to do just that. They hied themselves over to the Ritz-Carlton, where Larry sat down and turned out "Hands," which told the sad story of a manicurist who was sick and tired of being pawed by her customers. Good, but not good enough, they decided, and tried again. Before the afternoon was over, they had completed a plangent, affecting song about a downtrodden dance-hall hostess, "Ten Cents A Dance."

Of course, it's also possible the reason Ziegfeld complained so much was because he believed Rodgers and Hart were not giving him one hundred percent of their attention. On February 11, for instance, Dick did another WEAF radio program of Rodgers and Hart songs, including some from *Simple Simon*. Sometime during that same month, Larry and Dick also took time off to write four songs for a forthcoming movie version of the DeSylva, Brown, and Henderson musical *Follow Through*. All but one, "I'm Hard To Please," were dropped before the movie, starring Buddy Rogers and Nancy Carroll, was released in September.

On the final evening of the tryout of *Simple Simon*, in what Rodgers described as "an unprecedented show of friendliness," Ziegfeld invited Dick to sit with him in the audience as Ed Wynn trundled on with his piano bicycle and singer Lee Morse to introduce "Ten Cents A Dance." Unfortunately, and not for the first time, Morse was blasted. It wasn't merely that she could remember neither the tune nor the lyric: she was so drunk she nearly fell off the stage. They got her off, and as soon as the scene was finished, Ziegfeld roared backstage and fired her. Then he got on the phone to find a replacement. He was in luck: Moe the Gimp's girlfriend, Ruth Etting, had just been thrown out of work by the one-week flop of the *9:15 Revue*, in which she had introduced Harold Arlen's first big hit, "Get Happy." Ziegfeld told her to be on the next train to Boston.

Although she only had three days to learn her part and the songs—"I Still Believe In You," which the critics all liked, and "Ten Cents A Dance," which was Ziegfeld's favorite—Etting performed brilliantly. In fact she made "Ten Cents A Dance" so much her own that many believed it had been written especially for her. That "nice, simple hit" was to become one of Rodgers and Hart's biggest successes.

There was just one problem: Ziegfeld was so unhappy with the show he refused to pay Rodgers and Hart any royalties. Rodgers later claimed he wrested the money out of Ziegfeld by threatening him with action from the Dramatists Guild, of which he and Larry were both members. In view of Ziegfeld's well-known preference for being sued—he got more publicity that way—and the established fact that the Gershwins had to take that route to get paid for *Show Girl*, it may well be that Rodgers and Hart never got a penny in royalties or fees for *Simple Simon*. Even if they did, it wasn't much: the show ran only 135 performances, and by the time it closed Ziegfeld had yanked out three more of their songs and made a number of interpolations, one of them a Walter Donaldson–Gus Kahn number that was to become even more closely identified with Ruth Etting than "Ten Cents A Dance"— "Love Me Or Leave Me."

Six days after *Simple Simon* opened at the Ziegfeld Theatre, Richard Rodgers married his Dorothy in a simple family ceremony at the Feiner apartment, 270 Park Avenue. Dorothy's maid of honor was her Wellesley roommate Rosemary Klee; Dick's brother Mortimer was his best man; Larry and Herb Fields were ushers. Stephen S. Wise, chief rabbi of New York, officiated. Rabbi Wise was famed for a remark he had made after being told, at a family dinner party, that he ought to be sitting at the head of the table. "Wherever I sit is the head of the table," he replied.

Next day Cholly Knickerbocker reported in the *American* that the bride, who carried a bouquet of white calla lilies, had worn an ivory white satin gown cut in medieval style, with long sleeves made of family lace, and a tulle veil. After a honeymoon in Naples, Taormina, and the south of France, the couple would live in a nineteenth-floor terrace apartment at the Lombardy Hotel on East 56th Street.

The newlyweds sailed for Naples on the *Roma*; after a fortnight they rendezvoused in London with Larry, and rented the Regency town-house home of Zita James, daughter of Beatrice Guinness. Overlooking Regent's Park, beautifully furnished, 11 York Terrace came with three servants. The layout was ideal: Larry would take the top floor, Dick and Dorothy the rest. They met Larry at London's Victoria Station and on the way back told him about the house and the arrangements.

"Larry, there's just one thing," Dorothy told him. "You do have to be a bit careful about the hot water. There's plenty, but just don't waste it."

Next thing we knew, that very evening, the doorbell rang and it was a neighbor. She said there was a lot of hot water . . . running down the front of the house and going into the street. Larry had turned the water on to take a bath, started to read a book and forgot about the water. All the water was in York Terrace. He was terribly sad about that, full of apologies, but he would do the same thing the next day, you know. He was impossible to live with.

If Larry was disappointed to discover Cochran had hired playwright Benn Levy to develop the original Rodgers and Hart "idea" into a story, he went along with it, even joining the producer (while Dick and Dorothy were on their honeymoon) for a trip to Berlin—where Levy was working—to discuss the story line. If when it was delivered he privately agreed with Jessie Matthews, who pronounced it corny and the character she played a mess, he kept his opinion to himself, perhaps because he was genuinely fond of Cochran, going out of his way on other occasions to show him round New York, and presenting him with fine editions of rare books.

The score he and Dick put together for Cochran was brand new except for "Dancing On The Ceiling" (which, like "My Heart Stood Still," Cochran referred to as "one of my greatest hits") and another, called "The Color Of Her Eyes," of which Larry was particularly fond; it was one of his first attempts at the darkly comic approach he later perfected with "To Keep My Love Alive." There are two other quite beautiful songs in the score. It seems incredible that Rodgers and Hart did not persist with "When The Old World Was New," a little masterpiece of gentle regret, and the equally beautiful "Lovely Woman's Always Young" (which contains a couplet about love making Gibraltar crumble and tumble that predates a similar Ira Gershwin couplet in "Our Love Is Here To Stay" by a full decade), but inexplicably, they did not; until comparatively recently, neither song has ever even been recorded.

For their stars to sing together, they wrote "No Place But Home" and "Dear, Dear." Sonnie Hale would do one of his specialty dances (with a chair) to the music and lyrics of "Nobody Looks At The Man." Joyce Barbour, who had been in *Heads Up!*, was given the "hot" number "In The Cool Of The Evening" and set to sing the duet "If I Give In To You" with comic Albert Burdon. The big showpiece song, of course, was to be "Dancing On The Ceiling."

It can have come as no surprise to Larry and Dick that the star of their show, Jessie Matthews, was involved in a messy divorce. After all, the man in the case, Sonnie Hale, had been her co-star in *One Dam Thing After Another*. When Jessie came to America in 1929 with *Wake Up and Dream*, while Evelyn Laye, the woman whose husband she had "stolen," was playing the lead in *Bitter Sweet* a few blocks away, the tabloids had served it up to New Yorkers hot and juicy. In straiter-laced London, Jessie and Sonnie were perceived as the villains of the piece, and the judge who pronounced on the divorce was particularly hostile in his comments about Matthews.

In the face of all this adverse publicity, not to mention the blandishments of his current mistress, Ada May (she was one of that expatriate breed who specialized in playing American roles in London, her most recent being in *Follow Through*), Cochran was giving serious thought to dropping Jessie Matthews. When he suggested she step out to "give the dust time to settle"—and his mistress a starring part—Jessie refused point blank. Cochran then suggested dropping Sonnie Hale. "Sonnie and I have been through hell together. Please don't part us now. I wouldn't do the show without Sonnie, and I'm sure he feels the same way," Jessie told him.

Perhaps not altogether unaware that Hale and Matthews had an ironclad contract, Cochran decided to play it safe. He would go ahead with *Ever Green* but not until mid-October; and the tryout would be in Glasgow, Scotland—far enough to keep the tabloid reporters away. His decision left Larry and Dick high and dry. Just before they left for London they had signed a three-picture contract with First National, the first of which, written by Herb Fields and starring Marilyn Miller, was due to start in July. They had to get back to New York. Dick and Dorothy had an additional personal reason: after visiting a London doctor with the unlikely name of Beckett Ovary, Dorothy had received confirmation that she was pregnant. They packed their bags and headed down to Southampton.

By the end of 1929, a hundred million Americans were going to the movies every week. There was no question about it: talkies were the future. As if to confirm that supremacy, First National, now part of Warner Bros., had also purchased the music publishing houses of Remick, Harms, and Witmark. Their crowded production schedule for 1930 included a movie of the Rodgers and Hart show *Spring Is Here*, starring Inez Courtney and Lawrence Gray; Marilyn Miller in the 1920 Jerome Kern hit *Sally*, and another J. P. McEvoy story called *Showgirl in Hollywood*; Joe E. Brown in *Hold Everything!*; Jolson in *My Mammy*; a musical adaptation of *That Lady in Ermine*, and the screen version of *No, No, Nanette*; 1925's Kalmán-Stothart-Hammerstein operetta *Golden Dawn*; and many more. Also slated were several other Rodgers and Hart productions, including *Leathernecking*—formerly *Present Arms*—at RKO, and *Heads Up!* and *Follow Through* at Paramount, the latter probably filmed at the Astoria studio.

On July 7, Walter Winchell coyly broke the news to Broadwayites that Dorothy Rodgers was "knitting tiny garments." Her condition made it inadvisable for her to accompany Dick to California; so toward the end of July he took the train out to join Herb and Larry, who had gone ahead. Like everyone else, Dick had heard the stories about the capricious crassness of the illiterate pushcart peddlers who'd become Hollywood moguls.

> The first day in, I went to the studio to meet [Jack] Warner. He was just like the producers in the funny stories. He sprawled over the table and said "Vell, now you're here, you got to get to vork. And I don't vant none of your highbrow song-making. Musik vit guts ve got to have—songs vit real sediment like 'Stein Song' and 'Vit tears in my eyes I'm dencing.'"

Dick looked around for help: was this moron with the vaudeville Yiddish accent for real? Then he realized Larry and Herb and Jack Warner were rolling about, helpless with laughter; the whole thing was a set-up. Rodgers was not amused.

A few days after Dick's arrival, Jules Glaenzer threw a garden party at his sumptuous Beverly Hills home to welcome them to Hollywood. Among the eighty or so guests were Charlie Chaplin, Dolores del Rio, Joan Bennett, the Selznicks, Louella Parsons, Louis Bromfield, Jesse Lasky, Jack Warner and his son Lewis, director of the company's music publishing activities, Ben Lyon (who was to star in the movie they were making) and his wife Bebe Daniels, director Clarence Badger, and co-stars Thelma Todd and Inez Courtney.

When Rodgers and Hart reported for work at Burbank, it looked as if everything they'd heard about Hollywood was true. They were given a luxurious office with Oriental rugs, studio cars at their disposal, even a secretary standing by to take down their most casual remarks. At work and afterwards there was always someone around from the Broadway musical scene; current residents included Bert Kalmar and Harry Ruby, Irving Berlin, Sam Coslow, Joe Burke and Al Dubin, Vincent Youmans and Oscar Hammerstein, Richard Whiting, Joe Meyer, and many more. Everyone who could was working in Hollywood, even the ones who hated the place.

Larry and Dick got rid of the secretary and settled down to work on the Herb Fields script, *The Good Bad Girl*. The "story" concerned a riveter who accidentally tosses a bolt into the boudoir of an heiress; they fall in love when he tries to put out the fire. The first piece of bad news was that Marilyn Miller had (wisely, as it turned out) opted for the movie version of *Sunny*; Ona Munson, last seen on Broadway in the musical *Hold Everything!*, was assigned to make her second movie appearance in the title role opposite Ben Lyon, with Walter Pidgeon as Lyon's rival and Inez Courtney in support. Also cast were Thelma Todd, Tom Dugan, and Holmes Herbert. The director, Clarence Badger, was returning to musicals (he had directed the movies of Cole Porter's *Paris* and also *No, No, Nanette*) after a series of mediocre melodramas like *The Bad Man* and *Sweethearts and Wives*.

Writing the songs for the movie—now called *The Hot Heiress*—presented Larry and Dick with a set of problems they had encountered before: not only could their leading man not sing worth a damn, Ona Munson's range was as limited as her voice was insipid, which meant the melodies had to work within a fairly tight upper and lower register. The finished songs were a long way from their best; indeed, they might have done better to jettison all of them and start over. "Nobody Loves A Riveter," written with an insistent repetitive rhythm to match its theme, was as unmemorable as Larry's attempt to harness current slang phrases in "You're The Cats" and the very ordinary "Like Ordinary People Do"; rejigging "He Was Too Good To Me" to "He Looks So Good To Me" didn't improve it worth mentioning, either.

Fascinated by every aspect of the movies, Rodgers and Hart spent all their time finding out how things worked; they were surprised to discover that whenever a song was performed, the full orchestra had to be on set. Because so many films were made outdoors, cameramen worked from inside huge, soundproofed boxes. One day, Ben Lyon saw Larry and Dick exploring one of these, and knowing the microphone that Badger used to talk to the cameramen was live, he and Ona Munson began talking loudly about Rodgers and Hart, and what a nerve those New York people had coming out to Hollywood thinking they could write music for movies.

> We piled it on until suddenly Dick bounced out, absolutely livid, with Larry right behind him. Nothing we could say would convince them that it was all meant in fun. Dick was as mad as hell, and [even] Larry, who was a much sweeter guy, wasn't exactly sure we were telling the truth, either. Larry didn't talk to me for weeks.

When shooting was completed, Larry and Dick headed back to New York, arriving Monday, August 11; two days later they left for London, where Cochran was putting *Ever Green* into rehearsal. Almost immediately, Larry set Cochran's enthusiasm on fire by demonstrating how the show's many changes of scene could be enhanced by using a revolving stage; the producer immediately ordered a giant mechanism from Germany, stipulating it had to be in Glasgow by October 1, two weeks before the tryout began.

While Rodgers spent his leisure with his socialite friends, Larry Hart rushed about, seeing everything that was worth seeing: Shakespeare at the New; Wilde at the Lyric; Shaw at the Savoy; Molnar's *The Swan* at the St. James's; Bea Lillie and their original Sandy, Constance Carpenter, in *Charlot's Masquerade* at the brand-new Cambridge; Sophie Tucker and Jack Hulbert in *Follow a Star* at the Winter Garden. Highest on his list was Noël

Coward's new play, *Private Lives*; starring Coward and Gertrude Lawrence, with Laurence Olivier and Adrianne Allen in support, it opened at the Phoenix on September 24. On September 28, the New York newspapers noted he was sending postcards to them and to all his friends with rave reviews of the play. No doubt he attended Coward's parties. Perhaps he heard Coward tell how Gertrude Lawrence had cabled him after reading the play for the first time: YOUR PLAY IS DELIGHTFUL AND THERE'S NOTHING THAT CAN'T BE FIXED. To which Coward replied: THE ONLY THING TO BE FIXED WILL BE YOUR PERFORMANCE.

Early in October, badly missing his new wife, Dick left for Glasgow to attend the orchestra rehearsals; unlike American players, British pit orchestras rehearsed for days before the actual performance. After what seemed to him like a year, but actually was a week later, the company arrived at the cavernous pseudo-Gothic Central Hotel where, because it was an adjunct of the railroad station, porters could carry baggage directly through a special entrance to the hotel from the train platforms. When Larry arrived he refused to believe this and insisted—to the utter bewilderment of the Scots lackey attending him—on loading his luggage into a cab and having the equally puzzled driver take him around the corner to the front entrance.

The first problem that needed solving in Glasgow was the non-arrival of the revolving stage from Germany; frantic telegrams and considerable hair tearing were required before it arrived, just three days before opening. Harold Conway, then critic of the *London Daily Mail*, and in Glasgow to cover the show, recalled that when it was installed, the damned thing wouldn't work.

> Panic descended upon the entire company. It looked like it couldn't be used at all, which meant interminable stage waits [which] would disrupt the flow of the plot. To cover these, Larry had to add new lines; everybody had to spend hours learning them; the cast slept in the seats of the theatre and on the floor. It was October and freezing. Cochran caught cold and so did I. I came down with bronchitis and was put in hospital. . . . I'm still emotional when I talk about how Larry Hart devoted himself to me, a stranger. He came to the hospital and took a room near me, then sent off my dispatches to London for me. Somehow he managed to do some of his own work while looking after me. A marvelous man, I never knew anyone like him.

The revolving stage was fixed; the lines learned were unlearned again; rehearsals continued. Everything went well, proceeding to disprove the old theatrical saw that a good dress rehearsal means a bad opening night performance. Cochran had spared no expense, and the sets and costumes were dazzling: the Albert Hall in London, a Paris street fair, a fiesta in Spain. And as everyone had confidently expected, Jessie Matthews singing "Dancing On The Ceiling"—the revolving set worked like a charm—was a big hit.

The day after the Glasgow opening Cochran released Dick, who took the train south and sailed for New York and his Dorothy on the *Majestic*. Larry stayed on until the London opening at the brand new Adelphi Theatre on December 3. The show was a smash success that would run for over six months. Jessie Matthews was hailed as a bright new star; *Ever Green* would lead to a decade of movies in which she would become the queen of British musicals. The Rodgers and Hart songs were being played everywhere. Rubbing his hands with glee, Larry headed back to New York and work: Dick had called to say Herb Fields had a great idea for a new show. After that, they had to write the songs for another two movies to complete their Warner contract.

20

Hard Times on Broadway

hen Larry got back to New York, however, the Broadway picture didn't look so rosy; everyone was suffering from falling ticket sales and tight budgets. At $5.50 top, even the hit Cole Porter show *The New Yorkers* (book by Herb Fields), which had opened December 8, was struggling. Rudolf Friml and Sigmund Romberg both tried operettas: Friml with *Luana* for Arthur Hammerstein, and Romberg with *Nina Rosa* for the Shuberts. Thanks to the presence of the original Rio Rita, Ethelind Terry, the Romberg show managed a just-about-adequate 137 performances; *Luana* was a fiasco best described by Robert Littell in his New York *World* review: "Plentee grass skirtee. Plentee girlee swishee grass skirtee. Plenty nood girlee in swimming poolee. Plentee much Viennese waltzee by Rudolf Friml. Sounds mighty strange in middle of Hawaii. *Luana* same old stuffee if you askee. Aloha."

The Shuberts did *Hello, Paris*, starring Chic Sale; it folded after thirty-three showings. Two nights later, in partnership with Jed Harris, Billy Rose showcased his wife Fanny Brice—Mayor Jimmy Walker had married them at City Hall in February 1929—in a revue that had started life as *Corned Beef and Roses* but came to the 46th Street Theatre as *Sweet and Low*. Some of the song titles still bring a wince—"When A Pansy Was A Flower," for instance—but even with a score by Harry Warren and Ira Gershwin there wasn't enough to keep the show going more than 184 performances.

Lew Fields came back as producer of *The Vanderbilt Revue*, which opened at the theatre of that name on November 5. In spite of the presence of *Peggy-Ann*'s husky-voiced Lulu McConnell (who took over from Ruby Keeler Jolson when Al Jolson persuaded his wife to leave the show), the critics disapproved, and it lasted only thirteen performances. A by-product of this failure was Lew Fields's decision to retire—one more theatrical giant removed from the scene.

Next it was Ziegfeld's turn. On November 18, he brought in what he thought was a surefire crowd-pleaser: Fred Astaire and Marilyn Miller as a Salvation Army girl whose affections are trifled with on a bet (shades of *Guys and Dolls*) in *Smiles*. It looked good, with a book based on a Noël Coward idea partly written by Pulitzer Prize–winner Louis Bromfield, songs by Vincent Youmans, and lyrics by Harold Adamson and even Ring Lardner; but even with "Time On My Hands"—which, predictably, Ziegfeld had wanted to cut—it lasted only two months.

Dillingham was already finished. In December, Arthur Hammerstein was bankrupted by the failure of his nephew Oscar's show *Ballyhoo*, and lost the theatre so proudly named for his impresario father. Desperate for a hit, the Shuberts brought in *The Student Prince* and *Blossom Time*, both of which had been on the road non-stop since their original successes. The first lasted five weeks; the latter three and a half.

Hard though times were, it wasn't all gloom and doom. The Gershwins brought in one of their biggest hits for Aarons and Freedley. *Girl Crazy* featured Freedley's vaudeville discovery, the former Ethel Zimmermann, now Ethel Merman, electrifying audiences with her rendition of "I Got Rhythm," not to mention Ginger Rogers cooing "Embraceable You" and "But Not For Me." The production is still remembered for one of the most famous pit orchestras in history: among the personnel were Glenn Miller, Jack Teagarden, Gene Krupa, Red Nichols, and Benny Goodman.

Max Gordon, a new producer, had a moderate success with *Three's a Crowd*, which reunited the stars of *The Little Show*—Clifton Webb, Libby Holman, and Fred Allen—with its composers, Arthur Schwartz and Howard Dietz, whose "Something To Remember You By" (featuring newcomer Fred MacMurray) was a hit, as was an interpolation by John Green, "Body And Soul." Another hit in the show was Tamara Geva, who'd shone in Eddie Cantor's *Whoopee*. No one was more overjoyed than her agent. "I always said she was great," Doc Bender told anyone who'd listen. "Now I believe it myself."

<hr>

Doing the rounds of producers' offices following their return from London, Rodgers and Hart soon discovered their success there didn't mean a thing. The only way they were going to get a hearing was if they joined forces with a librettist and came up with something timely, something amusing, something new that would interest one of this new and tougher breed of producer. They set to work at once with Herb Fields.

Using some of their Hollywood experiences, they came up with a slip of a thing about Michael and Geraldine, who come to Hollywood from St. Paul,

Minnesota, determined to be silent-movie stars. Geraldine becomes a success, Michael doesn't. Then come the talkies, and the situation is reversed: Geraldine has a lisp (remember Marion Davies?), and now Michael is the hot property. Needless to say, love wins out in the end. First mooted as *Came the Dawn*, then *Come Across*, it finally appropriated Mary Pickford's nickname to become *America's Sweetheart*.

It is difficult to discuss the songs in *America's Sweetheart* because although most of the music was found in that famous Secaucus warehouse, the lyrics of all but nine have disappeared. The best-known survivor is "I've Got Five Dollars." A revised lyric was set to the melody of "Someone Should Tell Them," which was retitled "There's So Much More," and Larry had another shot at "I Want A Man." The songs that disappeared can hardly have been classics; even Rodgers couldn't remember any of them, although we might hazard a guess that "I'll Be A Star" was a song in the "You've Got That" mold. Many years later, Larry Adler recalled asking Larry Hart which of all his lyrics he was most proud of, and Larry told him it was one of these forgotten songs, "A Cat Can Look At A Queen."

"The sad thing is, all I can remember of it is the lines, 'Though it may be dark at night, A cat can see, More than you, More than me,'" Adler recalled. "But Larry said he considered that to be one of the funniest lyrics of all time. Not only the best thing he'd ever written, but the best thing anyone had ever written!" We have to assume there was an inside joke in there someplace—the title, perhaps?—that only Larry appreciated.

Now all they needed was a producer. They approached the Theatre Guild, who had sponsored a third *Garrick Gaieties* the preceding June featuring words and music by just about everyone in the business except Rodgers and Hart—Edward Eliscu, Johnny Mercer, Yip Harburg, Vernon Duke, Kay Swift, Ira Gershwin, even Henry Myers. No, thanks. Aarons and Freedley were just about holding on with *Girl Crazy* and had no intention of blowing their profits on another musical. What about Billy Rose, undergoing yet another metamorphosis as a producer? Perhaps not. Who else was there? Only Laurence Schwab and Frank Mandel.

Fortunately for Rodgers and Hart, Schwab and Mandel liked the idea and agreed to put up the money. As director Herb suggested his pal, Cole Porter intimate Monty Woolley, who had had a success with *The New Yorkers*. When auditions began, one of the applicants for the part of sweet Geraldine was a starlet who'd appeared in a forgettable 1929 Warner Bros. musical called *The Show of Shows*. She had a little round face, a breathy, nasal voice, and a formidable stage mother. Her name was Harriette Lake. Rodgers didn't like her at all, but Schwab became enamored with her. Needless to say, she got the part.

Jack Whiting was signed to play Michael; others cast included Inez Courtney, Gus Shy, and Jeanne Aubert, whose real-life husband had recently pleaded in court for an injunction to stop her acting ("The man is a dramatic critic," Dorothy Parker observed), as Premier Pictures' top star Denise Torel. Among the more successful additions to the proceedings were the Forman Sisters, Hilda, Louise, and Maxine, fresh from the Tennessee hills, who as "Georgia, Georgina, and Georgette" sang a satire on movie magazines in hillbilly style called "Sweet Geraldine."

During rehearsals, Dick became a father: a daughter, Mary, was born at Lenox Hill Hospital on January 10. Larry bought the cigars. On January 19, they began the tryouts in Pittsburgh, moving to Washington for the last five days in January. While they were there, Dick recommended the Forman Sisters as entertainment for a party. They had a good time and were paid as well. The following day one of them slipped him ten dollars commission.

After a final five-day shakedown at the Shubert Theatre in Newark, the show opened at the Broadhurst on February 10, 1931. As always on opening nights, the *Philadelphia Public Ledger* reported, Larry assaulted the phone

> with excited orders to have his suit pressed . . . attempting to do two things at once, continuing to telephone and pursuing a cake of soap in the bath tub. He dashes into a barber shop demanding that the frightened attendants do things with the razor in five minutes. He recalls twenty clever wires he had intended sending twenty members of the cast and associates, but the hands of the clock warn him that fleeting minutes play havoc with the best laid plans and he abandons his intention. Instead he hits upon the original idea of giving a list of names to the operator and a single message for the twenty. Dinner consists of one long and two short gulps. Three drinks give him sufficient strength to make the last lap to the theatre.

America's Sweetheart made no claims to being an intellectual feast. It had bad Hollywood jokes at every turn of the page: tycoon S. A. Dolan of Premier Pictures tells an aide to get hold of Gilbert and Sullivan and put them under contract, refers to Shakespeare as the Bird of Avon, suggests a movie of *Camille* with the title *Lovey Dovey*. Movie star Jeanne Aubert vamped "I Want A Man" and the rather more risqué sentiments of "A Lady Must Live." An ensemble of movie stars laments the fact they're no longer "Innocent Chorus Girls Of Yesterday."

Critical opinion was divided. Although Brooks Atkinson of the *Times* thought the wit clumsy and the humor foul, he said Rodgers and Hart had "acquitted themselves most creditably." Gabriel of the *American* thought the show "jovial, tuneful, lively," its songs "close to being the pleasantest Rodgers and Hart have composed." John Mason Brown of the *Post* liked Larry's "drolly rhymed lyrics." Sid Silverman of *Variety* singled out "I've Got

The cast of the 1916 Columbia Varsity Show, *Peace Pirates*. Larry Hart, as Mrs. Rockyford, is second from the left. The "minstrel" extreme left is Oscar Hammerstein II. *Author's Collection, courtesy Dorothy Hammerstein.*

Lorenz M. Hart, 1923, producer of "The Blond Beast." *Author's Collection.*

Rodgers, Hart, and Herbert Fields, 1927. *Vandamm Theatre Collection, New York Public Library for the Performing Arts (NYPL).*

Lew Fields, seated, with l. to r., Roy Webb, Alexander Leftwich, Herbert Fields, Rodgers, Hart, and Busby Berkeley, 1928. *Vandamm Theatre Collection, NYPL.*

Larry and Dick "acting up" a difference of opinion in *Makers of Melody*, 1929.
Courtesy of The Academy of Motion Picture Arts & Sciences (AMPAS).

Larry and Dick tell the lady reporter where they get their ideas from, in *Makers of Melody*, 1929. *Courtesy, AMPAS.*

Dick and Larry, and Larry's dog John, just back from London on the liner *Berengaria*, 1927. *Vandamm Theatre Collection, NYPL.*

Jimmy Durante, Howard Dietz, and Larry Hart in Hollywood, 1933. *Author's Collection.*

Alexander Gray and Inez Courtney in *Spring Is Here,* 1930. *Lynn Farrol Group, Inc. (LFG).*

Ben Lyon and Una Munson get ready to sing "You're The Cats" in *The Hot Heiress,* 1931. *British Film Institute (BFI).*

Maurice Chevalier pinches Jeanette MacDonald's cheek (for a change) in *The Merry Widow*, 1934. *BFI.*

George M. Cohan doesn't appear to be having much success romancing Claudette Colbert in *The Phantom President*, 1932. *BFI.*

Jessie Matthews in a scene from the British movie *Evergreen*, 1934. *LFG.*

Chester Conklin, Madge Evans, and Al Jolson in *Hallelujah, I'm A Bum!*, 1933. *Author's Collection.*

Director Edward
Sutherland and
cameraman Charles
Lang set up a scene for
W. C. Fields and Paul
Hurst (with cigar) on set
during filming of
Mississippi, 1935. *BFI.*

Helen Ford makes her entrance in
Dearest Enemy, 1925. *Author's Collection.*

Rodgers and Hart photographed on the set of *Love Me Tonight* at Paramount, 1933. *Vandamm Theatre Collection, NYPL.*

Babes in Arms rehearsal: watched by an adoring Mitzi Green, Rodgers worries over a point while George Balanchine, on chair, Dwight Deere Wiman, far right, and Larry Hart, the back of whose head is visible, look on. *Author's Collection.*

The creators of *I'd Rather Be Right*—l. to r., Sam Harris, Larry Hart, Richard Rodgers (at piano), Moss Hart, and George S. Kaufman—meet its star, George M. Cohan, far right. *Author's Collection.*

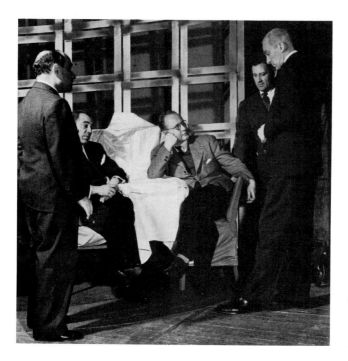

The creators of *Pal Joey* thrash out a point. L. to r., Larry Hart, Dick Rodgers, Robert Alton, John O'Hara, and George Abbott, 1940. *Mary Morris, NYPL.*

Larry and Dick
at work on *Pal
Joey*, 1940.
*Mary Morris,
NYPL.*

Larry and Dick at *By Jupiter* rehearsals, 1942. *Vandamm Theatre Collection, NYPL.*

Helen Ford accuses Edith Meiser in *Peggy-Ann,* 1927. *Author's Collection.*

Dick Rodgers and what looks like a badly hung-over Charles MacArthur in the front row at rehearsals of *Jumbo,* 1936. Sitting upper left is press agent Richard Maney, "the man who made a street walker out of Robert Benchley." *LFG.*

Ray Bolger and Tamara Geva rehearsing their "Slaughter on Tenth Avenue" ballet for *On Your Toes*, 1936. *Courtesy Mrs. Gwen Bolger.*

Eddie Bracken, left, and Hal LeRoy right, support Desi Arnaz for Marcy Westcott in *Too Many Girls*, 1939. *Author's Collection.*

Vivienne Segal makes her entrance at Robert Mulligan's nightclub in *Pal Joey*, 1940. *Vandamm photo, Theatre and Music Collection, Museum of the City of New York (TMC).*

Gene Kelly and the girls at Mike's Club, Chicago, in *Pal Joey*, 1940. *Vandamm photo, TMC.*

Ronald Graham and
Constance Moore during
a performance of *By
Jupiter*, 1942. *Author's
Collection.*

Dick Rodgers and
Oscar Hammerstein II
during a rehearsal of
Oklahoma!, 1943. *LFG.*

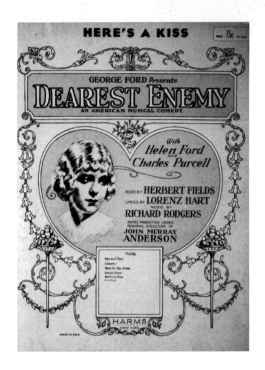

Sheet music from *Dearest Enemy*, 1925.

Poster from *A Connecticut Yankee*, 1943. *TMC.*

Five Dollars" and "We'll Be The Same" for praise and judged the rest "as sweet a group of topical lyrics as Hart has ever turned out. Maybe they're his best."

Charles Darnton of the *World* was less generous. The show had "more dirt than humor," he warned. "Smut leaves its mark on any book by Herbert Fields and lyrics by Lorenz Hart." The waspish George Jean Nathan felt much the same: "Where the book isn't covered with cobwebs it is, further- more, dully dirty," he sniffed. Excepting Dick's music from her attack, Dorothy Parker, subbing for Robert Benchley at *The New Yorker*, whaled away at Herb and Larry in fine style. Herb, she surmised, had "compiled his libretto with a pair of scissors," while Larry had produced "lyrics the rhymes of which are less internal than colonic," with "a peculiar, even for Broadway, nastiness of flavor."

Maybe the best thing was to be philosophical about it: the critics were tough on everything these days, especially musicals. A week later, for instance, Morris Green and Lewis Gensler brought Ted Healy to the Imper- ial in a revue with music by Gensler and lyrics by Owen Murphy and Robert Simon. Originally constructed around Ruby Keeler Jolson (who'd dropped out in Philadelphia, where *Variety* said her tap dancing "fell flat"), with uncredited directorial contributions by Oscar Hammerstein, it was called *The Gang's All Here*. "Who the hell cares?" snarled one of the critics.

On March 15, *The Hot Heiress* was put into national release. It was received apathetically; in a later interview even Larry and Dick admitted that it was "pretty bad, musically as well as pictorially." (Strangely enough, it was remade twice: in 1934 as *Happiness Ahead* starring Dick Powell—playing a window cleaner instead of a riveter—and yet again in 1941 as *Here Comes Happiness*.)

The failure of the movie was not a good augury for Larry Hart and Herb Fields. They preceded Dick to California at the end of the month to begin work on *Love of Michael*, their second Ben Lyon picture for Warner Bros. When they arrived, however, a cloud of gloom hung over the studio. Jack Warner's twenty-two-year old son Lewis had died when an infected tooth poisoned his system. Many of the sound stages were closed down. Warner Bros. was in the process of cutting its workforce by nine hundred people, and the front office planned to reduce the salaries of those remaining by 20 or 30 percent: everyone was retrenching.

1930 had produced some memorable musical moments: Jeanette Mac- Donald singing "Beyond The Blue Horizon" (by Hart's old collaborator Franke Harling) in *Monte Carlo*; the Marx Brothers performing "Hooray For

Captain Spaulding" in *Animal Crackers* (the fact that "Captain Spaulding" was the street name by which everyone's favorite Hollywood drug peddler was known adding a certain resonance to its lyrics); Marilyn Miller's "Who?" in *Sunny*; teenage itsy-bitsy Betty Grable leading the Busby Berkeley chorus (later Goldwyn) girls in *Whoopee*; and the amazing sequence leading up to Gershwin's "Rhapsody in Blue" in *King of Jazz*. But the exhibitors were all telling the same story: the public was becoming indifferent to musicals. So what Warner Bros. wanted to do was forget all about *Love of Michael* and buy back their contract.

Since it was clear they were in no position to bargain, Rodgers and Hart hired an attorney to work out the financial details and headed back east. For the first time the grim realities of the Depression came home to them. There were four and a half million men out of work. Average factory pay had dropped to less than seventeen dollars a week. Forty percent of the nation's farms were mortgaged. Everywhere the same grim reminders. Breadlines. Home relief offices. Apple Annie on one street corner, her husband on the other. And worst of all—they couldn't get anyone to back a show. Nobody wanted Rodgers and Hart. Except Billy Rose.

Rose had taken the Fanny Brice revue *Sweet and Low* out on the road; he brought it back in on May 19, shortly after the dedication of New York's— and at that time, the world's—highest structure, the Empire State Building. He gave it a new cast and a new score and called it *Billy Rose's Crazy Quilt*. Needing special material for his star, he prevailed upon Rodgers and Hart to write Fanny a song called "Rest Room Rose," which was doubtless a variation on an earlier Brice hit, "Second Hand Rose." No copy of it has ever been found, which suggests strongly that it was a long way from being a hit; the number that found favor with the public was "I Found A Million Dollar Baby In A Five And Ten Cent Store," which Fanny sang in top hat and tails. Forgotten though it may be, "Rest Room Rose" did manage to achieve unique status in the Rodgers and Hart canon: when the show reached Minneapolis, both it and "Million Dollar Baby" were banned for being "indecent and obscene."

June 3 saw the smash-hit opening of the new Schwartz and Dietz revue *The Band Wagon*, starring Fred and Adele Astaire in what would be their last show together. "It better be good," the cast sang as the audience took its seats. "It better be good and funny. It better be worth the money." It was. Astaire was brilliant ensemble with Tilly Losch or with his sister; the songs included "Dancing In The Dark" and "I Love Louisa"; the sketches were by George Kaufman and Howard Dietz. The critics loved it, and the show settled in for a profitable run.

Although he did not have a jealous bone in his body, one cannot help but wonder whether Larry Hart appreciated the irony of his former protégé's brilliant success, for just three days later *America's Sweetheart* closed; its star

Harriette Lake headed back for Hollywood. In July, *Ever Green* ended its run in London. The only income either Larry or Dick had now was from royalties on sheet music and records, the sales of which had been sharply reduced by the ever-growing popularity of radio.

If any of this bothered him, Larry gave no sign of it, continuing to maintain the lifestyle he had always enjoyed. The new season was imminent. Herb Fields signed with songwriter-turned-producer Ray Goetz, brother of Irving Berlin's first wife, to write the book for another Cole Porter show, *Star Dust*; Porter completed a dozen songs, but the show was never produced. Heywood Broun, the newspaper columnist, was putting together a revue on a shoestring (the budget was six thousand dollars) which he hoped would provide employment for a hundred of his actor friends. More interesting, Florenz Ziegfeld was mounting another edition—his last, it would transpire—of the *Follies*. As if to tempt the Fates, Earl Carroll was building a three-thousand-seat art deco theatre at 50th Street and Seventh Avenue to house a sumptuous new edition of his *Vanities*. The Shuberts, in spite of their shaky financial situation, were also trying another musical.

George and Ira Gershwin were preparing a new show with George Kaufman writing and directing. On October 31, the *New York Times* announced that Jerome Kern and Oscar Hammerstein II had nearly completed a musical adaptation of *Camille* for Helen Morgan; if this was so, they abandoned it and—for the moment, anyway—went separate ways. Kern teamed up with Otto Harbach to write a new show for Max Gordon, *The Cat and the Fiddle*. Oscar joined Sigmund Romberg to do *East Wind* for Schwab and Mandel. Amid all this activity, there was nothing for Rodgers and Hart.

Because Herb Fields had decided to accept a ready-money offer to write film scripts and had left for the coast, there was no hope of continuing their working arrangement. Dick and Larry worked on some ideas, but somehow nothing seemed right; and even if something had, there was still the problem of finding a producer with enough money to mount a show. However, as Hollywood's doors were still closed to them, they had to keep trying and hope that something would change their luck.

Something did: Maurice Chevalier.

21

═══◈═◈═◈═══

Hollywood Bound

O
n November 24, 1930, in association with Max Dreyfus, Rodgers and Hart announced the formation of a new music publishing house, its name a melding of theirs: Rodart Music Publishing Company. Its purpose was to exploit the existing catalogue and publish their future songs: they needed the money. The establishing of Rodart turned out to be their farewell to Broadway for four years.

Just a few days later they received a call from Mel Shauer, Larry's old camping and songwriting companion, who had gone out to California the preceding year and was now a producer at Paramount. Mel had met Maurice Chevalier in Paris, and they had become very friendly. When Chevalier came to work in Hollywood, Mel put up the suggestion that Rodgers and Hart be hired to write songs for him. The studio was not enthusiastic: the consensus was that Larry's lyrics were "too flip," perhaps too clever for the screen. "It was a struggle to sell Rodgers and Hart as a movie songwriting team," Shauer recalled. His persistence paid off, however; Jesse Lasky decided to give them a chance.

A short preamble here, necessary to understand the interplay of preceding events which resulted in Lasky's reaching out to rescue Rodgers and Hart. Tall, gentle, pince-nez'd, Jesse Lasky was an ex-newspaper reporter, cornet player, and vaudevillian who had once—long, long ago—tried to sign playwright William deMille to write a libretto; he ended up collaborating on comic sketches with deMille's younger brother Cecil.

Shmuel Gelbfisz, a refugee from Poland, had built up a successful retail glove business, and married Lasky's sister and former vaudeville partner Blanche. Tiring of gloves, he changed his name to Goldfish and proposed to his brother-in-law that they try to break into the movie business. One evening in 1913, Lasky and de Mille called in at the Lambs Club, where they encountered actor Dustin Farnum, brother of movie star William Farnum. They got talking, and someone mentioned the 1905 hit play *The Squaw Man*; Farnum said he thought it would make a great screenplay.

Next day Lasky Feature Play Company was capitalized at $25,000— $15,000 coming from the aggressive Goldfish, and $5,000 each—mostly borrowed—from Lasky and deMille. They acquired movie rights in *The Squaw Man* for $5,000 cash and $5,000 in notes, leaving $15,000 for de Mille to go to Arizona, set up a studio, and make the movie. Meanwhile Lasky and Goldfish worked feverishly to set up some kind of distribution deal.

This incredible gamble—two men who'd never produced a picture entrusting their entire capital to a twenty-five-dollar-a-week sketch-writer who'd never even directed traffic—paid off. Cecil deMille did not like the look of Arizona so proceeded as far west as the train would take him: to Los Angeles. There, in a sunny suburb, he set up his studio. The rest is Hollywood history.

The Squaw Man was immediately followed by another deMille feature, *Brewster's Millions*, which was an even bigger success. So much so that in 1916 Adolph Zukor, ex-furrier and motion picture pioneer—and father of Larry Hart's Camp Paradox friend Eugene—suggested that they merge with his company, Famous Players, the most famous of whom was Mary Pickford, a.k.a. America's Sweetheart; in short order Famous Players–Lasky was born. Zukor, a witty but calculating character, became president, Jesse Lasky vice-president. Nearly forty years would pass before Lasky addressed him as anything but "Mr. Zukor."

It was not long before Zukor decided to include Jesse's excitable brother-in-law out of the deal. "Mr. Lasky," Zukor is reported to have said, "Famous Players–Lasky is not big enough to hold Mr. Goldfish and me." By November 1916, Sam had entered into a partnership with the Selwyn brothers. He clipped off half of his own name and added half of the name of his partners to form the name of the new company—Goldwyn—and in due course adopted it as his own.

Paramount had begun life in 1914 as a distributing organization headed by a former Ogden, Utah, theatre owner named W. W. Hodkinson. Zukor pooled Famous Players–Lasky's interests with Paramount, the producers taking 35 percent, Hodkinson becoming president. Eventually Paramount tried to swallow Famous Players; some devious infighting by Zukor resulted in the $25 million purchase of Paramount's share of the business by Famous Players–Lasky.

B. P. Schulberg, father of novelist Budd, took over as head of production. By the time Rodgers and Hart appeared in *Makers of Melody* in 1929, Paramount had become the dominant force in the movie industry, with skyscraper offices on Times Square, and a nation-wide chain of eleven hundred theatres, each of which played an average of 175 features a year. The majority of these features were created at the Famous Players–Lasky studios on Marathon Avenue in Hollywood or the imposing pillared complex at Astoria.

⸻

Amiable and considerate, Lasky was one of Hollywood's better-liked production executives. But he had a problem, and the problem was Maurice Chevalier. At thirty-two, Chevalier was already something of a legend. Through Mel Shauer, he came to Hollywood in 1929 and signed with Paramount. He succeeded immediately in *Innocents of Paris*, singing a song that was to become a sort of signature tune, Richard Whiting and Leo Robin's "Louise."

He went directly into another success, *The Love Parade*, directed by Ernst Lubitsch and featuring Jeanette MacDonald in her screen debut, which won him a Best Actor nomination in the newly instituted Academy Awards— "Oscars." His next screen appearance was in the portmanteau 1930 spectacular *Paramount on Parade*, which featured practically every player under contract to the studio at the time. After that he co-starred with Claudette Colbert in a pedestrian nothing called *The Big Pond*, filmed at Astoria in both English and French.

Stopping only to do a one-night stand for Dillingham at the Fulton on March 30 with Eleanor Powell and Duke Ellington's Cotton Club Orchestra, Chevalier returned to Hollywood in *Playboy of Paris*, in which his co-star was Frances Dee. Even the Whiting-Robin song "My Ideal" couldn't save it. Reunited with Lubitsch (and Claudette Colbert) for *The Smiling Lieutenant*, a remake of Ludwig Berger's *Ein Waltzertraum*, Chevalier delivered another solid hit.

What the public wanted, however, was more Chevalier and MacDonald, so (although she detested him—she called him a "shameless bottom-pincher"—almost as much as he detested her) they were teamed again in *One Hour with You*, directed again by Lubitsch and George Cukor. Whichever director one gives the credit to, it was a sophisticated treat. Unfortunately, the producers were running out of ideas. It was at this point that Mel Shauer reminded Jesse Lasky about Rodgers and Hart. Lasky remembered them from *Makers of Melody* and offered them a contract.

Larry and Dick arrived in Hollywood in November 1931. At first Larry stayed with the Rodgers family in a Beverly Hills house rented from Elsie Janis. He was still impossible to live with, Dorothy Rodgers said:

> He decided, since it was our house and we were running it, that his contribution would be the liquor: he would buy it and he would dispense it. If we were having guests for dinner he would make the cocktails . . . so when our guests arrived,

Larry would go out in the pantry and there'd be one for him and one for the cook, and there'd be one for him and one for the cook, and pretty soon—no dinner!

He also loved to smoke big black cigars and he would think so hard about what he was working on that he would forget completely about the cigar and the fact that it was probably twelve inches long. One day he stood in front of the window looking out and thinking—and burning a huge hole in my curtain and not even knowing that he had done it. But you couldn't get mad at him somehow. He was so generous, so warm-hearted, so full of enthusiasm, that everybody loved him.

"Dick didn't like Hollywood at all," Mel Shauer added. "Larry accepted it matter-of-factly and proceeded to have a good time." Dick would go to parties because one met the right people; Larry because there'd be plenty of liquor, lots of people to talk to, and good dirty gossip. Thus both of them appeared shortly after their return on the guest list, faithfully reproduced by the *Hollywood Reporter*, for an opulent dinner dance given by director Edgar Selwyn and his wife Ruth at their Beverly Hills mansion for Mr. and Mrs. Nicholas Schenck. Held in a huge, peppermint-striped marquee raised over the tennis court, it was "the most elaborate as well as the largest attended party of the week," gushed the *Reporter*, adding, "There is always much celebrating by the Schencks and the Selwyns when the former come west and the latter go east."

"I was living alone in a little apartment in Hollywood," Mel Shauer said. "I didn't want an elaborate place. When I suggested to Larry that we take a house together, sharing the expenses, I thought of a small bungalow. But Larry rented a lavish house at 910 North Bedford Drive, a really large place with an Olympic sized swimming pool."

The house on North Bedford Drive was as untypical as its new tenant. Back in 1923, silent star Norman Kerry, a dashing leading man and occasional villain—and also heroic drinker—whose career ran from 1916's *Manhattan Madness* through 1931's *Air Eagles*, including such epics as *The Hunchback of Notre Dame* and *Phantom of the Opera*—purchased a very special, architecturally significant home that stood on the corner of Wilshire Boulevard and Berendo Street in Los Angeles.

Designed by architects Charles and Henry Greene of Pasadena, the house had a large, sloping roof, low-pitched gables, open sleeping porches (it was considered very healthy to sleep out of doors then), and exposed wooden structural elements. It was built in the Craftsman style, which had originated in England in the wake of the industrial revolution and was also known as the Arts and Crafts style, and featured handcrafted floors, doors, windows, and tiles in handsome woods, as well as handmade lighting fixtures and Tiffany glass. Designed to let in the maximum of fresh air, with windows placed in the front corresponding directly with windows in back so that when they were opened, fresh air would ventilate and cool the house, the architects also maximized the use of natural woods; the pieces were assembled like a puzzle, using wooden pegs instead of nails. These idiosyncracies may explain why its present owners have dubbed the house "Squeaky Hollow."

The same year Norman Kerry bought the Berendo house, he also purchased three lots in Beverly Hills, on the corner of Bedford Drive and Benedict Canyon. He called upon the Greenes to dismantle the downtown house and move it to this new location, building a new wall featuring Clinker bricks with inlaid Batchhelder tiles around it, and bringing in landscape artists from the Beverly Hills Nursery to design the gardens. A swimming pool sixty feet wide and twenty feet long was added on the south side. It must have been from Kerry, whose career had foundered with the coming of sound, that Larry Hart rented the house.

Mel insisted they split expenses for the house down the middle. Larry just said, "Yeah, yeah," and paid all the bills. Then he brought out Frieda, Teddy, the two servants, and Kiki the chow dog. "You see," Larry told Mel, "we've got too many here. We can't split expenses."

"He'd remember, almost to the penny, what his checking balance was, though you'd never think he bothered keeping track," Mel said. "Yet he'd forget his home phone number [OXford 2113] and have to ask me what it was, time and again."

Rodgers and Hart learned that the director assigned to their movie, which was to be called *Love Me Tonight* and scheduled to start shooting March 28, was George Cukor. A week later, Cukor was out, and producer Rouben Mamoulian, thirty-four, was in. It turned out to be an inspired choice. Born in Tiflis, Russia, of Armenian descent, Mamoulian was an owlish, exuberant, chain-smoking genius with a shock of black hair. He had worked with the Theatre Guild back east, where he had produced Du Bose Heyward's *Porgy*. In 1929, he had directed *Applause* at the Astoria studios, turning a routine sob story into a brilliant piece of movie making. His second movie, *City Streets*, was a gangster story, again told in deft cinematic terms. He had just completed *Dr. Jekyll and Mr. Hyde*, starring Fredric March, better known to Dick and Larry as Freddie Bickel from *The Melody Man*. Everyone said it was going to win the Academy Award, and it did.

"The first thing we did," said Larry, describing how they went about writing the movie, "was to study pictures, not on the sound set but in the cutting room."

Then, with Chevalier and Rouben Mamoulian, we developed for the first time dialogue with a sort of phony little half-rhyme, with a little music under it cut to the situation. We also put a portable soundtrack in an open field with an orchestra. We had a doctor coming to Jeanette MacDonald's room, and the sing-song conversation went something like this:

Now, my dear, remove your dress.
My what?
Your dress.
Is it necessary?
Very.

It isn't rhyme, it isn't anything like it; but it's screen talk and it isn't difficult if you know the medium. I'm a great believer in conversational rhythm. I think in terms of rhythmic dialogue. It's so easy, you can talk naturally. It's like peas rolling off a knife. Take the great screen actors and actresses, Bette Davis, Eddie Robinson, Jimmy Cagney, Spencer Tracy. They all talk in rhythm. And rhythm and movement are the life of the screen.

Larry's idea of using rhymed dialogue wasn't quite as new as he suggested; in a 1932 show with Jerome Kern called *Music in the Air*, Oscar Hammerstein had experimented with rhymed dialogue spoken over musical accompaniment. Rodgers and Hart were simply taking the idea to its logical destination—the screen.

For once—they would discover all too soon how rare it was—Larry and Dick had been teamed with a director who eagerly solicited their contribution and, more important, thought the same way they did. "One of the first things he insisted on," Rodgers recalled, "was that I compose all the background music, not simply the music for the songs. This was—still is—highly unusual."

Mamoulian's then-revolutionary techniques enabled Larry and Dick to work in a completely different way. Music and lyrics would be interwoven into the action. Instead of filming the performers from a static position while they sang or danced, the camera would move with them, among them, past them, over them. Songs would be pre-recorded so that they could concentrate on acting them. The songs sprang forth almost effortlessly: "That's The Song Of Paree," "Isn't It Romantic?," "Lover," "Mimi," "Love Me Tonight." One song, "Give Me Just A Moment," was deleted from the screenplay before shooting commenced. Another, "The Man For Me," never reached the screen either, although it was a good deal better than many songs which did that year.

One day as Rodgers and Hart were working in their cramped Paramount cubicle, they were surprised to receive a visit from Chevalier, dressed from head to foot in his favorite color, blue. He asked to hear the songs, and Dick played them while Larry did his usual performance on the lyrics. Chevalier listened expressionlessly; when they were finished he left without a word. They were stunned. If Chevalier didn't like the songs, they would be fired off the picture in a shot, Mamoulian or no Mamoulian.

Both of them had a sleepless night. Next morning Chevalier breezed into the office again. "Boys," he said, putting his arms around them, "I just had to come back and tell you. I couldn't sleep a wink last night because I was so excited about your wonderful songs."

Chevalier became very friendly with both Dick and Larry and visited their homes. He was not a success at the Harts', where the formidable Big Mary was now in residence. The reason for his unpopularity was that Maurice was renowned as a tightwad who never picked up a check and never left a tip: they said, although he didn't smoke, if you offered him a cigarette, he would take it and put it in his pocket. After spending the weekend with the Harts, during which time he was treated royally, Maurice got ready to leave. He didn't even say thank you to Big Mary. As he went out the door she slipped him a dollar as a tip. To her everlasting disgust, he took it.

Love Me Tonight, which began shooting, after a short delay, on April 4, was—still is—a small miracle of movie making. In its opening sequence of Paris awakening, the camera picks up a road mender digging, a housewife sweeping her stoop, a cobbler and his assistant banging nails into shoes, smoke belching from chimneys, shutters being thrown wide, tablecloths being shaken, sounds blending rhythmically (and uncannily reminiscent of morning on Catfish Row in *Porgy*) as the camera swoops into Maurice (Chevalier) the tailor's room. He dresses, singing, and hits the street calling "How are you?" to neighbors on the way to his shop.

The screenplay was inventive; it was bold; it was witty; it was clever. Rodgers's music was fresh and lyrical, notably in the famous extended "Isn't It Romantic?" sequence, where he switches effortlessly from jaunty to martial to lilting, holding back the sweet *zigeuner* crescendo of the melody until it lifts the camera—and the audience—on "love, perchance," to Jeanette MacDonald on her balcony. Larry's lyrics were unfailingly apposite and witty; in "Lover," which Jeanette sings as she drives through the woods in her carriage, key end-of-line rhymes like "woe" and "hay" become her commands to the horse. The history of the line "Like two children playing in the—hey!" suggests Larry tried to get a slightly risqué phrase past Mrs. Grundy—or in this instance, movie censor Will Hays-("in the hay" being slang for "in bed together") but didn't: "rolling" was altered to "playing."

Maurice Chevalier clearly knew exactly how to perform for the camera; the camera in turn loved him, even when he was laying on the Gallic charm with a shovel. For the first time, Jeanette MacDonald was able to hint at the comedienne inside the Iron Butterfly persona, but even so, as one critic remarked, it was difficult to know why Chevalier preferred her to Myrna Loy. Loy, rescued at last from playing Chinese princesses, not to mention the daughter of Fu Manchu, stole every scene she appeared in as the man-mad Countess Valentine. "Don't you ever think of anything besides men?" Jeanette asks exasperatedly. "Oh, yes," answers Myrna, wide eyed. "School-boys." On another occasion Charlie Ruggles rushes into a room where she is languishing on a chaise longue. "Could you go for a doctor?" he asks her. "Sure!" she replies, her lethargy disappearing instantly. "Bring him on."

The supporting cast was as classy as the top billing: as well as Loy, there were Charlie Ruggles, Charles Butterworth, C. Aubrey Smith (singing!), and the three aunts—Elizabeth Patterson, Ethel Griffies, and Blanche Frederici—who pop up from time to time like Macbeth's three witches. Other felicities included singing butlers and chambermaids, and an extended, "choreographed" deer hunt featuring a slo-mo ballet. All combined to make *Love Me Tonight* unique.

When Hart wasn't working on rewrites of his lyrics—he composed completely new verses and refrains for the published versions of "Lover" and "Isn't It Romantic?"—he wandered around the lot, watching other productions being filmed, always curious, always ready to stop to gossip, smoke a cigar, have a drink. Among the old New York friends he found at Paramount was Henry Myers, who had parlayed his collaboration with Oscar Hammerstein and Otto Harbach on the book of the Kalmar-Ruby show *Good Boy* into a two-year contract with the studio.

> We had met earlier at a party, where he had arrived ahead of me, and when I entered the big reception room he hurried over to me, flung his arms around me, and exclaimed: "You fabulous sonofabitch!" He asked me almost instantly what were the terms of that first contract of mine, and I told him: $500 a week for the first three months; then they had an option on me at $650 a week for nine months; then $750 a week for a year. I expressed my opinion that this did not compare with princely salaries I had been hearing about, but he said, positively: "No! This is what they all really get." After a moment he added: "I wish I was sure of it for the rest of my life."

Henry was working on a film which had come about when studio head B. P. Schulberg sent round a staff memo saying that since the 1932 Olympics were to be held in Los Angeles, he urgently wanted a story built around the event. Joe Mankiewicz, then a staff writer at Paramount, sent in a goofy satire called *On Your Marks*, which the studio bought for twenty-five hundred dollars, and which his brother Herman was assigned to supervise. With W. C. Fields signed to make his Paramount feature debut after a four-year absence from the movies, Herman tapped Henry Myers, "a marvelously daffy and very funny man," according to Joe Mankiewicz, to collaborate. The story they came up with was called *Million Dollar Legs*.

Shooting began; chaos ensued. Director Edward Cline kept insisting Fields stick to the script; Fields could no more do that than stay off the gin.

In the end, Cline bowed to the inevitable, and Fields ad-libbed everything. Myers and Mankiewicz stuck in every bit of crazy business they could invent, remember, or steal from anyone who happened to be passing. Anarchy was the norm. When B. P. Schulberg accused the writers of being disloyal to the studio, Henry Myers's response was to compose an honest-to-God Gregorian chant and march into Schulberg's office singing, "Let us all give our loyalty and best co-operation/ To Paramount Publix Pictures, a Delaware corporation." Another day Mankiewicz rushed into the front office to offer the marketing boys a new slogan: "If it's a Paramount Picture, you don't have to stand on line."

The predictably outrageous *Million Dollar Legs* was set in the mythical kingdom of Klopstokia, where everyone was sports-crazy; apart from Fields it featured Jack Oakie, Andy Clyde, and Hugh Herbert. There was even a fleeting appearance by Betty Grable. The part of Mata Machree, "the greatest woman spy of all time, a woman no man can resist," was played by Lyda Roberti. One scene required her to vamp the entire Cabinet of "President" Fields. Ralph Rainger wrote her a song to go with the scene but couldn't come up with the right words. It just so happened Larry Hart, whose brother Teddy also had a bit part in the movie, had wandered onto the set. Rainger—or someone—asked Larry to help out, whereupon, Myers recalled:

> Larry wrote the lyric, anonymously, Lyda Roberti sang the song most effectively, and Larry furtively told me about it. "But don't let Dick know!" he enjoined. "He'd kill me!"
>
> My only familiarity with that song comes from having seen the finished picture. A few of the lines that have stuck with me are:
>
> I'm terrific when I get hot
> I'm just a passionate forget-me-not,
>
> and
>
> I'm amazing when I get mean
> I'm just a lady made of gelatine.

At least four other musical numbers were written, possibly more; all the music and lyrics are credited to Rainger. Myers said one number, the Klopstokian national anthem, was actually the verse of *The Love Parade* recited backwards. The one he credits Larry with writing was probably "It's Terrific When I Get Hot," registered as an unpublished work with the Library of Congress May 18, 1932, by Paramount's publishing arm, Famous Music; however, there is no copy of the song in the Library's collections.

This wasn't the only instance of Hart's writing a lyric for anyone who needed one; in fact, all the other writers dropped into his office all the time, seeking advice, looking for ideas. So heavy did this traffic become that one of Larry's acquaintances chided him for being so profligate with his own talent. Didn't he know the people he was helping were taking him for a ride, exploiting him without compunction, coining money at his expense? "Oh, come on," Larry said impatiently. "They're only words."

That first weekend in April, after which *Love Me Tonight* was to begin principal photography, Dick and Dorothy Rodgers threw a buffet supper at their Beverly Hills home. Among the guests were Groucho Marx, Herman and Joe Mankiewicz, Ernst Lubitsch, Miriam Hopkins, David and Irene Selznick, Inez Courtney, Myrna Loy and Arthur Hornblow, Fredric March and Florence Eldridge, Billie Dove, Howard Hughes, and many other famous faces. Maurice Chevalier and Jeanette MacDonald sang, Bert Kalmar did card tricks, Dick played Rodgers and Hart. And Larry? The newspapers reported that, as usual, he was "all over the place."

In fact, he was all over the place all the time. "The house on [North] Bedford [Drive] was great for entertaining, and entertain Larry did," Mel Shauer said. It became the place where the Hollywood famous congregated, or freeloaded. Any new actor who came to town quickly got the word. "Go to Larry Hart's. Everyone's welcome."

Mel Shauer's main worry—as his producer—was that Larry's penchant for having a good time would interfere with his studio work, and indeed, there were one or two occasions when lyrics were not ready when they were needed. But then, late at night, after everyone had left, Larry would sit down in his shorts, light up a cigar, and set to work. And next morning he would charge into Mel's room at some ungodly hour and shake him awake, saying, "Get up, get up! I want you to hear what I wrote!"

Directors, stars, producers, writers, everyone came to Larry's. And a lot of unknowns, hopefuls, all of them handsome, most of them broke, came as well. Larry didn't care: the more the merrier, as far as he was concerned. In the summer of 1932, the Olympic Games opened in Los Angeles. Mel Shauer knew the coach of the Olympic team and others. They invited the entire Olympic swimming team to a party. All the movie stars came as well. It was the party of the year.

Studio screenings of *Love Me Tonight* impressed the front office sufficiently for Rodgers and Hart to be offered another contract. This time they were to write a score for a movie based on a George F. Worts novel, *The Phantom President*, which would be the first starring vehicle of "the Man Who Owned Broadway," George M. Cohan. Paramount felt they had effected a major coup in getting him. The Yankee Doodle Boy had flirted a number of times with the idea of appearing in talkies but—perhaps instinctively realizing his style wouldn't work on film—had hitherto always backed out. His only firm commitment had been with United Artists in 1928, not to star, but to write, produce, and direct a movie for Al Jolson with Cohan songs. The movie was never made.

Neither Larry nor Dick had ever met Cohan, although of course they knew—as everyone in show business knew—who he was. Fifty-three years of age, Cohan had been everything there was to be in the theatre: boy violinist, dancing bootblack, songplugger, vaudevillian, playwright, songwriter, stage director, actor, dancer, producer, theatre owner; but the significant words in all this were "had been." His star was waning, and everyone knew it except Cohan.

It transpired he had signed with Paramount on the understanding that he would write his own material—as he had done throughout his career. It must have been galling for someone as vain, hard-headed, stubborn, and impetuous as Cohan to learn the bitter truth of Sam Goldwyn's oft-quoted saw to the effect that a verbal contract isn't worth the paper it's written on. And more galling still to hear that he was going to have to perform songs by two college-boy composers who'd still been in diapers when he was headlining on Broadway in *Little Johnny Jones*.

Cohan didn't like Rodgers, and he didn't like Hart. In fact, he didn't like anyone. He seemed to think the studio people weren't treating him with proper deference, and he responded with disdain. It was obvious from the start he regretted having agreed to make the picture and wanted everyone to know it. "Those fellows didn't know anything about me," he told a biographer. "Lot of them had never heard of me and didn't care about being told. They treated me like a man from another world. On the level, kid, Hollywood to me represents the most amazing exhibition of incompetence and ego that you can find anywhere in the civilized world."

Jimmy Durante's biographer Gene Fowler suggested Cohan was dubious about the picture long before he came west; and matters did not improve when he drove up to the studio on his first day and the doorman wouldn't let him in—because he had never heard of him! Fowler also indicated that, in repayment of a favor Durante's former partner Lou Clayton had done for him in New York, Cohan determined to "throw" the picture to Durante.

With director and former child actor Norman Taurog in charge, and Claudette Colbert providing the glamor, *The Phantom President* began shooting on Monday, July 18. The story was a simple one: "Doc" Varney, who runs a medicine show, just happens to be the exact double of Theodore K.

Blair, presidential candidate. Trouble with Blair is, he's stuffy; his party's leaders—one of them Sidney Toler, looking like Charlie Chan long before he ever actually played him—prevail upon Doc to campaign in Blair's stead, with predictable complications where Blair's sweetheart Felicia Hammond (Colbert) is concerned.

Rodgers and Hart wrote half a dozen numbers for the film, as well as quite a lot of rhythmic dialogue and special material for Durante, on which the actor "collaborated" with Larry. In the opening sequence (one of the movie's few inventive moments) "The Country Needs A Man" was a dialogue between four portraits: Jefferson, Washington, Lincoln, and Teddy Roosevelt. Alan Mowbray, famed later for his portly Englishmen but slimmer then, played Washington. The Lincoln part was taken by Charles Middleton, still remembered fondly as Ming the Merciless of Mongo in the Flash Gordon serials.

The party leaders find the man the country needs. He is, of course, Doc Varney (Cohan), whom we first meet as his partner Curly Cooney (Durante) spiels the crowd into Doc's medicine show with a series of Durante specialties. Hammy beyond belief, these are but curtain raisers to Cohan's "Somebody Ought To Wave A Flag"—performed first in blackface, then without—accompanied by a demonstration of the stiff-legged, strutting, back-flip-off-the-proscenium Cohan style of dancing so perfectly caught by James Cagney playing Cohan in *Yankee Doodle Dandy*.

Cohan was utterly unconvincing as a romantic lead; playing against wood, Colbert—herself not the most expressive of actresses—was unable to do much opposite him. She is supposed to be confused because one minute her suitor is charismatic (when he's not Blair but Varney) and next boring (when he's not Varney but Blair), although moviegoers probably could not detect any difference. She drives to a secluded glade: cue romantic music.

Taurog clearly did not think Cohan could carry off a ballad, so Rodgers and Hart came up with a song the birds, bees, and frogs could "sing." Originally the camera was to hold on Colbert while a lady frog sang Colbert's "thoughts" and then on Cohan while a male frog "replied." Then birds and bees—clunky models that wouldn't have fooled a ten-year-old kid even then—would adjure him to "Give Her A Kiss." In the final print, the verse was cut, leaving the whole scene looking like an afterthought. Colbert is stranded; Cohan merely looks uncomfortable. With absolutely no onscreen chemistry between them, the story flounders, and no amount of lunatic mugging by Durante—often nearer brute than cute—can rescue it.

When the movie was completed at the end of the first week in September, Cohan took the first train back to Broadway. "When I left," he said, "I thanked them for a million laughs. I didn't say goodbye to any of the executives. Couldn't find them. They were away on weekends. . . . Parties? I didn't see any. I went to somebody's house for dinner one night and that was about all. If you want to ask me why in hell I ever went out there I can only say that maybe I was laughed into it. My lawyers did it, I guess, Cap O'Brien and Jesse Lasky and a few others. Hereafter I'll get my sunshine at the ball park."

Shortly after the completion of *The Phantom President* tragedy struck Dick and Dorothy Rodgers. Dorothy—pregnant for the second time—went into premature labor and was rushed to Cedars of Lebanon Hospital. The child, a girl, lived only a few minutes. As soon as his wife was well enough, Rodgers terminated the lease on their furnished house on North Elm Drive, and they, too, returned to New York.

22

A Jolson Story

By the end of 1932, American industry was operating at less than half its 1929 volume, the total amount paid in wages 60 percent less. For every four automobiles that had been produced in that year, only one was made now. Construction of buildings was often halted so abruptly that naked girders were left rusting in the open. Foreign trade slumped, blue chip stocks fell, crop prices plummeted. In the wake of the collapse of the First National Bank of Beverly Hills, salaries throughout the movie industry were cut between 10 and 30 percent.

There were now more than thirteen million unemployed—one in four of the national labor force. Of these, nobody knew how many were wandering the country; one million, perhaps even two. The Southern Pacific had given up on ejecting vagabonds, most of them young men between sixteen and twenty-five, from its freight cars. Jobless and homeless, men spent their days scavenging or begging, building makeshift shelters in parks or empty city lots. In June, the Republican convention at Chicago renominated Herbert Hoover; two weeks later, the Democrats nominated Franklin D. Roosevelt, who in his acceptance speech promised the country a "New Deal."

Twelve million homes—two out of every five—now had radios. In May, NBC put on a new comedy show featuring Jack Benny (although it would be some time before he developed the cheap, vain, egotistic boob persona for which he became famous). The same month, Ed Wynn made his debut as the Texaco Fireman. This was not just a radio character he created; the idiosyncratic Wynn actually dressed up as a fireman. "As far as Ed was concerned," George Burns said, "radio was just a stage show being broadcast. He played to the studio audience. During the broadcast he wore makeup and even changed his costume as many as six or seven times." In June, Irving Berlin was on the Ed Sullivan show. Will Rogers, Rudy Vallee, Burns and Allen, Eddie Cantor, Joe Penner, Jack Pearl—essentially repeating their vaudeville acts on the air—were all making it big. And would make it bigger.

In July, an era came to an end. Florenz Ziegfeld's Broadway fortunes had sunk so low that the preceding year he had accepted financing for a new show from the only place he could get it: gangsters Dutch Schultz and Waxey Gordon. The revue *Hot-Cha!*—Schultz and Gordon wanted to call it *Laid in Mexico*—featured Lupe Velez, the "Mexican Spitfire" of the movies; in support were Bert Lahr and Buddy Rogers.

It opened March 8, an event completely overshadowed by headlines about the kidnapping of Charles Lindbergh's baby son two days earlier. Velez was a disaster, frequently drunk, always erratic. As *Hot-Cha!* foundered, Ziegfeld became obsessed with what he felt was his failing virility. He began taking pills and hormone treatments; his palatial home became the scene of a series of weekend orgies, with chorus girls ferried in relays to Hastings-on-Hudson in Ziegfeld's Rolls-Royce.

A weekly radio show and a successful revival of *Show Boat* in May somewhat revived the producer's fortunes but took a heavy toll on his health. When in a June heat wave ticket sales dropped alarmingly, every member of the cast and Ziegfeld's staff had to take pay cuts. Early in July, Ziegfeld contracted pleurisy, and his wife took him out to California. After a brief rally, he was transferred to Cedars of Lebanon Hospital, where he died, July 22, 1932, a million dollars in debt.

Times were hard in the movie industry as well. Box office receipts were half of what they had been in the halcyon days of 1931. At MGM, Louis B. Mayer, weeping enough crocodile tears to win an Oscar, persuaded everyone at the studio to follow his example and take a 50 percent pay cut, although forty-dollar-a-week secretaries and $1.50-an-hour studio technicians complained justifiably that his taking a cut from a million to half a million a year wasn't sacrifice on quite the same scale.

At Paramount, too, overexpansion and the pressure of falling attendance led to a vicious power struggle for control of the studio. Sam Katz of Balaban and Katz, the Chicago theatre chain, and Sidney Kent, a self-made ex-stoker from Lincoln, Nebraska, whom Zukor had put in charge of distribution in the early twenties, moved in on the company through its subsidiary, Publix. In the battle which followed, only Zukor of the original partners survived. Overextended and outmaneuvered (or "outkatzed," as studio wits had it), Jesse Lasky was bankrupted and forced out; Ben Schulberg—deeply enmeshed in an extramarital affair with Sylvia Sidney—was summarily dismissed.

In such an uncertain climate, it seemed unlikely Rodgers and Hart would get a further commitment from Paramount, so Larry briefly rejoined his partner in New York, only to find there was nothing for them there either. So, even as *Love Me Tonight* opened at the Rivoli in New York, the newspapers reported Larry and Dick were back together in Hollywood (Dorothy Rodgers had remained in New York) helping with recordings of *Phantom President* and "huddling on the subject of Al Jolson's next production." How this came about requires a short preface which might be entitled "The Banana Bag Contract."

At the end of 1928, not long after he married Ruby Keeler and found himself sittin' on top of the world ("*Abie's Irish Rose* comes true!" shouted the tabloids), Jolson was sunbathing on the rooftop of his Palm Springs home with Joe Schenck, president of United Artists. Joe and his Russian-born brother Nicholas—the man for whom those lavish Selwyn parties were thrown in Hollywood—had been top executives with the Marcus Loew organization, which eventually became the parent company of MGM; Nick was now its president. Joe had cut loose in 1917 to become an independent producer of films starring his wife Norma Talmadge and her sisters Constance and Natalie, as well as Roscoe "Fatty" Arbuckle and Buster Keaton. Chairman of United Artists since its founding, in 1924, he was one of those who had stubbornly insisted in the early days that sound was just a passing fad.

During their sunbathing session Schenck made Jolson an unheard-of offer: two million dollars if Al would star in four UA pictures. Needless to say, Al agreed, and a contract was drawn up on their paper lunch sack: hence the title of the story. There was just one small problem: Jolson was still under contract to Warner Bros. Schenck agreed to keep their deal a secret until Jolson was free.

Shortly afterwards, the first of Jolson's three remaining pictures for Warner was premiered. Originally released as *Little Pal* (the title song was a blatant—and hugely inferior—rip-off of "Sonny Boy"), and changed to *Say It with Songs* for the New York opening on August 6, 1929, it was so bad it was unceremoniously withdrawn within forty-eight hours. Jolson's next, *Mammy*, finished shooting just before the Wall Street crash, in which the singer lost a mere million and a half dollars. He accepted his loss as philosophically as the failure of the movie the following March; in fact, he even offered to give Warner Bros. their money back rather than make the third picture, but the studio wanted to go ahead. In spite of its early use of Technicolor, *Big Boy* became Jolson's third flop in a row.

Meanwhile, Joe Schenck's executives were predicting—like everyone else in the business—that 1931 was going to be a bad year for musicals; the Jolson projects were put on hold. While he waited, Al went back to Broadway to do a show—for the first time not wearing blackface—for the Shuberts, *The Wonder Bar*. Opening March 17 with a $6.60 top, it ran in profit for a while; but toward the end of April ticket sales began to slide, and—after a series of non-appearances by Jolson—it closed on Saturday, May 30, a ten-week disaster. Piqued—and with nothing much else to do—Jolson took the show on a road tour, beginning in Newark on September 18 and winding up in San Francisco at the end of March 1932. He reported from there to United Artists.

The property Schenck had originally intended for Jolson was the 1929 stage hit *Sons o' Guns*, which United Artists planned to film dual-language, German and American. This idea had been abandoned because its Great War theme was now too dated; instead Jolson's first movie for UA would be a Ben Hecht original, *The New Yorker*, about the effects of the Depression on the hobos in Central Park, with a screenplay by Ziegfeld's old stand-by, William Anthony McGuire. Knowing McGuire was a drunk, Jolson vetoed him, but approved Schenck's second choice, ex-Harvard, ex-Columbia, ex–*New York Times* playwright S. N. Behrman.

By June 1, a draft screenplay had been completed. It revolved around Bumper, "mayor" of the Central Park hobos, and his friend John Hastings, Mayor of New York. Bumper is carefree until he rescues a girl from drowning; she has amnesia. Unaware she is Hastings's girlfriend, Bumper falls in love with her and goes straight, gets a job in a bank, where he finally learns the truth about the girl, reunites the lovers, and returns to his carefree existence in the Park. Jolson was underwhelmed; the inclusion in the script of a character named Egghead w. spouts a very direct variety of socialism—in one scene he refers to the m nted cops in the Park as "Hoover's Cossacks"—did nothing to nthuse i. . Was this the vehicle that was supposed to revive his flagging career?

The movie was scheduled to begin at 10:00 a.m. on Thursday, July 7, under the new title *Happy Go Lucky*, with Roland Young as Mayor Hastings, Madge Evans as his girlfriend June Marcher, silent-movie comedian Harry Langdon as Egghead, and Edgar Connor as Acorn, Bumper's black buddy. Irving Caesar was signed to write the songs. Lewis Milestone, originally slated to direct, was still busy on the Joan Crawford movie *Rain*; Schenck brought in the exotic Argentinian Harry d'Abbadie D'Arrast. Within hours of starting, Jolson balked, demanding changes. Shooting was abandoned the same morning.

"They had me playing almost a silent character," Jolson told a reporter later. "All I said for half the picture was 'yes' and 'no.' I had no dialogue. Now that isn't what people expect of me. I know what kind of stuff I can do on the stage. That's the kind of dialogue I want now." But that was only part of the problem; a hint of the on-set tension may be had from the fact that at the same time D'Arrast went on record as saying he loved the story but thought Fred Astaire would be much better for the lead.

Schenck's response was to take D'Arrast off the picture and bring in Viennese-born Chester Erskine, a Broadway producer and director on his first Hollywood assignment as Milestone's assistant on *Rain*. Production recommenced July 21. By August 26, Dick Rodgers and Larry Hart had been brought in. To their dismay, however, they discovered the picture was a mess.

Irving Caesar was paid off and Rodgers and Hart assigned to write the score, with Larry—who was particularly enthused, because he admired Jolson tremendously—providing rhythmic dialogue. Writing furiously against the clock and the shooting schedule, he and Dick produced what might almost be described as a semi-operatic score, with recitative dialogue and song so closely intertwined that only five individual melodies emerge from the entire soundtrack: the title song "Hallelujah, I'm A Bum!," "I Gotta Get Back To New York"—one of the most undeservedly neglected songs ever written about that city—"What Do You Want With Money?," "You Are Too Beautiful," and "I'd Do It Again."

With another new title, *Heart of New York*, filming was resumed in late September—most of the Central Park locations were shot at the Riviera Country Club in Pacific Palisades—and continued until mid-October, when Jolson flew to New York to sign a thirteen-week radio contract (at five thousand dollars a week) with General Motors. On October 15, Schenck called him back: they were in trouble. Perhaps sensing disaster, Roland Young had become "ill" and could no longer continue. Extensive reshooting was scheduled, with Frank Morgan now playing the part of Mayor Hastings; Schenck promised Jolson it would be completed by November 14, which gave them three weeks to rewrite the script, the rhythmic dialogue, and the songs, then rehearse and reshoot the film.

Just how hectic it was is apparent in a letter Rodgers wrote to his wife in New York in October, 1932. Apart from writing and preparing manuscripts, he told her, his day was a continuous round of conferences—about rehearsals, about orchestrations, and no doubt a hundred and one other things. After a hasty half-hour dinner, eaten alone at the Brown Derby, it was back to the studio. When he got there, Jolson wouldn't work because he wanted to go to the fights. They made a deal; Dick would go to the fights with him if Jolson agreed to an hour's rehearsal afterwards. After the fights and an hour at Jolson's place, Dick then went on "to meet the boys at Milestone's house to hear the final dialogue scenes." He got home at two a.m.

After yet another title—*The Optimist*—was discarded, they settled finally on *Hallelujah, I'm A Bum!* The movie ran over its deadline, being completed around the end of November. Then someone pointed out that a lot of the slang would be incomprehensible to people outside the United States, so further scenes had to be reshot. As if that were not enough, the distributors told Schenck he could never get away with releasing the film in England, where "bum" had a totally different connotation. This meant further expensive re-recordings of the title song as "Hallelujah, I'm A Tramp!" The final cost of the production was a staggering $1.2 million—an unheard-of sum for that time.

With nothing further to keep him in Hollywood, Rodgers headed back east. Hart stayed on, confident they would get more movie work. *Love Me Tonight* had premiered in New York on August 28 to widespread critical praise and satisfactory box office returns. The following month, *The Phantom President* was released. "A delight," said the *Daily News*. "A crackerjack show," enthused the *Times*. "Perfectly swell picture," added the *Post*. The auguries were good.

Of course, work was not the whole reason Larry elected to remain in Hollywood. He stayed on because it suited him. He loved the parties, the stories, the scandal, the studio gossip about the still-unexplained Labor Day suicide of Jean Harlow's husband Paul Bern, about Marlene Dietrich's torrid affair with Claudette Colbert, about Joan Crawford's blue movies. In a world where every variation of sexual activity was not only possible but practiced, Larry's personal hang-ups must have seemed considerably less oppressive.

At the end of November, his confidence proved to have been justified. In Dick Rodgers's account of events, what happened was that Irving Thalberg, the boy genius producer at MGM, called. He was already a Hollywood legend: the frail and sickly son of a middle-class Brooklyn lace importer who at age twenty-four had become Louis B. Mayer's right-hand man. Married in 1927 to Norma Shearer, Thalberg was responsible for much of MGM's artistic success, although his name rarely appeared onscreen. "Credit you give yourself isn't worth having," he said. As powerful—and in his own way, as tyrannical—as Mayer, Thalberg concealed his despotism behind a disarming and elegant manner. There was at the studio an unofficial society known as the Waiting for Irving Club; the settee in his waiting room was known as the Million Dollar Bench. "On a clear day," George Kaufman once said, "you can see Thalberg."

With a year behind them that included *Grand Hotel, Tarzan of the Apes*, and *Red-Headed Woman*, Thalberg and Mayer were poaching the top names of competing studios, building up the grandiose claim emblazoned across the MGM studio's portals: "More stars than there are in Heaven." One of their more recent signings had been Jeanette MacDonald. Because he was vastly impressed with their work in *Love Me Tonight*, Rodgers related, Thalberg offered Rodgers and Hart a choice of two assignments in which she might star. The first was a musical adaptation of the Thorne Smith novel *Topper*; the second, Hungarian Janos Vaszary's play about a banker whose wish comes true when he marries an angel. They hesitated not a moment; the fantasy seemed a far likelier prospect for the kind of screenplay they wanted to do.

Other evidence, however, suggests that the deal Thalberg offered Larry and Dick was a straightforward one-year contract as staff writers on the studio payroll, assignable to any production the studio chose. Whichever is true, they accepted. On December 12, Louella Parsons announced in her *Los Angeles Examiner* column that Dick and Dorothy Rodgers were on their way back to Hollywood.

Just a fortnight later, Irving Thalberg had a massive heart attack; by New Year's Day, rumors had spread through the studio that he would never again hold the power he had once had. It was under this cloud of uncertainty that Rodgers and Hart buckled down to work on the screen adaptation of the Vaszary play with another Hollywood neophyte, Moss Hart. Co-author with George Kaufman of the successful 1930 Broadway play and later movie *Once In A Lifetime*, twenty-nine-year-old Moss Hart was not related to Larry ("except by mutual consternation," as a newspaperman later observed).

Moss Hart, who had grown up in the Bronx with what he called "the grim smell of actual want always at the end of my nose," was also one of the world's busiest bachelors. He maintained this status until his middle forties by telling every woman he went out with—and they were legion—that he would never marry because he was still mourning his first and only love, a schoolteacher who had died. He delivered this fiction so emotionally that almost everyone believed it. George Kaufman summed up its effect one evening when he saw his partner coming into a restaurant with his latest conquest. "Take note," he said. "Here comes Moss Hart with the future Miss Smith."

Moss Hart was working on a revue called *The March of Time*, a shelved musical project dating back to 1930 that producer Harry Rapf was trying to revive, when on January 9 he was transferred to the Rodgers and Hart project. They hit it off immediately; in short order, with scriptwriter Howard Emmett Rogers, they formed the "Rodgers-Rogers and Hart-Hart MGM Luncheon Club" and toasted Larry and Dick's successes: *Hallelujah, I'm a Bum!* was released that day, and two days earlier Paul Whiteman had included "Lover" in one of his Carnegie Hall concerts.

According to Rodgers, Moss Hart was "an intense, Mephistophelean fel-

low, fairly bursting with ideas." He must have been; in a little over two
months they completed both the story—now titled *I Married an Angel*—and
the score. This consisted of at least seven songs, some of them extended
numbers in which Larry again employed the rhythmic dialogue technique
that had proved so successful in *Love Me Tonight*. Samuel Marx, then head of
MGM's Story Department, remembered Dick calling to tell him they'd just
finished the title song; he went over to their office in the series of clapboard
bungalows the studio grandiosely called its Music Department, a cubbyhole
"barely large enough to accomodate Dick, Larry, and a scratched-up baby
grand," to hear it. The date was March 10.

> They had begun to play "I Married An Angel" when we heard a terrific rumbling
> going on underneath us and Larry said "What's that?"—a question that none of us
> could answer, by the way. The rumbling continued, and then the whole room
> began to shake—so we knew what it was. The three of us realized that little office
> was far down a corridor, and the corridor was kind of like trying to run a kind of
> gauntlet to get out front when the whole earth is shaking—there's nothing more
> terrifying than the whole earth which you've counted on to be quiet, suddenly
> gyrating, which this was doing.
>   As a matter of fact this old wooden building, the Music Department of MGM,
> was coming apart at the seams, and the piano literally moved away from Dick
> Rodgers and then rolled back. We were now contemplating getting out of the
> window, but the window faced on the road, and whoever was responsible for
> guarding the treasures of MGM had carefully nailed that window shut!
>   Larry Hart rose to the occasion. He scrambled up on the piano, threw the piano
> stool through the window, and dived after it. It sounded from the noise like the
> whole building was coming down, although we found out later it was only a chim-
> ney; whereupon Dick departed the same way, through the window, and when it
> continued I followed him out. We all rolled out in a lot of broken glass—fortu-
> nately we weren't cut—and an automobile had come along the road that ran
> alongside the building, and the driver leaned out and said, "What the hell is going
> on here?"
>   Apparently you don't feel earthquakes in a moving automobile. He just saw
> these three guys come diving out of a window, and he couldn't figure out what had
> happened.

That very same day, the heads of the major studios were having what
turned out to be the last in a series of so-called but not very "secret" crisis
meetings. The industry was in chaos in the wake of President Roosevelt's
March 5 decision to close the banks. Universal had suspended all contracts
during the "national emergency." Fox employees were told they would not
be paid until the banks reopened. Columbia and United Artists shut down.
More than two thousand workers had been laid off to join Los Angeles's
forty thousand unemployed. The Academy suggested a 50 percent pay cut
across the board, which was largely hot air, since all salaries were paid in cash

and there wasn't any cash to be had. The screaming match that Louis B. Mayer, Jack Warner, Columbia's Harry Cohn, and Universal's Merian C. Cooper were having in the conference room on the top floor of the Hollywood Roosevelt hotel ended abruptly as the entire room began to shake, and a massive wheeled safe slid across the room and went right through the wall.

The following week, the producers and the unions agreed on an eight-day 50 percent pay cut; there were a few wildcat strikes, but in the main production returned to normal. Shortly afterwards, the writers and actors would form their own unions, but in the meantime it was business as usual. Larry and Dick continued their work to the accompaniment of the hammering of repairmen.

With Thalberg out of action until July, Louis B. Mayer took over all production at MGM. He became involved in abortive discussions with Darryl F. Zanuck, then at Warner; these fell through, and Zanuck joined Joe Schenck in a new production company called 20th Century. Now Mayer turned to his son-in-law David Selznick, offering him the autonomy at MGM the producer couldn't get at RKO. Selznick accepted, although his complete freedom from Thalberg's supervision and first call on the studio's stars were seen as nepotism. "The son-in-law also rises," cynics sneered, as Selznick plunged into a production schedule that included two Dickens adaptations, the play *Dinner at Eight*, with George Cukor set to direct, a film to star radio comic Jack Pearl, and a musical for Joan Crawford. These last two would become minor milestones in the Hollywood career of Rodgers and Hart; but not quite yet.

Immediately after *I Married an Angel*, or perhaps even at the same time they were working on it, Rodgers and Hart were asked to provide songs for two other movies. The first was the Marion Davies B-feature *Peg o' My Heart*. (MGM's prestige movies were rated A, whereas those made for mass consumption were a notch down the alphabet in both budget and casting.) The movie, which featured Onslow Stevens and Alan Mowbray, was an adaptation (the third) of the 1912 Broadway hit that made Laurette Taylor a star. Such music as was in the finished print was credited to Herbert Stothart, although Arthur Freed and Nacio Herb Brown also contributed a number called "I'll Remember Only You" for it; the two songs Larry and Dick wrote, "Tell Me I Know How To Love" and "When You're Falling In Love With The Irish," were not used.

Almost simultaneously, they were asked to contribute a song to a new movie Selznick—a boyhood schoolmate of Dick Rodgers—was producing for William Randolph Hearst's Cosmopolitan Pictures. One of many projects* he was juggling simultaneously during his first months with the studio—Selznick's perfectionism, his eye for detail, his incredible energy, his toughness and ambition (and his fondness for writing memos) had made him as much a legend as Thalberg—it was a story written by Arthur Caesar, called first *Three Friends* and later *East Side*, about two childhood pals (the third friend in the original story was their priest) who grow up on opposite sides of the law. The lyric of the song Larry and Dick turned in, "It's Just That Kind Of A Play," suggests it might have been intended to run behind the titles, or for a backstage scene. It was never used.

Selznick had selected his close friend—and some said alter ego—Oliver H. P. Garrett to write the screenplay, aided by Joe Mankiewicz, newly returned to the studio after an RKO loanout. As they expanded the original story to fit its stars—Clark Gable, Myrna Loy, and William Powell—the name of the movie became *Manhattan Melodrama* and the scene for the song changed to a Harlem nightclub. Selznick liked the tune, so Larry took another shot at the lyric, which now became "The Bad In Ev'ry Man." It was sung on film by blonde Shirley Ross fitted out in a black wig, "sepia" body makeup, and a slinky sequinned Dolly Tree gown; the tune was also reprised, instrumentally, in a prison sequence.

Directed by W. S. "Woody" Van Dyke and completed, without credit, by Jack Conway—Mankiewicz says it was typical of Selznick to use more than one director on a picture, as he was "trying to destroy the power and concept of the director as the one who makes the film"—and shot in twenty-four days at a cost of $355,000 between March 12 and April 3, 1934, *Manhattan Melodrama* was destined, when it was released the following May, to reinforce Gable's tough-guy image and to revive Powell's flagging career. The happy chemistry between Powell and Loy in the picture resulted in their being immediately paired in the hugely successful "Thin Man" series (also directed by Van Dyke) that followed.

Although the movie was a success ($415,000 profit), "The Bad In Every Man" made no impression at all on the public, even when *Manhattan Melodrama* got a new lease on life as "the film that caught Dillinger." Public Enemy No. 1, bank robber John Dillinger, was shot dead by the FBI on July 22, 1934, in Chicago after watching the picture at the Biograph Theatre on North Lincoln Avenue; studio publicists dreamed up the story that he was unable to resist going to see it because it was about a gangster who went to the electric chair.

Of the score Rodgers and Hart wrote for MGM Production No. 1116,

*I have examined the Rodgers and Hart songs for Selznick productions as closely as possible to the order in which they appear to have been written, using as my guide the dates on which they were registered for copyright; see Appendix for the chronological order in which the movies were shot.

however, only the title song, "I Married An Angel" (its first three notes lifted from the opening of the foundation stone–laying scene in *Hallelujah, I'm a Bum!*), is widely known; the others disappeared when Louis B. Mayer abruptly canceled the project. Exactly why he did so remains unclear. Rodgers said Mayer arbitrarily decreed that fantasies were uncommercial and no one could budge him. Another story suggests that the Catholic League of Decency intervened; when they made representations to Mayer that the portrayal onscreen of a mortal man going to bed with an angel was unacceptable, he canceled the movie. Either way, the project was a dead duck.

So, too, it transpired, was *Hallelujah, I'm a Bum!* When it opened in New York on February 9, the critics were less than kind, and in spite of a personal onstage appearance by Jolson, moviegoers didn't want to know. Although there were many bright and inventive moments—the scene showing the laying of the foundation stone (in which Rodgers makes a cameo appearance as a photographer), the montages of the kangaroo court, with their musical echoes of "Three Blind Mice," the millions-to-zero dialogue that takes the camera from the doors of the bank to Jolson's desk (and features Larry as a bank teller)—they were not enough to save the picture. Perhaps most of all, it just wasn't the Al Jolson everyone wanted up there on the screen. Whatever the reason, *Hallelujah, I'm a Bum!* failed miserably. According to Rodgers, even Hoover—voted out of office in Franklin Delano Roosevelt's landslide November victory—was more popular.

In March 1933, Warners released *42nd Street*, the most famous, most slyly knowing, and most authentic backstage musical of them all, the one in which Warner Baxter tells Ruby Keeler, "Sawyer, you're going out there a youngster, but you've got to come back a star!" Produced by Darryl F. Zanuck and directed by Mervyn LeRoy, with a quite brilliant score by Al Dubin and Harry Warren that included "Young And Healthy," "Shuffle Off To Buffalo," and "You're Getting To Be A Habit With Me," it made major stars of Keeler and Dick Powell, while Busby Berkeley became the hottest dance director in Hollywood.

The triumph of *42nd Street* generated an almost instantaneous revival of studio interest in musicals. Had they been free, Rodgers and Hart might well have been able to write their own ticket in Hollywood. But they were already under contract, and there wasn't a thing they could do about it. "So Larry and I accepted our sentence," Rodgers said. "One year at soft labor at MGM."

23

Goldwyn's Folly

arly in 1932, Samuel Goldwyn, with ten years of independent production now behind him, decided to produce a movie of *The Brothers Karamazov* by Fyodor Dostoevsky. He came to this decision after seeing the photograph of an actress named Anna Sten in an advertisement for a German film based on the book, *Der Morder Dimitri Karamazov*, then playing in New York. Inquiries revealed that Russian-born Anna's real name was Anjuschka Stenskaja Sudakewitsch; she was reported to be twenty-four years of age (she was actually born in 1900), a former student of Stanislavsky, briefly married to Soviet cinematographer Anatoli Golovnia, and a protégée of director Fyodor Ozep, who had shortened her name and directed her in the film.

Goldwyn had a copy of the film rushed out to him in Hollywood and watched it over and over, totally smitten. He immediately dispatched his (and much later Rodgers and Hammerstein's) publicity chief Lynn Farnol to Europe to sign her, determined she would now star in a Goldwyn version of the Dostoevsky novel. Eschewing preamble, Farnol offered Anna a four year contract: fifteen hundred dollars a week for forty weeks the first year, two thousand a week for forty weeks the second. Sten and her astonished manager-husband, the former razor-blade salesman Eugene Frenke—she had by this time left Ozep—grabbed it. There was just one minor problem: the lady spoke not a single word of English.

By the time Sten got to New York, Farnol had taught her enough for her to be able to say "I lof you" to the reporters before setting off for Hollywood. There she discovered that Goldwyn had fallen out with her prospective co-star Ronald Colman, and that the Karamazov project was shelved. Instead, Goldwyn announced, she would star in a film based on Emil Zola's *Nana*.

From the moment he first heard (he didn't *read* the book, of course; someone told him the story) how Nana raises herself out of the gutter to become the toast of Paris, Goldwyn was hooked: this doomed-girl story was exactly the right showcase for Sten's acting and singing. She would be the greatest female star in the history of movies, a wonderful amalgam of Garbo and Dietrich.

Set to write the script was scenarist George Oppenheimer, author of the 1932 "Dorothy Parker" play *Here Today*, starring Ruth Gordon in a part so clearly identifiable as Dorothy Parker that it led Parker to reply, when asked why she never wrote her autobiography, that she was afraid George Oppenheimer and Ruth Gordon would sue her. Twenty-seven-year-old Fred Kohlmar, Goldwyn's executive assistant, took on production duties.

A whole year later, during which time Anna Sten received a daily diet of four hours of English tuition a day and three motion pictures a week (and via Ozep, point-blank refused to accept any of the pay cuts forced on everyone else in the industry during the midsummer panic of 1933), shooting began. The director was not Josef von Sternberg, as Goldwyn had hoped, but Paris-born George Fitzmaurice, who had directed Pola Negri in *The Cheat*, Valentino in *Son of the Sheik*, and more recently Greta Garbo in *Mata Hari* and *As You Desire Me*. Delayed and harrassed at every stage by Sten's agent-husband with his autocratic brinkmanship, self-awarded 'doctorate' and imagined entrepreneurial skills, this version was scrapped; the film was totally remade by Dorothy Arzner.

In the meantime, Goldwyn had approached Cole Porter to write Anna's songs. YOU ARE THE ONE MAN IN THE WORLD TO WRITE TWO SONGS TO BE SUNG BY ANNA STEN IN HER FIRST PICTURE FOR ME, he cabled Porter in Paris. Although they were friends, Porter turned down the commission; he was writing a new show for Charles Cochran in London, *Nymph Errant*, which would star Gertrude Lawrence. Goldwyn then asked MGM to lend him Rodgers and Hart. Whether they were asked for two songs, as Porter had been, cannot now be established. It is at least possible that "The Night Was Made For Dancing (And You Were Made For Love)" might also have been written for *Nana*, but there is only one song in the picture.

Even if it was hardly what Rodgers and Hart were hoping for, a commission to write a song for a major Goldwyn production was not to be dismissed lightly. Dick went home to Angelo Drive and composed a lilting minor-key melody with a nice Mittel-European feel; Larry came up with the title "That's Love," and a gently rueful lyric to go with it. They rushed down to United Artists and played it for their old friend Alfred Newman, who had led the pit orchestra in *Heads Up!* and was now Goldwyn's musical director.

Newman loved it; they took it up to Goldwyn. He raved. Nothing would do but that he immediately phone Anna Sten to tell her it was the best song he'd ever heard. His parting remark to Larry and Dick was, "Boys, I thank you from the bottom of my heart!" In fact, Goldwyn was so enthused about the song, he had them convinced they had another "Lover" on their hands. One day he asked Dick to play the tune for Frances Marion, one of the top Hollywood screenwriters, a tiny but hugely talented woman who was one of Goldwyn's favorite people. They had been close friends since the days of Sam's first, heady crush on Mabel Normand.

When he finished playing, Rodgers looked up to see Frances Marion standing with her eyes closed, as if in a trance. When Goldwyn asked her what she thought of the song, she slowly opened her eyes and said she had never heard anything so Parisian in her life. Without hesitation, Goldwyn turned around to face his musical director. "Newman!" he decreed. "In the orchestra, eight French horns!'"

Hart, too, had his Goldwyn story. At the same time he was filming *Nana*, Goldwyn was producing a multi-million-dollar extravaganza starring Eddie Cantor—at $270,000 a year the highest-paid actor in the movies–not inappropriately called *Kid Millions*. Based on an original screenplay by Arthur Sheekman, Nat Perrin, and Nunnally Johnson, it was a nonsense about a *nebbish* who inherits seventy-seven million dollars from the archaeologist father he never knew, and the misadventures that happen as various crooks try to separate him from his money. While the film was still shooting, Goldwyn found himself unhappy about the ending. He called Larry, asking him if he would come in and do a quick rewrite. Larry told him he couldn't, he was sick; Goldwyn wouldn't take no for an answer. No matter how much Larry protested he was really ill, Goldwyn wouldn't give up.

"Sam, Sam," Larry protested, "You've got to believe me, I can't get out of bed. I'm sick, don't you understand? I'm at death's door!"

"Look here, Larry," Goldwyn said sternly, "don't be such a kleptomaniac."

Back to *The March of Time*, but now without Moss Hart, who had returned to Broadway to work with Irving Berlin on a new revue, *As Thousands Cheer*. Cobbled together as *Broadway to Hollywood*, a backstage musical starring ten-year-old Mickey Rooney (in his movie debut), Frank Morgan, and Alice Brady, it managed to successfully incorporate some of the earlier color sequences of *The March of Time* and make a decent showing at the box office on its early 1933 release, encouraging producer Rapf to try another musical.

Production supervisor at MGM, ex–vaudeville booker Harry Rapf was famous for discovering (Howard Dietz also claimed credit) chorus girl Lucille Le Sueur at the New York Winter Garden in 1924, making her first his mistress, and then Joan Crawford. He was a favorite butt of cruel jokes because of his long, thin nose and his legendary malapropisms. Discussing a story idea, he once asked, "Take out the essentials and what have you got?" He turned down a picture about the Queen of Rumania because he didn't like "mythical kingdom stories," and when the life of the Virgin Mary was mooted, he asked if it could be done in modern dress. One time he completely threw Thalberg by telling him, "I woke up last night with a terrific idea for a movie—only I didn't like it." This was the man to whom Rodgers and Hart were assigned to write the songs for a glitzy new movie spectacular.

Like most of Rapf's work, it was a retread. Its genesis was in a 1929 MGM musical called *The Hollywood Revue*, which featured just about everyone on the lot excepting Garbo. Although it was no more than a collection of songs and sketches, the picture was nominated for an Oscar. Right now, Rapf was desperate to find vehicles for two of the radio stars he had brought to Hollywood: Jack "the Baron" Pearl and Ed "the Perfect Fool" Wynn. Why not feature them in another *Hollywood Revue*, suggested MGM publicity chief Howard Dietz. Collaborating with producer-playwright Arthur Kober (married to Lillian Hellman, he had once produced Henry Myers's play *Me*), Dietz concocted a story line on which to peg appearances not only by Pearl but by as many MGM stars as could be rounded up. It would be called *Hollywood Revue of 1933*.

Rapf had special material—mostly Rodgers and Hart songs—written for the guest stars, who would be introduced by other performers, including Robert Young, Jimmy Durante, and Laurel and Hardy. While they were waiting for the stars, who were working on other pictures, he shot the other scenes he would require. Even the most dedicated Hollywood detectives have never properly established what happened next; all that is known for certain is that three dance directors (Seymour Felix, Dave Gould, and George Hale), at least five directors (Edmund Goulding, Allan Dwan, Russell Mack, Richard Boleslawsky, and Roy Rowland, none of whom was prepared to take the credit, or blame), and cinematographer James Wong Howe were somehow involved behind the cameras, with Durante, Pearl, Lupe Velez (fresh, if that is the word, from Ziegfeld's *Hot-Cha!*), Charles Butterworth, Polly Moran, Eddie Quillan, and June Clyde in front of them. Appearing as themselves were Shirley Ross, Harry Barris, Frances Williams (singing the title song), Robert Young, Ted Healy and his Stooges (before they went trio as Curly, Larry, and Moe), and Mickey Mouse.

It was around this time that agent Myron Selznick got his client Joe Mankiewicz a ten-week loanout guarantee deal with Sam Marx at MGM. When he turned up for work, Marx sent him over to Harry Rapf. Rapf called in director Eddie Goulding to outline the story they wanted Mankiewicz to script. "Goulding was a tremendous ham," Mankiewicz records, "and he started by telling me, 'We fade in on the Earth, and we pull back and it's spinning on Karl Dane's finger, which leads into a sketch between Dane and his partner. Then Joan Crawford sings 'Blue [Black] Diamond' and Jean

Harlow croaks 'Make Me A Star." All of this was gibberish to me, and as I sat
there listening, I thought, 'They're kidding me. This is a put-on.' When
Goulding finished, I laughed and said, 'O.K., fellows, you've had your fun.
Now tell me what the story is.' Rapf, who was a very explosive man, got up
and screamed at me, 'Get out, you fake. They told me you were a good
writer. Get out, you don't know your business,' as he drove me from his
office."

Mankiewicz reported back to story editor Marx. "Oh, boy, this is terrible,"
Marx said, and told him to report for reassignment Monday morning. On
Monday, Marx told him he was the luckiest guy who ever lived: David
Selznick wanted him. Some luck. Shooting back to back with Rapf's *Hollywood Revue* was another picture, which Selznick described as "a horror that I
produced called *Meet the Baron*, starring Jack Pearl. . . . I have never been a
devotee of radio comics. I had never heard either [Ed] Wynn or Pearl on the
radio. You will recall that Rapf had been my benefactor in the early days at
Metro, and so, when he appealed to me to take one of them over (because he
had initiated both, and felt he could only handle one at a time), I agreed and
he gave me Pearl. I made the picture with a loathing for it, and it was a terrible flop."

Jack Pearl was unlikely movie material. Born on Rivington Street on the
Lower East Side, he had begun his career as a German dialect comedian in
Gus Edwards's vaudeville show, then worked as a songplugger, as a burlesque comedian, and eventually on Broadway in Shubert extravaganzas like
A Night in Paris and *Artists and Models*. He graduated from the final, 1931
Ziegfeld *Follies* via the "Ziegfeld Follies of the Air" on radio to his own show,
sponsored by Lucky Strike. On the radio Pearl played Baron von Münchausen—the biggest liar in the world. For instance: he was in the Arctic trying to talk to the Eskimos, but the words froze in his mouth before he could
get them out.

"Oh, yeah?" straight man Cliff "Charlie" Hall would say. "So how did you
know what they were saying?"

"Dot's easy, ve took der converzation into der ikloo und thawed it out."

"Boy, Baron," Charlie ventured, "that's pretty hard to believe."

And the Baron would ask, in a very offended voice, "Vus you dere, Sharlie?"

This catchphrase had passed into the vernacular; anytime anyone doubted
anything, the other person would ask, "Vus you dere, Sharlie?"

An army of gag-writers (Mankiewicz was not one of them: he ducked the
project and was instead loaned out by Marx to RKO) put together Selznick's
vehicle for Pearl—so many writers, studio wags had it, the producer was
thinking of putting their names against a credit title that showed thousands
of marching men. A number of things about *The Big Liar* (the title was
changed during production to *Meet the Baron*) suggest a close relationship—

perhaps even an incestuous one—between it and *Hollywood Party:* Rodgers and Hart wrote at least two songs for it; Arthur Kober was one of the principal writers; and on at least one occasion Pearl apparently segued directly from one set to the other without even changing his costume, bringing his "African" bearers with him, not to mention Ted Healy and his Stooges.

With the aid of some creative guesswork, the Rodgers and Hart songs for what became *Hollywood Party* provide further clues to this relationship (as well as a chilling illustration of the value Rapf—and Selznick—placed on their work). They wrote a total of twenty songs, only four of which were used in the final print: the title song—two different songs performed as one—"Hello," and "Reincarnation." Everything else was discarded.

"Yes, Me," "The Mahster's Coming," and perhaps also "Give A Man A Job" were written for Durante in *Meet the Baron* but were not used. "You Are," intended for Zasu Pitts and Jack Pearl, was also recycled for *Hollywood Revue* (as it then was) and again dropped. "The Pots" (possibly written for *I Married an Angel*) and "Baby Stars" were clearly designed as big production numbers, but for which movie is less clear. "Fly Away To Ioway," intended for Durante and Polly Moran (who ended up singing "I've Had My Moments"), was a satirical riposte by Larry and Dick at the huge "Shuffle Off To Buffalo" production number in *42nd Street.*

There are two versions of "You've Got That." One was to be sung by Lupe Velez and Charlie Butterworth in their comedy seduction scene—and cut—but who was to perform the director-star version? "I'm One Of The Boys" could have been intended for Marlene Dietrich, who was making a point of cross-dressing in all her movies, or even Marion Davies. Most unbelievable of all, as Joe Mankiewicz indicated, Joan Crawford (in blackface?) was to be the "Black Diamond," with her "ruby lips and iv'ry teeth and ebony cheeks."

But for whom was "I'm A Queen In My Own Domain" written? And the impassioned "Burning"? There are further mysteries, none bigger than the neo-operatic "My Friend The Night," a quite extraordinary song that defies allocation. Perhaps it was intended to be spoken soliloquy-fashion; it is difficult to imagine anyone at MGM who could—or would—have sung it. What about the "Party Waltz," later salvaged for *Jumbo*? We'll never know, any more than we'll ever know whether "Keep Away From The Moonlight" was written for Bing Crosby (its first five notes are an ascending musical phrase identical to that of the Johnny Burke–Jimmy Monaco song "Too Romantic" he sang in *Road to Singapore* seven years later).

The last and most famous casualty of *Hollywood Party* is the Jean Harlow song. Taking his cue from *Make Me a Star*, the retitled 1932 movie version of *Merton of the Movies*, Hart wrote yet another lyric to the tune used in *Manhattan Melodrama*. This time it was a plaintive "Ten Cents A Dance"–style song called "Prayer" about a stenographer who wants to get into the movies. As far as can be ascertained, Harlow never came near the set, so, like everything else they'd written, "Prayer" was dropped in favor of the Walter Donaldson songs, "I've Had My Moments" (lyric by Gus Kahn) and "Feelin' High" (lyric by Dietz). Also interpolated—in the Mickey Mouse sequence—was another Nacio Herb Brown–Arthur Freed number, "Hot Choc'late Soldiers."

At this juncture, to Harry Rapf's astonished consternation, Louis B. Mayer decided to sneak-preview the unfinished material, overruling all Rapf's protests that the movie wasn't finished with the assertion that a live audience would provide valuable insights into what the picture needed. Arrangements were made to show it at a San Bernardino theatre, while in panicked haste the studio's film editors patched together some sort of story line from the available material.

Came the day: in the big red studio car chartered from the Pacific Electric Company taking the studio executives down to San Bernardino, Louis B Mayer was reduced to a stony rage when his bridge partner, test director Felix Feist, ruined his game. Worse was to follow. The preview audience couldn't make head nor tail of the story—hardly surprising, in the circumstances. Watching it today, it is difficult to believe Mayer sat through it himself. The comedy was cretinous, the musical interludes at best shoddy, the dancing elephantine, and the plot abysmal.

Jimmy Durante is the star of the movie *Schnarzan the Conqueror*. At a sneak trailer, an exhibitor tells his producer Schnarzan's stunts with moth-eaten stuffed lions are pathetic; unless Schnarzan starts wrestling some real animals, forget it, and the same goes for his biggest movie rival, Liondora. Learning that Baron Münchausen is arriving from the Congo with some really wild beasts, Durante decides to throw a huge party—cue the title song, a lead balloon with a pathetically cut-price Busby Berkeley–style dance routine—to impress Münchausen so much he'll sell Schnarzan the lions.

Liondora—played by Broadway song-and-dance man George Givot—crashes the party disguised as a duke, determined to get them himself. His idea is to bamboozle oil zillionaire Charlie Butterworth into buying them, then (to the tune of "Temptation") romance Charlie's wife, played by Polly Moran, into making Liondora a present of them. Add Lupe Velez in a dress Cher would think twice about, Durante chewing up the scenery in a number called "Reincarnation" as "Adam" and "Paul Revere's horse" (a sequence set in the French Revolution with Schnozzola as Marie Antoinette was mercifully cut), and take it from there.

The San Bernardino audience booed, they hooted, they walked out. Those who stuck it out to the bitter end (Durante's real-life wife Jean Olsen Durante wakes him up, and he realizes it was all a dream brought on by reading *Tarzan the Untamed*) wanted their money back. The theatre manager accused the MGM executives of trying to foist off a collection of rushes on him. Mayer stalked out, seething with rage. On the trolley he furiously dismissed Rapf's explanations that the movie was unfinished, that the star roles had not yet been filmed. "Don't spend another cent on it," he rasped angrily. "Patch it up the best you can and let it go."

"Rapf went off to sit, separate and apart," wrote Samuel Marx, who was there, "pretending to gaze out at the dark landscape, averting his eyes even from his sympathetic friends who were aboard, which was just as well, for none could find words to cheer him. Later, all pretense gone, he began to cry."

24

Night Madness

*R*odgers and Hart had precious little time to be cast down by the cavalier treatment accorded to their songs in *Hollywood Party*. Almost simultaneously they were assigned to yet another Selznick movie, *Dancing Lady*, scheduled to go into production in June starring Joan Crawford. The property had been moving around from producer to producer ever since the studio had purchased the novel (written by James Warner Bellah, better known for his westerns) in April 1932.

After a series of flops that included *Rain*, Crawford was not convinced a musical would be good for her. Selznick knew how to get around that problem. "I don't know if you can play this part, anyway," he told her. "It's kind of tarty. I think it's more Jean Harlow's style." Crawford bristled. She'd been playing hookers before Harlow even knew what a hooker was, she told Selznick; and so the hooker was hooked.

With Crawford set, Selznick assigned veteran director and former silent movie actor Robert Z[igler] Leonard—fresh from *Peg o' My Heart*—to the project, and early in June began thinking about getting someone to write the music the picture would need. As with everything else he did, Selznick had specific ideas in mind. A story about Dimitri Tiomkin's later experience with Selznick may explain why Dick Rodgers came to view "King David" (as he far from admiringly called his onetime schoolmate) with something less than admiration.

When Selznick engaged Tiomkin to write the music for *Duel in the Sun* in 1946, he told him he intended to approve all the musical themes for the picture. Shortly thereafter, Tiomkin received a long memo instructing him to write eleven themes; these were to include a Spanish scene, a love theme, a desire theme, and an orgasm theme.

"Love themes, desire themes, Spanish themes I can write," Tiomkin told the producer. "But orgasm? How do you write an orgasm?"

"Try!" Selznick urged him. "I want a really good *shtupp*."

Selznick knew what he wanted for *Dancing Lady*, too: the title song, an incidental song for the opening sequence, a torchy number for Joan, three full-scale production numbers, and something jazzy and hot for the big finale of the show-within-a-movie that makes Janie Barlow (Crawford) a star. He handed the job of writing the finale to Rodgers and Hart, who produced an up-tempo number called "That's The Rhythm Of The Day." When they called Selznick to tell him it was ready, Rodgers said, they were obliged—his verb—to perform the number before "King David" and everyone who was on the set. Although this was clearly not to their taste, they did as the King commanded, finishing to applause and compliments and Selznick's profuse thanks.

Satisfied with what he thought was a good reception for their song, Dick took the next day off to play tennis. In the middle of the game he was called to the telephone. It was Selznick. He wanted to tell Dick how much he loved the song, how much everyone who had heard it loved it. It would make a terrific finale for the movie. There was just one thing he wanted to ask him.

"Sure, David," Rodgers said. "What's that?"

"Uh . . . could you make it a little better?"

Rodgers explained carefully that you couldn't make a song "better." It was either good or bad. You either liked it or, if not, you threw it out and wrote another. Selznick was puzzled that the song couldn't be improved; later he told scenarist Allen Rivkin he considered it "monotonous" (Rodgers insisted this was because he'd heard the middle eight rehearsed on the set so many times) and had reluctantly decided not to have anything else by Rodgers and Hart in the picture. This meant that their two shots at a title song, "Dancing Lady"—same title but different words and music—were also consigned to limbo.

One night Rivkin was at a party at the home of Fox writer Leonard Spigelgass when a melody someone was playing at the piano caught his attention. The pianist was Burton Lane, and the song was "Everything I Have Is Yours." When Selznick heard it, he agreed with Rivkin that it was exactly right for the torchy ballad spot—Joan doesn't actually sing it, merely hums along—and asked Lane and his lyricist Harold Adamson what else they had. They came up with a couple more songs, one of which was "Heigh Ho, The Gang's All Here." They were signed, but even then, Selznick hedged his bet, hiring Nacio Herb Brown and Arthur Freed to write "Life Is A Merry-Go-Round" and Jimmy McHugh and Dorothy Fields to do the title song. With that out of the way, he concentrated on casting.

Crawford wanted Gable. A year or two earlier, during the filming of *Pos-*

sessed, Louis B. Mayer had broken up a torrid romance between them after complaints from Gable's wealthy socialite wife Rhea Langham. As she later admitted, Crawford still "had a case on Gable," but the King wasn't interested. He turned the project down flat; only when Mayer and Selznick promised him a vacation following completion did he accede, and at that reluctantly. Selznick had Robert Montgomery in mind for the part of the Long Island millionaire who is Gable's rival, but Montgomery was committed; at Crawford's urging, he cast Franchot Tone, her current lover.

Dancing Lady went into production June 10, 1933; after a total of sixty-five days in front of the camera (fifteen days over schedule), and at a cost of $923,000, the movie wrapped on October 20 with a medium two-shot of Gable and Crawford in their final—and only—kiss. It opened at the Capitol, New York, on December 1, 1933, and went on to play all of Loew's 135 theatres around the country, producing a profit of $744,000 over and above the cost of production.

Selznick had a solid success. No doubt relishing the irony of playing a tough little hoofer who with the assistance of a millionaire "patron" makes her way from the chorus in a burlesque house to stardom, Crawford had the smash hit that would put her back at the top. Gable, as usual playing the tough lug with the heart of gold, was loaned out for his pains to Columbia for a "Poverty Row" picture that turned out to be his biggest break: *It Happened One Night*. Fred Astaire, who made his debut in *Dancing Lady*, moved on to RKO, where he got lucky when dancer Dorothy Jordan, who decided she'd rather be married to studio chief Merian C. Cooper than do "The Carioca" on top of seven white pianos, was replaced by Ginger Rogers.

Dancing Lady also featured the debuts of Robert Benchley, Ted Healy and his Stooges, and—unbilled in a small role as a hoity-toity Southern gal—a peroxide-blonde unknown named Eunice Quedens who later changed her name to Eve Arden. Burton Lane and Harold Adamson had their first big success. Everyone was pleased, except Rodgers and Hart. Their song, "That's The Rhythm Of The Day," sung by Nelson Eddy (another debut), and the pretty waltz theme Dick wrote for the sequence made no impression. They did the only thing left for them to do: they resigned.

———————

Richard Rodgers and Lorenz Hart ironed out their MGM difficulties and withdrew the resignation they handed in just before they hopped off on a two weeks' vacation. The songwriting team was dissatisfied with the assignments they had been receiving and felt they could not do themselves or the studio justice with them. The team will write the musical score and additional music for the Lehar operetta "Merry Widow" on their return.

That brief item in the *Hollywood Reporter* of December 10, 1933, indicates just how disenchanted with Hollywood Larry and Dick had become. Determined to either get a decent assignment or get out, they took their grievances to Irving Thalberg. But the Boy Wonder had one or two problems of his own.

After convalescence in the south of France, Thalberg had returned to MGM anxious to restore his reputation and position at the studio. In New York, after conferring with Nick Schenck, he agreed to head a unit which would embark upon a more limited program of prestige pictures, rather than the forty to fifty films a year he had previously produced. Almost immediately, he discovered he was running into obstacles he had never encountered before: not only did he have to contend with Selznick's freedom to use any MGM star, the other producers on the lot were also competitors. When he began to look for properties for Garbo and Joan Crawford, with whom he planned to make two films, Mayer told him Crawford was tied up for at least two years; the same was true of Gable and Harlow, whom Thalberg wanted for *China Seas*. Clearly he was no longer the golden boy of old.

As if that were not bad enough, Nick Schenck was helping finance his brother Joe's 20th Century company; Mayer, too, poured in money because of his son-in-law Bill Goetz's involvement. In addition, Schenck and Mayer gave 20th Century access to MGM's incomparable roster of stars. Thalberg now had to compete with the hungry, talented Darryl F. Zanuck—soon to become head of the even larger 20th Century-Fox—as well.

After abandoning the idea of filming Michael Arlen's famous novel *The Green Hat*, Thalberg decided his first project would be a sophisticated comedy-drama about adultery among the rich, written and to be directed by Eddie Goulding. Starring Thalberg's wife Norma Shearer, *Riptide* would feature Herbert Marshall, Robert Montgomery, and—in the most unlikely of movie debuts—Mrs. Patrick Campbell, once London's dazzling Eliza Doolittle, the Covent Garden flower girl of George Bernard Shaw's *Pygmalion*.

It was at this juncture that Rodgers and Hart trooped in. Their arrival sparked an idea in Thalberg's mind. He was teaming the recently signed Maurice Chevalier with Jeanette MacDonald in a big-budget remake of *The Merry Widow* directed by Ernst Lubitsch and written by Samson Raphaelson, who had scripted *One Hour with You*, and Ernst Vajda, adapter of *The Love Parade*. He was assigning Cedric Gibbons, unchallenged as the most influential production designer in the industry, art director Frederic Hope, and cinematographer Oliver Marsh to the picture to give it a look of "unparamountable" opulence. Who better to "enhance" the Franz Lehár score than Rodgers and Hart, whose work on *Love Me Tonight* he had so admired? He offered them the picture.

What Thalberg didn't tell them—although he of course knew, even as the deal was struck—was that the studio had just offered Jerome Kern and Oscar Hammerstein a commission to write a score for Jeanette MacDonald's next picture, *Champagne and Orchids*, in which it was planned she would star not with Chevalier but Nelson Eddy. This little piece of double dealing clearly defines his—and the studio's—perception of Rodgers and Hart. Froth, bubble, wit, naughtiness—send for Larry and Dick. But the lush, deeper, romantic stuff? Jerome Kern or Sigmund Romberg every time.

"Actually," Rodgers wrote of Thalberg, "he didn't sign 'us,' he signed Larry, since Franz Lehár had written a pretty fine score without any help from me." This is yet another example of Rodgers's disingenuity; he knew perfectly well he and Larry wrote at least three new songs for the movie, although they were not used: "It Must Be Love," "A Widow Is A Lady," and "Dolores." He went on to detail Larry's unhappiness while working with Lubitsch. In sharp contrast to Mel Shauer, who said Larry and the director were great friends, Rodgers said Lubitsch was autocratic, with a "decidedly Teutonic approach"; he insisted Larry be present at all meetings and that all his lyrics be submitted on neatly typewritten sheets. "He took it as a personal affront when Larry would show up late," Rodgers wrote, "and fumble through his pockets for scraps of paper on which he had scribbled the lyrics."

There were still one or two other little problems to be ironed out, not the least of which was that Chevalier was unhappy playing the same part—himself—over and over, especially opposite MacDonald, especially directed by Lubitsch. He stormed into Thalberg's office demanding that Grace Moore be cast as the widow. Thalberg refused. "After plenty of arguments and bickerings one way or another," the *Hollywood Reporter* told its readers February 15, "MGM has set Jeanette MacDonald as the widow in 'The Merry Widow' opposite Maurice Chevalier. The French star fought to the last ditch to keep Jeanette out, but had to bow to the logic of the situation as was outlined by both Irving Thalberg and Ernst Lubitsch." It's not hard to imagine the kind of "logic" they employed: on March 18, a chastened Chevalier hastily issued a statement denying he'd had any intention of "high-hatting" MacDonald and Lubitsch.

Principal photography was completed by midsummer; once *Widow* was finished, Rodgers and Hart were again reduced to kicking their heels, fiddling around with song or show ideas, hanging out with the other occupants of the Writers' Building and the Music Department. Goldwyn's much-heralded *Nana* had opened at New York's Radio City Music Hall on February 1 to an unprecedented wave of publicity. During the first week it broke all existing records at the box office, but the paying public decided that Anna Sten had been oversold. After a few days the lines disappeared and never re-formed. To be fair: Anna was not at all bad, and she was very beautiful.

Although she sang "That's Love" much better than Garbo would have, even her most ardent admirers—and none was more ardent than Sam Goldwyn—had to admit that the film was a turgid disaster. Any hopes Rodgers and Hart might have had that their song was, indeed, another "Lover" were quickly dashed.

While Dick played tennis and lived what he described as the life of a retired banker, Larry's erratic, disorganized life grew ever more chaotic. He had what one of his friends called "night madness." "Larry was respectable until ten-thirty," he said. "He would take his mother to a show, or to dinner, or a party. Then they would go home and he'd wait till she went to bed, and he'd be out with his own crowd. Those crazy nights. It was like he never wanted to go to sleep."

"He was a stay-up-late, a jack-in-the-box, restless, nervous, impatient, buzzing into every corner like a bumble bee," Leonard Sillman said, "the most generous and gentle of geniuses."

> Scott Fitzgerald called him the poet laureate of America and nothing that has happened to the American popular lyric since his death can alter that epitaph, but Larry did not look or act the poet. He was barely five feet high and chunky all the way. The one distinguishing feature of his intense homely face was his burning black eyes. He had five o'clock shadow every day at noon and smoked about two dozen cigars per diem. He talked in machine gun bursts and held his cigar like a gun. He could stay up all night talking like a Lindy character and then go home with a hangover and over two cups of coffee write a lyric like "My Funny Valentine."

Frances Manson, a scenarist with Columbia, saw a lot of Larry while he was in Hollywood; she went out with him often, and said that from time to time he asked her to marry him, but she was never tempted. His lifestyle in general and his appetite for liquor in particular bothered her: she was afraid she might wind up drinking as much as he did.

One night they went to the Clover Club, a gambling place on the Strip. Larry bet like a madman, money pouring out of every pocket as he placed his bets. No matter how much he won, back on to the tables it would go. Next day he called Frances and asked her whether he had lost a lot of money the preceding night because he hadn't got a cent. He had absolutely no recollection of having won at all.

Mel Shauer told of a dinner party where Fanny Brice and Larry "sat at the table trying to top each other with funny stories and reminiscences." Around midnight Larry insisted they all go to the New Yorker club on Hollywood Boulevard. At closing time—Larry never went home until closing time—he called the headwaiter over and insisted on paying everyone's check. He had never seen any of the forty or fifty people in the place before, and he'd never seen any of them again. They didn't even know who he was, just some crazy little guy who paid their bills.

Larry became a regular at the New Yorker, where he went often with friends (or freeloaders, depending on your point of view). The club, which was at the Christie Hotel, featured female impersonator Jean Malin, whose act Larry loved to see. He'd take a whole crowd along with him, a sprinkling of big names like Tallulah Bankhead, a group of struggling but handsome young actors like Tyrone Power, and always Doc Bender, whom Irving Eisman called "Larry Hart's Rasputin," and who in turn brought along his new "protegés," the campy, outrageous Rocky Twins.

They had come into Larry's orbit through his acquaintanceship with Leonard Sillman, a young producer who had put together a revue at the Pasadena Playhouse the preceding May called *Low and Behold*, with "new faces" who hadn't yet made their names—young Tyrone Power, still "Junior" then, and comedienne Eunice Quedens (Eve Arden) among them. Larry persuaded Sillman to give Teddy a part. "I think it always weighed heavily on Larry's conscience that he himself had come into the world so loaded down with gifts from the Gods, whereas his stage-struck kid brother seemed to have little more to offer the world of the theatre than a somewhat comic speech impediment," Sillman wrote. "In spite of his own fame, his influence, [Larry] could not get steady work for Teddy."

Among Sillman's other "finds" were Norwegian-born identical twins Paal and Leif Rocky. They were dancers; they were young, handsome, and rich. "Talent they had not," Sillman recalled,

> but they were fairly bursting with boyish charm. They had been headliners in European vaudeville, had come to Hollywood to conquer new worlds, had taken over a huge villa in Beverly Hills, and proceeded to give the most gilded parties since the great days of the wild, or silent era, when a party *was* a party. The Rocky parties were always climaxed by a mass visit to their bedroom where the twins would display their jewelry. They had diamonds by the peck and gold ornaments by the pound, given to them (they said) by the crowned heads of Europe.

According to Sillman, the Rocky Twins gave their lavish parties in the hope of getting parts. "Everyone came to the parties," he said, "but nobody gave them contracts." Like Malin, the twins performed in drag, their act a burlesque of the famous Dolly Sisters. Irving Eisman, who described them as "wild" and "brazen," said Bender was in love with them. Larry just thought they were wonderful and took them everywhere, enjoying their tremendous notoriety. They hit the bars, they hit the clubs, they went to parties at Malin's home or Lew Cody's beach house in Malibu. "They weren't orgies," a dancer who went often with Larry said. "Nobody got raped, for

God's sake. If someone showed up and was shocked by what was going on, he left."

Had Larry Hart finally quit dabbling his toes in the water and slid into the pool? He was certainly in the right place for it. He was seen more and more with beautiful boys, to whom he gave expensive gifts of gold cigarette cases and watches. "It was easy for him to be led astray," Edith Meiser said. "He *wanted* to be led astray. And there was nothing you could do about it. You couldn't ride herd on Larry Hart. Not in Hollywood."

In spite of his lack of affection for the task, Larry created some sparkling lyrics, with precisely the right Ruritanian flavor, for *The Merry Widow*: the lilting "Melody of Spring" and a bravura rendering of of "Maxim's," where—in Larry's lubricious version—the girls are dreams. Even his limpid, if entirely unnecessary, new lyric for the famous title waltz is exactly right for the context. One song, "Tonight Will Teach Me To Forget," is credited to Gus Kahn; no one seems to know how he became involved.

Featuring Edward Everett Horton, Una Merkel, and Sterling Holloway— reunited briefly with his *Garrick Gaieties* chums—*The Merry Widow* was shot in eighty eight days (exactly the same time it took Frank Lloyd to film the Gable-Laughton *Mutiny on the Bounty*) for a total cost of $1,605,000, which included a simultaneously filmed French language version. It was a perfect example of what MGM called a prestige picture. Although it never broke even at the box office, it was and remains one of the most beautifully photographed, effervescently performed, and stylishly mounted of all thirties musicals.

It also marked the end of the career of Rodgers and Hart at MGM— except for one final surprise. In May, *Manhattan Melodrama* went into release. As already noted, "The Bad In Ev'ry Man"—formerly "Prayer" and "It's Just That Kind of a Play"—failed to make any impression on the public. The story goes that one day Larry bumped into MGM's music publisher, his old friend Jack Robbins. They were talking about the picture, and Robbins said, "You know, Larry, that's a really good tune you boys have got there. I'd be glad to get behind it, but it needs a commercial lyric." Stung by this remark—it was a point of pride with him that he always wrote lyrics to fit the scene or the personality of the performer—Larry retorted, "Oh, yeah, I suppose what you'd like me to write is something corny like 'Blue Moon.'"

"Yeah," Robbins breathed, 'Blue Moon!'"

Rodgers's version is less romantic. He said that the following year, when he and Larry were in California working on *Mississippi*, Jack Robbins sent them a wire suggesting they write a "commercial" lyric for the melody and let him publish it. "Sometime within the next three months, a new lyric was written. . . . When Larry showed the lyrics to Jack Robbins, Robbins personally suggested changes in the last three lines of the chorus."

Now we know why Larry never liked it.

25

✦ ✦ ✦

Yesterday's Men

*I*n April 1934, on what was to be his final day at MGM, Dick Rodgers called in to see Irving Thalberg. "Larry and I are leaving today," he said, "and I just wanted to say goodbye." As Thalberg looked up uncomprehendingly, Dick realized that not only did the Boy Wonder not know what he was talking about, he did not have the faintest idea who he was. It was, he thought, a fitting end to their MGM years.

The Broadway they returned to was vastly different from the thriving theatrical mainstream it had been four years earlier. Now luxurious movie palaces like Radio City Music Hall and the Roxy—"The Most Magnificent Motion Picture House in Any City, or Any Country, Anywhere!"—offered patrons not only a movie but also a lavish stage show for less than the price of a balcony seat in the theatre. Ritzy "clubs" and "cabaret restaurants" like Billy Rose's Casino de Paree—"Believe it or not! For $2 you get the Revue, the Dinner, the Dancing, the Girl in the Fish Bowl, and a hundred other novelties!"—also competed. On Broadway at 48th Street was the Hollywood, where $1.50 bought dinner and a floor show featuring Rudy Vallee and his "Connecticut Yankees," singer Alice Faye, and "50 Gorgeous Girls 50." At the Paradise, a block further up, Paul Whiteman and his orchestra offered an "International Revue with 50 World's Loveliest Girls 50," plus Jack Fulton, Ramona, the Rhythm Boys, and Peggy Healy for the same price, and "never a cover charge."

George M. Cohan's fortunes had revived with a big hit in Eugene O'Neill's *Ah, Wilderness*. The stage version of Erskine Caldwell's steamy *Tobacco Road*, slated at first by the critics, had settled in for what would be a record run. Audrey Christie and Bruce Macfarlane starred in *Sailor Beware!* Walter Huston scored a personal triumph in *Dodsworth*, adapted from the Sinclair Lewis best-seller by Sidney Howard. Among the musicals, the Moss

Hart–Irving Berlin revue *As Thousands Cheer*, with Marilyn Miller and Clifton Webb singing "Easter Parade" and Ethel Waters cooking up a torrid "Heat Wave," was still at the Music Box. Jerome Kern's lovely score was keeping *Roberta* going; everyone was singing "Smoke Gets In Your Eyes," in spite of the curiously unsingable lyric Otto Harbach had put to it.

Broadway veterans had gaped in astonishment when the Shuberts—the *Shuberts???*—announced they would produce a new *Ziegfeld Follies*. Prompted by a legion of creditors, Ziegfeld's executor sued: the Shuberts had no right even to use the Ziegfeld name. Wrong: Mr. Lee produced a contract with Ziegfeld's widow, Billie Burke, who had sold him all rights in the Ziegfeld name and properties for one thousand dollars plus 3 percent of the gross. Now the creditors closed in: what right did Billie have to negotiate with the Shuberts or anybody else? Ziegfeld had signed an agreement with her, Mrs. Ziegfeld replied, giving her all rights in all his properties in exchange for loans she had made to him. And where was this agreement? She had mislaid it, Billie fluttered. So go sue a penniless widow. The show went on, "produced by Mrs. Florenz Ziegfeld," and scored a hit. A little while later the Shuberts sold the name *Ziegfeld Follies* to MGM for twenty-five thousand dollars. The creditors screamed, but they never got a penny.

Dillingham's final production—he died later that year—was a New York version of the revue at which Larry Hart had met the Rocky Twins in California. *New Faces* introduced Imogene Coca and Henry Fonda (who had been understudying Humphrey Bogart) to Broadway audiences. When its run was over, Fonda won the lead in *The Farmer Takes a Wife* in Hollywood and so began his long film career. The luck of the draw, or in this case the toss: at the auditions for *New Faces* there had been a young man whom everyone had wanted but whose physical similarity to his good friend Fonda created a problem. They flipped a coin to see which of them would appear: Fonda won. James Stewart had to wait a little longer for his big break.

As the 1933–34 season drew to a gloomy end, Dwight Deere Wiman's production of *Champagne Sec*, an Americanized version of *Die Fledermaus* starring Kitty Carlisle, Peggy Wood, and Helen Ford, had registered only 113 performances, and even Bill "Bojangles" Robinson couldn't keep Lew Leslie's *Blackbirds* revue going longer than three weeks. Only one new musical opened on Broadway between March and June: an operetta called *Caviar*.

Caviar had special significance for Larry Hart; during the run of *America's Sweetheart* he had become greatly enamored of a young singer at the Metropolitan Opera named Nanette Guilford. Nanette fitted beautifully into the Rodgers-Hart-Fields world. Delivered by Dick's father, Dr. William Rodgers, and a niece of Lew Fields's former partner Joe Weber, she had gone straight from finishing school to the opera, where her remarkable voice earned her the sobriquet "Baby Star of the Met." Recently divorced, she had decided to do a musical; she was in rehearsals for *Caviar* when Larry returned to New York. They began seeing each other again.

Calling itself a "romantic musical comedy," *Caviar* owed a lot of its plot to that old clunker *The Count of Luxembourg*. The producer and librettists were unknowns. The music was by someone called Harden Church, and lyrics were by Edward Heyman, who had done the words for "You Oughta Be In Pictures," currently being sung in the *Follies* by Jane Froman. The director was Clifford Brooke, who had had a couple of hits before and during the Great War and an almost unbroken string of flops ever since. Larry Hart sat through a rehearsal at the Forrest on West 49th Street, then sent Nanette a card: "Only you can make a swan out of this turkey." She recalled:

> The director was absolutely hopeless, every night was different. We were at our wits' end. I talked to Larry about it. He told me not to worry. "It'll be all right, baby," he said. He always called me baby. "I know it's the most unprofessional thing in the world to do, but we'll do it. You get the cast together, bring them in to the theatre at nine thirty tomorrow and I'll direct." Well, you can imagine! Larry Hart coming in to direct! Everyone was so excited. We all sneaked into the theatre next morning, and Larry started running us through. Of course, you know what happened. Clifford Brooke walked in! He didn't need to ask what was going on. He just glared at Larry. Larry got out of there somehow, I don't know how. He laughed! He laughed and laughed about it! But he was right. It was a turkey.

Mel Shauer painted a vivid little word picture of Larry and Nanette together. They had been to the opera, he remembered, and went on afterwards to Reuben's, an all-night restaurant at the corner of Madison and 59th Street. Larry was wearing evening clothes and top hat, and Mel's impression was that even with the top hat on, Larry was only slightly higher than Nanette's head. Another time, as she and Larry were leaving the opera, a couple of rowdies made fun of the diminutive Hart. Although he affected to ignore them, he never wore a top hat again.

Nanette was beautiful, Shauer said, but like most prima donnas of that day, buxom. He always believed that if Nanette had been more petite, if they had made less of an odd pair, Larry would have married her. In fact, Nanette said, Larry did ask her to marry him; but after her "miserable" experience with violinist Max Rosen, she had decided never to marry again, and after a time they drifted apart.

Rodgers and Hart had hardly got used to being back in New York when they had to turn right around and go back to Hollywood. For once, the old pals'

act was working in their favor: Dick received a call from Arthur Hornblow
Jr., with whom they had become friendly while he was romancing Myrna
Loy on *Love Me Tonight* (she became Mrs. Hornblow in 1936) and later
when he was production supervisor for Sam Goldwyn.

Hornblow, now a producer at Paramount, had been searching for a vehicle
in which to feature the studio's singing discovery Lanny Ross, a young tenor
who had appeared (with Harriette Lake, now calling herself Ann Sothern) in
1933's *Melody in Spring*, and more recently with radio star Joe Penner
(famous for his catchphrase "You naaaahsty man!"), Jack Oakie, and Lyda
Roberti in *College Rhythm*. Hornblow thought he had come up with exactly
the right story. By one of those delightful coincidences with which movie
history is littered, it was based on an early adaptation by Herbert Fields.

In 1929, looking back over his shoulder as usual, Fields got together with
reporter-screenwriter Claude Binyon (reputedly the man who wrote that
most famous of all *Variety* headlines on the day of the stock market crash:
WALL STREET LAYS AN EGG) on an adaptation of a forgotten 1923 Booth Tarkington play, *Magnolia*, itself in turn adapted from Tarkington's novel *The
Fighting Coward*, about a young Southerner (Leo Carrillo in the ill-fated
play) who is considered a coward because he refuses to fight a duel. Eventually he gains a "killer" reputation, and redeems himself down on the old
plantation.

Paramount had already filmed it as *River of Romance*, an early musical starring Buddy Rogers, Mary Brian, and Wallace Beery. This was the vehicle
Hornblow had decided on for Lanny Ross, and he wanted a score by
Rodgers and Hart. With a new script by Francis Martin and Jack Cunningham, and a new title, *Mississippi*, it would also feature Joan Bennett as the
love interest and W. C. Fields in the comedy part of the riverboat owner
originally played by Beery.

Rodgers and Hart reported to Marathon Avenue in mid-July and set to
work. Once again, they aimed toward creating music and lyrics which would
integrate with what moviegoers would see on the screen. Thanks to the sur-
vival of a demonstration disk of their score, it is possible, even with a synopsis as bare as the one above, to follow the way they worked. An aside: on the
record, Dick Rodgers sings and plays all the songs except a few lines in the
opening track ("No Bottom"); it is clear to a careful listener that from the
lines "By the sandbar/ Quarter Twain here you are" the song is sung by
someone else. It seems most unlikely it could have been anyone but Larry.
And if it was, he certainly sings well enough to belie all those stories about
his having a range of about three notes and a voice like a raven.

Of course, Rodgers and Hart must have been aware, from the moment
they began, of the dangers of inviting comparison with Kern and Hammerstein's *Show Boat*; they wrote accordingly. The first song they completed was
"Down By The River" (its first five notes are a musical phrase from the verse

of "How About It?"). Next came a second ballad for Lanny Ross, "Soon (Maybe Not Tomorrow)." These were followed quickly by "No Bottom"—featuring the badinage between the leadsman on the riverboat, a deckhand, and the Fields character, Commodore Jackson (not used in the movie, where they were replaced by banjo phrases)—and the title-music song, "Roll, Mississippi."

Three more numbers completed the score: one for a street scene in which passers-by discuss the reputation of "that laughin' sneerin' snarlin' fiend, notorious Colonel Blake"; a comedy song for W. C. Fields, "The Steely Glint In My Eye"; and a "Mexican" number called "Pablo, You Are My Heart" (which—musically, at least—is a clearly recognizable prototype of "Johnny One Note"). Dick and Larry also wrote a number called "I Keep On Singing," which Dick, at least, thought charming. Paramount clearly did not, and it was dropped like these last three during filming.

While they were working on *Mississippi*, Rodgers and Hart noted in one of the Hollywood trade papers that RKO, which had just finished shooting the Fred Astaire–Ginger Rogers *The Gay Divorcée*, a film that would establish them as inseparable for the next decade, was looking for something new for the team. Larry and Dick began thinking of ideas while they completed their Paramount assignment. They wrote "a two-page outline and two or three songs" about a former vaudeville song-and-dance man who gets mixed up with the world of ballet and a temperamental ballerina before returning to the sweet kid he really loves. They even had a title, *On Your Toes*, which suggests it was here, at the Beverly Hills Hotel, that the song of the same name was written. They took it to Pandro Berman at RKO; he liked it and promised to show it to Astaire.

Fred Astaire was intrigued; although he'd done shows with just about everyone else, he had never worked with Rodgers and Hart. He came over to the hotel, read the outline, and listened to the songs, all of which he apparently liked. But not enough; Rodgers records that Astaire turned it down because he felt his public would not accept him in a role that didn't allow him to appear in top hat, white tie, and tails. Maybe it made even less of an impression than that; Astaire didn't mention the project, the meeting, or even Rodgers and Hart in his autobiography.

By the end of August, Larry and Dick had completed the *Mississippi* score. In a letter to Dorothy that month, Dick delivered himself of the judgment that it was the first work he'd done since *Love Me Tonight* that had satisfied him. He, sagely, expressed qualms about Lanny Ross, whose screen presence was proving nearer to an absence; maybe a sympathetic director might be able to improve his performance.

Perhaps so; but Ross never got the chance. Larry and Dick were hardly unpacked when there was yet another studio revolution. Paramount Publix, which had been declared bankrupt in 1933, was reorganized as Paramount

Pictures, Inc., with Barney Balaban, the former partner of Sam Katz (who moved on to MGM), becoming president, in the process kicking Adolph Zukor all the way upstairs to the figurehead office of chairman of the board. After Balaban sat through a rough cut of *Mississippi* he gave Lanny Ross the thumbs-down, replacing him with another tenor who was certain to be much bigger box office: Bing Crosby.

Crosby was in the middle of a brilliant career. He had started in the 1920s as one of the Rhythm Boys, a trio who sang with Paul Whiteman's orchestra; he made his Hollywood debut in 1930 as a featured singer with Whiteman's orchestra in *King of Jazz*, then appeared in a few Mack Sennett shorts. So far, he had scored in a couple of lightweight features, notably in MGM's *Going Hollywood*, where he had introduced "Temptation," and in an adaptation of Barrie's *The Admirable Crichton* produced by Benjamin "Barney" Glazer called *We're Not Dressing*, in which Bing co-starred with Carole Lombard, Ethel Merman, Burns and Allen, and a very young Ray Milland. His most recent appearances had been opposite Kitty Carlisle, first in another Glazer production called *She Loves Me Not* (which featured "Love In Bloom" and "Cocktails For Two") and *Here Is My Heart*, co-written by Larry Hart's old pal Edwin Justus Mayer, in which Crosby warbled "Love Is Just Around the Corner" and "June In January."

What was exciting about the casting was that Crosby was one of the nation's favorite radio performers; he had been a top attraction since 1931. One of the greatest plugs any song could get was to have Crosby sing it; his legion of radio fans bought all his records. In addition, since the repeal of Prohibition, there had been a nation-wide resurgence of bars, saloons, and cocktail lounges; most of these had installed jukeboxes, swelling record sales even more.

Consequently Rodgers and Hart were devastated when Crosby professed himself unhappy with the songs they had written for *Mississippi*. Balaban agreed with Bing and told Hornblow to commission a completely fresh score—by someone else. Arthur Hornblow dug in his heels; if the Rodgers and Hart songs went, so did he. Balaban relented, and Crosby went along, but not all the way. Behind Bing's easy-going public persona there was a tough, shrewd, determined negotiator who knew his pictures were playing a major role in keeping the studio solvent. He still wanted changes, and insisted—despite furious opposition from Larry and Dick—on interpolating the Stephen Foster classic "The Old Folks At Home" into the movie.

Rodgers and Hart returned to New York with the situation still unresolved; shortly after they got back, Hornblow called them and asked them for another ballad Crosby could sing backed by a gaggle of Southern belles. In short order, they wrote "It's Easy To Remember," made a demo disk, and mailed it out to the Coast. But no matter how much they continued to protest, Hornblow wouldn't—or couldn't—drop the Stephen Foster ballad.

To rub salt into the wound, when the picture was released Rodgers and Hart discovered to their chagrin that Paramount had credited the song to them.

Summer was almost over, and the new Broadway season shaping up. It was time to get down to some serious business: Larry was broke—he was always broke, no matter how much he earned, because he spent money as fast as he got his hands on it. Although Dick was a lot more careful, with Dorothy Rodgers expecting another baby and Mary in kindergarten, he needed work, too. But producers were hardly beating a path to their door; Larry and Dick were chastened to discover they were looked on as yesterday's men. Oh yeah, Rodgers and Hart, that collegiate stuff. It doesn't go any more.

They looked back on their year with little or no affection. Apart from "Rhythm," a piece of special material written for Bea Lillie that had gone into an André Charlot revue called *Please* in London at the end of 1933, their sum total of work for the year was "Blue Moon," a hit; "That's Love" in Goldwyn's *Nana*—a disaster; a Gaumont-British movie of *Evergreen*, (the original title telescoped into one word) starring Jessie Matthews and Sonnie Hale, with all but three of their songs dropped; the May-released *Hollywood Party*, likewise; the lyrics for the Chevalier-MacDonald *The Merry Widow*, released in October—which Larry wasn't particularly anxious anyone should know he had written. As for the songs for *Mississippi*, which they were pleased with, no one would hear them until next year.

Rodgers and Hart did some more work on *On Your Toes* and tossed some other ideas around, but nothing jelled. Toward Christmas, Larry and an old collaborator, the one-legged Laurence Stallings, down on his luck after a spell in Hollywood, decided to put together a revue, which seemed to be the kind of entertainment producers were looking for. They had a "bicycle"—an idea—for something with a South American flavor; they even wrote a song for it called "Muchacha," which Rodart published at the beginning of 1935.

In that same December, with Doc Bender back in tow, Larry embarked on the *Monarch of Bermuda* for a Caribbean cruise. While he was away, he jauntily told reporters, he would be working on a "new musical for production next spring." Perhaps he had decided to develop the South American idea on his working vacation; more probably, in the words of the old song, he was painting the clouds with sunshine.

26

Billy Rose's Jumbo

arry Kaufman—no relation to George—was a garment-district graduate who had gone into the ticket brokerage business, then joined Lee Shubert as a sort of assistant producer and—more important—scout. It was he who had brought Lee Shubert together with Florenz Ziegfeld's widow for the successful *Follies* of 1934, as well as steering the revue *Life Begins at 8:40* their way. One day early in 1935, he bumped into Larry Hart, or Dick Rodgers, or both of them, and asked—as he asked everyone—what they were working on.

This and that, they replied. They had written the opening number, "What Are You Doing In Here?" for a one-performance benefit show called *The Post-Depression Gaieties*, presented by Marc Connelly at the end of February. "Blue Moon" was being played everywhere; Bing Crosby had been heard plugging some of their songs from the forthcoming *Mississippi* on his CBS radio program February 5; and there was talk in the papers that the "revue with a threat of a plot" they were working on with Laurence Stallings would follow Max Gordon's elaborate and expensively mounted *The Great Waltz*—a Strauss family operetta written by Moss Hart—into the Rockefellers' Center Theatre.

When Kaufman remarked that Lee Shubert was looking for something for Ray Bolger, who had starred in *Life Begins at 8:40*, Larry and Dick told him they had the perfect vehicle, and gave him a brief synopsis of their vaude-ville-hoofer-in-the-Russian-ballet idea. Kaufman liked what he heard and set up a meeting.

On the appointed day, Larry and Dick went up to Shubert's apartment—his office was too small to hold a piano—and started to do some of the songs they had written. They had already been warned that holding an audition for Shubert was a chastening experience, and that was exactly what it turned out

to be. With his leathery skin, prominent nose and slicked-down black hair—
Gypsy Rose Lee once described him as "a fascinating cross between a
wooden Indian and a hooded cobra"—Lee Shubert's impassive, beady-eyed
stare could make the heart of even the stoutest supplicant falter. Chastening
was the word: as Dick played and Larry sang, Shubert did something astonishing even by his bizarre standards: he fell fast asleep!

More than somewhat discouraged by this demonstration of Shubert's lack
of enthusiasm, Larry and Dick slunk away. A day or two later they got a call
from the Shubert office. Lee Shubert—probably prodded by Harry Kaufman—loved the story and the songs; Bolger—who'd worked with Larry and
Dick in *Heads Up!*—did, too. Shubert was willing to advance them a thousand dollars for a one-year option; they accepted with alacrity.

On March 15, just ten days after the birth of Dick Rodgers's second
daughter, Linda—delivered at Lenox Hill, as her sister had been, by Dr.
Mortimer Rodgers—Larry Hart told a newspaper reporter that Lee Shubert
had approved the first act of *On Your Toes*. Larry also said that he and Dick
were working on a second project for Shubert: *Going Places and Doing Sings*.
Whether this was the project they discussed with Laurence Stallings (no
show of this title was ever staged), or an earlier title for Shubert's *Something
Gay*, which premiered April 29 starring Tallulah Bankhead and Walter Pidgeon, and to which Larry and Dick contributed "You're So Lovely And I'm
So Lonely," is no longer apparent.

At the beginning of April, *Mississippi* was released to mixed reviews.
Crosby, perhaps because he was thirty or forty pounds overweight and
trussed up in a corset to hold in what he called his "bay window," looked
uncomfortable, and platinum blonde Joan Bennett merely vapid. The critical
consensus was that W. C. Fields—especially in his "five aces" scene—had got
the best of whatever the picture had. Of its songs, only "It's Easy To
Remember" would last.

———※———

Still waiting for Lee Shubert to give them a start date on the new show,
Rodgers and Hart got a call from Billy Rose. Had they heard about his big
new project? Yes, indeed: who had not? Broadway rumor had it that
financier and show angel John Hay "Jock" Whitney was putting up a quarter
of a million dollars for Billy—these days seeing himself more and more as a
combination of Ziegfeld and P. T. Barnum—to produce a gigantic spectacular, part indoor circus, part vaudeville, part famous-name revue.

Rose had been working on the project for months; everyone was talking
about it. The show would have girl aerialists, horse ballets, tightrope walk-
ers, lion tamers. He had taken over the old Hippodrome Theatre at 44th and

Sixth Avenue and was going to gut and totally refurbish the interior to house his extravaganza, which would feature Jimmy Durante and the Paul Whiteman Orchestra. Ben Hecht and Charlie MacArthur would write the book, and Rose wanted Rodgers and Hart to do the score.

At the time Rose approached them, Larry and Dick were at something of an impasse. The deeper Larry got into the first draft of their vaudeville-hoofer-in-the-ballet story, the more complicated and demanding it became. They decided to enlist the expertise of George Abbott, whose skilful direction of *Three Men on a Horse*, in which Larry's brother Teddy had just scored his first hit, had so impressed them. The tall, sharp-featured, thirty-eight-year-old Abbott was an inspired choice. As an actor, writer, collaborator, fixer, and director, his record of successes stretched from *Dulcy* and *Hell Bent fer Heaven* to *Twentieth Century* and *Room Service*. "He told us what he liked and what he disliked about the script," Dick said, "and what he said made so much sense that we asked him not only to rewrite the book but to direct the production."

At around the same time Doc Bender, who haunted the dance rehearsal studios on 54th Street, told Larry about a Russian-born choreographer he had seen. Born Gyorgy Balanchivadze, he had worked with Diaghilev in the late twenties. The preceding year, after a season with the Ballets Russes de Monte Carlo, he had co-founded with Lincoln Kirstein the School of American Ballet, changing his name to George Balanchine. Soon after their first meeting, Larry entreated him to consider staging the dances for the new show; the thirty-one-year-old Balanchine enthusiastically agreed.

Meanwhile, it looked as if Lee Shubert was never going to produce *On Your Toes*. They took the project to Dwight Deere Wiman, who liked it and wanted to go ahead immediately, but until Shubert's option ran out there wasn't much they could do but wait and get on with the Billy Rose extravaganza.

"To acquaint myself with musical comedy routine," George Abbott recorded, "Rodgers and Hart thought that it might be a good idea for me to break in by doing another musical first, and they suggested *Jumbo*, for which they had done the songs and which Billy Rose was to produce. . . . I went to see Billy Rose, and it was agreed."

Ben Hecht and Charles MacArthur, who had written the book, were an almost legendary New York writing team. After an apprenticeship of newspaper work, furiously writing *Smart Set* stories and no-pay flops, they had a big success in 1928 with *The Front Page*. Hecht was a New Yorker, short, fast-talking, brash, tough, and abrasive; he once told a Hollywood studio head who presumed to advise him on how to write a screenplay, "Look, imbecile, I'm here at your request to write a script for your goddamn movie. I'm a writer and you're not. What's more I'm a writer who knows his business. So I don't want any of your goddamn horning in." The bewildered executive retreated; no one had ever dared speak to him like that before.

MacArthur, born in Scranton, was the son of a minister, tall, well dressed, handsome, effortlessly charming; people were attracted to him as if by magnetic force. But beneath the outward air of calm reserve was a sense of humor as outrageous and unfettered as Hecht's. Once in Hollywood he hired a plane and bombed the homes of several producers he didn't like with empty whiskey bottles. Another time, kept waiting by banker Otto Kahn in the latter's palatial library, MacArthur amused himself by taking down priceless first editions from the shelves and inscribing them with such sentiments as "To my friend Otto, without whose help this could not have been written, Socrates."

For all their expertise, Hecht and MacArthur had never collaborated on a musical comedy book before. What they eventually came up with was little more than a variation on the Romeo and Juliet theme, a story of rivalry between the children of rival circus owners named Considine and Mulligan, which bore considerably more resemblance to the love affair between the children of rival vaudeville magnates John Considine and Alexander Pantages, with which the tabloids were currently entertaining the masses, than it did to Shakespeare. If it wasn't quite as sophisticated a story as Rodgers and Hart had hoped, it was enough to be going on with; their initial list of ideas illustrates how they approached running up the score:

> Blue Day
>
> Barnum & Bailey
>
> You Who Are Lovely
>
> Barnum I & II
>
> Diavolo I & II
>
> Spring Song II
>
> Women!
>
> Laugh
>
> No Job in Omaha

Their final tally was a dozen songs, not all of which have survived intact. "Blue Day" can be easily identified as an early version of the winsome "Little Girl Blue." "You Who Are Lovely" may have been a retread of "You Are So Lovely"; it was not used. Neither was "Spring Song." "Women!" and "Laugh" were both specialty numbers for Durante. The Barnum, Diavolo, and Omaha ideas were developed as, respectively, "Memories of Madison Square," "Diavolo," and "The Song Of The Roustabouts."

To these were added a "Party Waltz" originally intended for *Hollywood Party*, which had never got past the planning stage; with Larry's new lyric it became "Over And Over Again." Another piece, "The Circus Is On Parade," was written especially for the Whiteman orchestra. Five new songs were created.

The first, a swirling waltz called "The Most Beautiful Girl In The World," was also used for "The Circus Wedding" (sung by Durante) with a different lyric; it surfaced many years later in the movie version of *Jumbo* starring Durante, Doris Day, and Martha Raye. Another, "My Romance," features some of the most elegantly wistful lyrics Larry Hart ever put on paper; it is, quite simply, one of the best songs Rodgers and Hart ever wrote, although Dick always said it was one of the hardest to "get"—it took him almost a day and a half!

Even more surprisingly, or so the story goes, there was one tune of Dick's Larry so detested that he point-blank refused to write a lyric for it. This may account for the existence of that rarity among Hart mementos, a "dummy" lyric jotted down at first hearing of the melody in order to fix the rhyming scheme.

> There's a girl next door
> She's an awful bore
> It really makes you sore
> To see her . . . [repeat]
> By and by
> Perhaps she'll die
> Perhaps she'll croak next summer
> Her old man's a plumber
> She's much dumber . . .

Finally cornered (locked in the men's room, according to one newspaper story), Larry wrote a lyric to Dick's "obnoxious" melody—inspiration provided by a recent stay at Colligan's Stockton Inn in New Jersey—about the small hotel and the wishing well that would make the song a world favorite, although not quite yet.

As rehearsals of *Jumbo* proceeded, it was clear some of the songs would have to be cut. A cute serenade called "The More I See Of Other Girls," fashioned to be sung by Durante to Rosie the elephant was one of them; "There's A Small Hotel" was another.

The attrition common to most stage productions was tripled in *Jumbo*. As usual, Billy Rose's ambition far exceeded his grasp. *Time* magazine reported he was raiding the circuses of the world for talent, from aerialists to zookeepers. The cast included twenty principals, Paul Whiteman and his orchestra, nine showgirls, sixteen dancers, seventeen "Allen K. Foster Girls," thirty members of Henderson's Singing Razorbacks, and Big Rosie as the eponymous elephant. The only "name" in the production was Durante, who played Claudius B. Bowers.

Billy Rose's press agent Dick Maney—the man who made Benchley a streetwalker—released reams of stories to the columns. Some were routine: Ella Logan, a big success in the preceding season's *Calling All Stars*, had been cast in the lead opposite Welsh-born Donald Novis, a young movie actor and radio singer who had appeared in Paramount's 1932 *The Big Broadcast*. Others were bizarre, like the one about Billy scouring Africa for animals bigger than elephants, because the Hippodrome stage was so big they would be using elephants to simulate dog acts. Another Maney story indicated that for the climax of the show an elephant would be shot from a cannon.

A huge sign was erected on the Sixth Avenue frontage: SHHHH! *JUMBO* IS IN REHEARSAL! The planned August opening date was abandoned; there were more delays as John Murray Anderson, responsible for staging the production, wrestled with the enormous problems of getting animals and actors to perform together on cue while trying to fit scenery, lighting, and costumes into a viable whole. The costs mounted and mounted, well past Whitney's original investment to an eventual $340,000—at that time the most ever spent on a musical.

"This show will either break Jock Whitney," Billy told everyone, "or make me." Every time he bumped into the increasingly worried Whitney he would grin and say, "Hiya, sucker!" The more Whitney bucked, the more often Billy said it. Then one day, when Billy was interviewing actors, a man of exactly his own height and build stepped forward. "What do you do?" Billy asked. "I'm a midget," the man said. Billy exploded with anger; he never did realize it was a set-up, Jock Whitney's revenge for being called a sucker so often.

In September, Gloria Grafton, a graduate of *The Second Little Show* and a protegée of Cole Porter and Noël Coward, replaced Ella Logan. Rehearsals went on and on and on; twenty-one principals worked in a church on West 48th Street, while the chorus—nine showgirls, seventeen aerialists, sixteen dancers, and thirty-two male singers—worked at the Manhattan Opera House alongside the thirty-some circus acts. As the Gershwins' *Porgy and Bess* opened, as Moss Hart and Cole Porter brought in *Jubilee*, and as the Shubert revue *At Home Abroad* found its audiences at the Winter Garden, postponements pushed the opening of *Jumbo* past Labor Day and nearer and nearer to Thanksgiving. Billy Rose put defiant ads in the newspapers: "I'll be a dirty name if I'll open *Jumbo* before it's ready!" Every day, dozens of theatre people dropped in to watch rehearsals; so many of them that Charlie MacArthur darkly predicted: "Billy isn't going to open the show until everyone in town has seen it."

A final, final opening date was set: November 16. Billy Rose even signed (and of course published) an affidavit to that effect. Unable to continue with *On Your Toes* while George Abbott was directing *Jumbo*, Rodgers and Hart accepted what can only be described as an extraordinary commission: to write an original musical comedy for CBS radio producer Howard Brenner once a week, every week. Brenner had signed Helen Morgan, "star of Ziegfeld's *Show Boat*," brash, cheerful comedian-singer Ken Murray from Earl Carroll's "Hysterical Historical Revue" *Sketch Book*, and Romney Brent, whose association with Larry and Dick went all the way back to the first *Garrick Gaieties*.

The word "extraordinary" is not out of place here. Rodgers and Hart were not renowned for following in anyone else's footsteps, yet by going into radio they were doing so with a vengeance. Between February and May 1934, George Gershwin had appeared in a twice-weekly series of fifteen-minute shows called "Music by Gershwin" sponsored by Feen-A-Mint, a laxative chewing-gum; then (while writing *Porgy and Bess*) the following September, he had done a half-hour weekly show with the same title and sponsor. During the same season, Ivory Soap sponsored a thirty-nine-week NBC Red Network radio series called "The Gibson Family," a musical situation comedy that featured original songs composed every week by Arthur Schwartz and Howard Dietz. It was on the occasion of signing for this grueling timetable that Schwartz was asked whether writing so many new songs would not take an awful lot out of him. "Yes," he replied. "But not as much as it will take out of Bach, Beethoven, and Brahms."

The story for the first of the Rodgers and Hart half-hour shows, written by someone called Peter Dixon, was about a modern-day crooner (Murray) who is transported by scientist "Fourth Dimension" Bates (Brent) back to the time of Cleopatra (Morgan). Larry and Dick provided two songs (with plenty of reprises), "A Little Of You On Toast" and "Please Make Me Be Good," together with some patter and other special material. After the show, listeners were promised "thirty minutes of dancing, gaiety, and entertainment in the company of Lois Long and Freddie Rich and his forty-piece orchestra." The show was broadcast live over the CBS network on October 22 under the catch-all banner "Let's Have Fun." It was an unmitigated disaster.

Working only with an aircheck of the broadcast it is difficult to decide whether Murray was under-rehearsed, merely incompetent, or blind drunk. He missed all his cues for "A Little Of You On Toast," and not only blew the lyrics but convincingly demonstrated he either couldn't remember or had never learned the melody. Adding insult to injury, he proceeded to do exactly the same with "Please Make Me Be Good." Helen Morgan (whom the announcer introduced as "Cleo-*pay*-tra") and the chorus did what they could to repair the damage, but the disarray was compounded by a further series of muffed lines and cues during the "dancing, gaiety, and entertainment" that followed, as Murray, Helen Morgan, and the ensemble sang a ragbag of current hits, including Cole Porter's "Begin the Beguine," which Miss Long contrived to rhyme with "benign."

The performance ended with Murray promising another original musical comedy and more new Rodgers and Hart songs next week. Dick Ford would be Chris Columbus telling Queen Isabella he is going to make her a movie star—America's Sweetheart! Murray, Morgan, and the ensemble trailered one of the "new" songs that would be in next week's show; it turned out to be a partial rewrite of *Hollywood Party* reject "You've Got That."

The documentation of early radio is extremely sketchy, and it is difficult to confirm whether—although it seems highly unlikely—"Let's Have Fun" was broadcast the following week, or indeed ever again. If it was, no record has been preserved; Rodgers and Hart probably remained in hiding until Saturday night, November 16, when the Hippodrome marquee blazed out BIGGER THAN A SHOW, BETTER THAN A CIRCUS! *Jumbo* was open for business.

27

"The Saddest Man I Ever Knew"

"Well, they finally got *Jumbo* into the Hippodrome," observed *The New Yorker*. "Now all that remains is to complete the Triborough Bridge and enforce the sanctions against Italy."

They got the bridge finished, but the sanctions against Italy and Il Duce, the Italian dictator Benito Mussolini, whose army had invaded Abyssinia in October 1935, were never enforced. This was a grim reminder of the resurgence of fascism in Europe; even now German chancellor Adolf Hitler was preparing to annex the Saarland. Most Americans, however, preferred to believe events in Europe were none of their concern; they were too busy hoping the Depression was really nearing its end. For those who had any sort of income, however, it was a good time. A complete pre-theatre dinner at the Russian Eagle restaurant in the Sherry-Netherland Hotel cost exactly $2.50, a new tuxedo at Finchley's on Fifth Avenue $60, a 1937 Packard 120 $945. Or, if one's pocket didn't stretch that far, a Dodge four-door touring sedan with built-in trunk sold for just $640.

Jumbo was in business; and so, now, were Rodgers and Hart. Just a month after the Billy Rose show opened, they agreed to write songs for a new Jock Whitney picture with a Spanish locale. Although they were already deeply embroiled in putting *On Your Toes* together, they were able to take on the commission because by the beginning of the year the score of the show was virtually complete. Following the withdrawal of Marilyn Miller, originally announced to star, Dick and Larry went down to Atlantic City to write three new songs, which they did, so they claimed, over one weekend.

In addition to the title song and "There's A Small Hotel," salvaged from *Jumbo* (and cleverly cross-referenced in the *pas de deux* of the spoof ballet), Rodgers and Hart created a wonderful number—inspired by one of Roosevelt's radio "Fireside Chats," in which FDR had spoken of the aspirations of the average man—that lampooned all the latest fads of the seriously rich: jam-packed supper clubs, all-night parties, infidelity, "nerves," psychoanalysis, and even abortion ("the modus operandi") were clearly "Too Good For The Average Man." One other number, "Olympic Games," was registered with ASCAP, but no trace of it has ever been found.

To these were added "Two A Day For Keith" and "Questions and Answers" (The Three B's: Bach, Beethoven, and Brahms), plus a couple of bright, lighthearted love songs, "It's Got To Be Love" (it couldn't be tonsillitis) and "The Heart Is Quicker Than The Eye." This is the one in which the singer ruefully reflects that due to her mother's warning her not to mix men and hard liquor, she was "made on lemonade." There were two splendid ballads: "Quiet Night" and the perennially popular "Glad To Be Unhappy," so often thought to be Larry Hart's most autobiographical lyric. I once asked Richard Rodgers if Larry was glad to be unhappy. "He most emphatically was not!" he replied forcefully. "In fact he was very unhappy about it."

The book of the show—credited to Rodgers, Hart, and George Abbott (but mostly written by Abbott) and rejigged following Marilyn Miller's defection to accommodate the real balletic abilities of Russian-born Tamara Geva, who had danced for Diaghilev in the 1920s—was a slight enough thing. Junior, son of vaudevillians Phil and Lil Dolan, forsakes his hoofer heritage and goes to Knickerbocker University to study music; there he meets and falls in love with co-ed Frankie Fayne. Their fellow student Sidney Cohn is writing a jazz ballet, "Slaughter On Tenth Avenue"; Frankie's friend Peggy Porterfield tries to sell it to a Russian ballet company. The prima ballerina, Vera Barnova, is attracted to Junior and wants to do the ballet (and thus snare Junior), but Sergei Alexandrovitch, the head of the company, says no. Complications ensue, with Junior plunged unreadily—and ruinously—into a performance of the "La Princesse Zenobia" ballet.

Misunderstandings between Junior and Frankie follow. "Can a good man be in love with two women at the same time?" he asks Peggy. "Only if he's *very* good," she tells him. Meanwhile Vera's other swain, ballet dancer Morrosine, is in trouble with gangsters, to whom he owes money. Peggy confronts Sergei and threatens to remove her million-dollar patronage unless "Slaughter On Tenth Avenue" is performed. Junior is chosen to dance it over Morrosine, who pays a hit man to bump off "Juniorvich Dolanski" during the performance. Warned that a gangster is in the audience, the exhausted Junior has to keep on dancing till the cops arrive. Of course, true love conquers, and it all works out just fine.

"The Slaughter on Tenth Avenue ballet was Larry's idea," George Balanchine said. "He was so full of wonderful ideas and worked very hard. What no one realized about Larry was that he was so quick; his mind worked so fast that in a few minutes whatever you needed—a suggestion, a lyric—he would do it very fast."

Just as they were ready to begin rehearsals, George Abbott delivered a bombshell: he quit the show. Rodgers said Abbott just wanted to get away to Palm Beach and play golf. Abbott suggests it was Dwight Deere Wiman's procrastinations that caused the split. Wiman, he said, was "another talented amateur who had money and taste but lacked the drive and know-how that comes from having to scratch your way to the top. He kept postponing. I grew impatient, asked them to get another director, and took off for Palm Beach."

With replacement director Worthington Miner—who had had a success the preceding year with the melodrama *Blind Alley*—rehearsals began in February. Ray Bolger was cast as Junior; Doris Carson, a graduate of Kern's *The Cat and the Fiddle*, was Frankie, Luella Gear was Peggy, Monty Woolley (making his acting debut in a role originally intended for Gregory Ratoff) Sergei, and Tamara Geva, last seen in the revue *Three's a Crowd*, the ballerina. They were scheduled to begin tryouts in Boston March 21, but even before rehearsals were over, it was clear they were in trouble: the journeyman Miner was no George Abbott.

"They began to bombard me with appeals to come to the rescue," Abbott recorded. "Wiman wired me frantically, and then Dick Rodgers got on the phone and told me bluntly, 'It's your show, and it's your obligation to come and protect it.'" Abbott couldn't argue with that; he took the next train north.

Abbott saw the show, but to Rodgers's surprise he made no notes, nor did he say anything about it after they left the theatre. To his further surprise, when he suggested it might be a good idea if they all went back to the hotel to talk things over, Abbott demurred. Clearly he knew exactly what was wrong and exactly how to fix it. "There's nothing to talk over," he told Rodgers. "Let's get some girls and go dancing."

In fact, Abbott had found *On Your Toes* in better shape than he had expected, but "the book was in a mess because the story line had been destroyed by experimenting, and the actors were out of hand." Abbott laced into them ruthlessly, forcing them to play their parts instead of fighting for material. As for the book, "I straightened [it] out by the simple device of putting it back the way I had written it in the first place. In this arbitrary and unpleasant fashion I brought things under control and achieved order and even smoothness by the New York opening."

Hart's pal Doc Bender was especially in evidence during rehearsals. In his own strange way Bender was completely stagestruck, and he aspired to being something of a dancer, perhaps because it meant he could go backstage and gossip with other young male dancers. Which is what he was doing one afternoon when Balanchine, awaiting the overdue entrance of the "Nubian slaves" for the Princess Zenobia ballet, yelled out "Ver de hell is de zlaves?" This prompted Larry to sing out, to the tune of "There's A Small Hotel":

> Look behind the curtain
> You can see six slaves and Bender
> Bender's on the ender
> Lucky Bender!

After two and a half weeks in Boston, the show opened at the Imperial Theatre in New York on Saturday night, April 11, and was an immediate success. The critical reaction was strongly positive, and *On Your Toes* settled in for a run that would last most of the year; no doubt it was a considerable source of satisfaction to the "forgotten" Rodgers and Hart that of all the musicals that opened during the 1935–36 season, the two longest runners were *Jumbo* and *On Your Toes*.

Now they turned to the movie assignment, which was for a new company called Pioneer Pictures, whose origins require a little explanation. When he first joined RKO in 1932, Merian C. Cooper had seen an early demonstration of the new Technicolor process and enthusiastically recommended it to David Selznick, who in turn tried to get the RKO board to take it up; but they declined. When Walt Disney started to use it with great success for his cartoons, Cooper became a heavy investor and persuaded the Whitneys to join forces with him in a new company—Pioneer Pictures—which would make and market movies using the new process.

The preceding February, Paramount had released *The Trail of the Lonesome Pine*, directed by Walter Wanger and shot in the new three-color process. Starring Fred MacMurray, Henry Fonda, and Sylvia Sidney, this creaky hill-billy melodrama attracted large audiences and received much favorable comment because of the "realism" of its outdoor photography. The picture for which Rodgers and Hart had signed was to be the very first musical to be filmed in Technicolor; this was doubtless what sparked their interest.

In keeping with President Roosevelt's new initiative, the positive support

of North American–South American relationships which was known as the
"Good Neighbor" policy, the movie was set—as so many would be in the
next decade—south of the border. A Bostonian dance teacher (Charles
Collins) is shanghaied by pirates "somewhere along the coast of Mexico"
and, after some swashbuckling adventures, wins the hand of Serafina (Steffi
Duna), daughter of the *alcalde*. It cannot have been any coincidence that the
director, Lloyd Corrigan, had already photographed a Technicolor short
with a "Mexican" theme called *La Cucaracha*, some scenes of which would be
incorporated into the new movie.

Rodgers and Hart produced two songs for *The Dancing Pirate*. "Are You
My Love?" was sung in the movie by Duna, and "When You're Dancing
The Waltz" was performed by Duna and Collins. Whether "Muchacha"—
minimally rewritten (probably as a result of the Harry Warren and Al Dubin
song with the same title featured in a current movie, *In Caliente*) as "Little
Dolores"—was also offered for the project is not certain: *Pirate* is the only
thing they were working on at this time in which it might have fitted. Not
that it mattered one way or the other. The picture—"The first dancing
musical romance in 100% NEW Technicolor!"—was released on June 17,
1936; in spite of its novelty value, it was a total failure. It appeared that as far
as the movies were concerned, Rodgers and Hart songs were jinxed.

By the time *The Dancing Pirate* came out, however, they had other preoc-
cupations. At the end of May, Dwight Deere Wiman, thrilled by the huge
and continuing success of *On Your Toes*, signed Larry and Dick to write
another show for the 1936–37 season. That was some distance away: meanwhile, if South America was so popular, why not do a musical set there?
They got together with George Abbott, and on May 31 a newspaper item
announced Abbott would write and produce, and Rodgers and Hart write
music and lyrics for, a show with "a Mexican locale" based on Porter Emerson Brown's 1920 play *The Bad Man*, with William Gaxton as Pancho Lopez.

This old warhorse—"Bad Man" Lopez was a thinly disguised Pancho
Villa—had already been filmed twice (and would be four times more). The
most recent incarnation was a mediocre 1930 Warner Bros. production starring Walter Huston. How the trio felt they were going to be able to make a
musical out of a story which ends in the hero's being killed by the Rangers is
not on record, but it didn't matter, anyway: by the beginning of July they
had abandoned the proposition. "George Abbott has a better idea (his own),"
it was reported, "and they will be working on it soon."

In the event, Abbott elected to produce and direct a new play by Sam and
Bella Spewack, *Boy Meets Girl*, very clearly based on the antics of the Hecht-
MacArthur writing team, to do with a wild-eyed scheme to make a movie
star out of a waitress's illegitimate baby. Meanwhile Rodgers and Hart came
up with an idea of their own.

It hit them while they were taking a walk in Central Park and noticed some youngsters making up games with their own rules. And so that undying musical chestnut—"Hey, kids, why don't we do the show right here in the barn?"—was born. To save themselves from being sent to a work farm, the children of vaudeville performers–mere babes in arms—put on a show to raise money for a youth center; it flops. Second time around, however, when they "stage" an interview with a famous French transatlantic aviator, the ensuing publicity results in success.

From the start, they decided *Babes in Arms* was going to be a Rodgers and Hart—and nobody else—show. They no longer wanted to be seen merely as songwriters who could run up a collection of numbers to hang on some glitzy production, but as musical dramatists, unique in that they would write book, lyrics, and music together. They would get Balanchine to work on the choreography again. And for *Babes in Arms* they would make another dramatic departure from established practice: their show would have no big names; it would be their own *New Faces*, with an ensemble corps of young, fresh, talented kids.

On July 26, a newspaper report noted that producer Alex Yokel, Richard Rodgers, and Lorenz Hart had a musical show "well under way." They were auditioning for a cast of fifty young people, no one over the age of sixteen, with a view to starting rehearsals in September. This can only have been *Babes in Arms*, and the time scale indicates that a November or December opening was originally planned. Even if the report was incorrect and the producer was Dwight Deere Wiman, it still suggests that the show ran into difficulties. The fact that Jock Whitney is known to have later invested money in Wiman's production reinforces the probability; whatever the reason, Alex Yokel dropped out, and *Babes in Arms* was not produced until the spring of the following year.

It was at this juncture that Rodgers and Hart once again decided to do something completely different and totally unexpected. Early in September, just a week or two before the sudden death of Irving Thalberg in Hollywood, Paul Whiteman and his orchestra recorded "Slaughter on Tenth Avenue" for Victor. During the session, Whiteman asked Rodgers, as he had done on earlier occasions, to write something special for him.

Paul "Pops" Whiteman was approaching his twentieth anniversary as a bandleader. Among the growing number of dance band musicians and singers—the "swing" era was just around the corner—who owed much of their success to the smooth and savvy operator who first showcased their talents were the Dorsey brothers, Bix Beiderbecke, Bing Crosby, Jack and Charlie Teagarden, Johnny Mercer, Joe Venuti, Frankie Trumbauer, and

Mildred Bailey. Hardly the world's greatest musician, not much of a conductor, more showman than leader, Whiteman was above all a salesman—especially to radio sponsors and hotel managers. He had commissioned and launched Gershwin's "Rhapsody In Blue" and performed his *An American in Paris*. Anything Pops Whiteman showcased was sure to be noticed.

Rodgers told his partner about Whiteman's request, and almost at once one of them—"and I can't recall whether it was Larry's idea or mine," Rodgers said disingenuously—came up with the idea of a "symphonic narrative" set to music, the soliloquy of a train announcer at Grand Central Station. They would call it "All Points West."

It's a simple little melodrama, opening with the announcer calling out the track and the time of departure of the Great Lakes Express, bound for Albany, Syracuse, Rochester, Buffalo, Cleveland, Toledo, Detroit, Kalamazoo, and (he begins to sing) "all points west!" He comments on the people who are boarding the train: a salesman, a young man, soldiers, college girls, a criminal being taken to Sing Sing by an armed guard, a honeymoon couple, all heading west. As he watches, he keeps wishing he, too, could get on that train and go. The prisoner escapes; the cops fire at him, and kill the announcer by accident. Thus he gets his wish: he has "gone west" in the most literal sense.

Rodgers and Hart's "Symphonic Narrative" was first performed at the Academy of Music in Philadelphia by Paul Whiteman and the Philadelphia Orchestra on Friday and Saturday, November 27 and 28. The following Tuesday, Whiteman conducted a performance at the Hippodrome in New York, dark since *Jumbo* had closed there the preceding April.

Around Thanksgiving, Bea Lillie, who was in rehearsals for a new revue, *The Show Is On*, asked Hart if he would again write some special material for her, to which he agreed with alacrity turning out a rewrite of "Rhythm," which Bea Lillie had sung in the 1933 revue *Please*, and coming up with a second chorus that gave her a superb chance—which she grabbed with both hands—to take a swipe at the Broadway show-stoppers of the moment, most notably Ethel Merman's brass-throated rendition of the Gershwins' "I Got Rhythm" in *Girl Crazy*, not to mention "Sonny Boy," "Rose Of Washington Square," and Hart's own "On Your Toes." Another piece performed by Lillie and credited to Rodgers and Hart is "Go, Go, Benuti," in an "Al Fleegle arrangement"—Al Fleegle being none other than Lillie's pianist-accompanist, Reginald Gardiner. Lillie also briefly revived "I've Got Five Dollars," but dropped it soon after the show's Christmas Day opening. A number Larry drafted called "The Desert Lullaby" may well have been intended for her co-star Bert Lahr, but if so, it was never used.

Around the same time Rodgers and Hart were among the many Broadway composers—Vincent Youmans, Arthur Schwartz, Albert von Tilzer, and

W.C. Handy among them—who were invited to write songs for a Radio City concert organized by Leonard Sillman, Helen Hayes, and Ruth Gordon to aid victims of the great floods in the Mississippi Valley. What their contribution was does not appear to be on the record.

On Sunday, January 3, 1937, "All Points West" was broadcast on network radio. Crooner and bandleader Rudy Vallee—who had recently had a big hit with "The Most Beautiful Girl In The World"—made it a feature of his floor show on the Astor Roof and announced he had bought rights to film it starring himself as the announcer, but the movie was never made.

Just a few weeks after the radio broadcast, *Babes in Arms* came back to life, and the young cast was reassembled: an electrifying new tap-dancing act—eighteen-year-old Fayard and twelve-year-old Harold, the Nicholas Brothers—and a young dancer named Duke McHale were the first. For their heroine, Larry and Dick chose child movie star Mitzi Green, now sixteen; they had seen her do her famous impressions of George Arliss and Edna May Oliver many times in Hollywood. One by one they brought in other young people they had noticed or heard about: a fine baritone named Alfred Drake; Wynn Murray, a choirgirl from Scranton who had come to New York via Hollywood, and who looked and sang like a junior Kate Smith; Ray Heatherton, who had shone in the 1930 *Garrick Gaieties*, at twenty-five the oldest member of the cast. By the time the *Herald Tribune* reported on March 2 that casting was complete, the show was already in rehearsal.

The score for *Babes in Arms* is one of the richest of the entire Rodgers and Hart canon. By the time they got to the Shubert Theatre in Boston to begin tryouts on March 31, 1937, half a dozen of the songs were already done, among them the flowing, affecting "Where Or When" and the dyspeptic "I Wish I Were In Love Again." One of their best songs from this score—notable for being that rarity, a popular song containing a six-syllable rhyme: "laughable / unphotographable"—has over the years become one of their most-performed ballads. "My Funny Valentine," a lament to a lame-duck February 14 lover, was so clearly destined for success the name of one of the characters in the show was changed to match it.

As they expanded the show, so the specialty numbers came. One was a wicked spoof of the "singing cowboy" song; having roamed over the range "where seldom is heard an intelligent word," the singer extols the joys of city living, "Way Out West" (on West End Avenue). Another, for Wynn Murray, deriving its melodic construction and rhyming scheme from the discarded *Mississippi* number "Pablo You Are My Heart," became a bedtime story about "Johnny One Note," the show-off opera star who could outsing anything: boat whistles, steam whistles, train whistles, cop whistles, thunderclaps, lions in the zoo—even Niagara.

It was a score abounding with musical riches. "All Dark People" was a specialty number for the Nicholas Brothers; "Imagine" began and ended the interpolated ballet. A duet for Grace McDonald and Rolly Pickert, "I Wish I Were In Love Again," brilliantly played ebullience against disillusion. Rodgers's musical versatility was never greater; Hart's acerbic wit never sharper. But it was during rehearsals, writing for Mitzi Green, that they hit their best form. In the defiantly brazen confessions of the nonconformist lady who's called a "tramp" because she won't do what the "smart set" does, Larry joyfully thumbed his nose at the pretensions of "society" ladies who went up to Harlem in ermine, ringsided at prizefights, patronized—the exact word—the opera, and similar conceits.

Their eleven-song score complete, the show set to open at the Shubert in New York on Wednesday, April 14, Larry Hart relaxed by giving a few interviews. Their approach to the task of writing *Babes in Arms* had been the same as always, he told a reporter from *Young America*. "We map out the plot. Then Dick may have a catchy tune idea. He picks it out on the piano—I listen and suddenly an idea for a lyric comes. That happens often. On the other hand, I may think of a couple of verses that will fit into the show. I write them out and say them over to Dick. He sits down at the piano and improvises. I stick my oar in sometimes and before we know it, we have the tune to hang the verses on. It's like that—simple!"

In spite of rave opening-night reviews, the ticket agencies showed little interest in *Babes in Arms*, maybe because of the verdict rendered by *Variety's* out-of-town correspondent: "No nudity, no show girls, no plush or gold plate may mean no sale," a put-down not dissimilar to one that would be pinned to the first Rodgers and Hammerstein show some years later. During April and May, receipts were just about the break-even mark, sometimes below it. In June, Wiman cut fifty cents off the top ticket, but sales continued to slide. Then all at once, as if by divine intervention, every competing show on Broadway folded. On July 17, *Babes in Arms* became the only musical on Broadway. The following week takings jumped 50 percent; after that, the show never looked back.

When one newspaper asked for their response to the accusation that "Here In My Arms" owed a lot to "Nobody Knows The Trouble I've Seen," or that "Where or When" was remarkably like certain passages of *La Bohème*, Rodgers and Hart demonstrated what they called "the telephone stunt." This was done by asking for the heckler's phone number, then substituting the note C for 1, D for 2, and so on, rapidly transposing it into a tune. It could be elaborated for purposes of relaxation, Larry said, by adding the exchange as well. They gave as an example the telephone number of police headquarters: SP7-3100, which turned out to be a waltz. Q.E.D.: if you could write tunes from phone numbers, why would you need to steal, from Puccini or anyone else? And did they ever quarrel? Never, said Dick. "We have discussions, but they are clinical, you might say." "Yeah," Larry commented darkly. "Clinical—with lancets."

Another interviewer revealed—gasp!—that Larry collected dirty books. "He will pay anything for a rare edition of a salacious story," the reporter breathed. "His library shelves are filled with erotic books." What about his love life, asked a lady from *Popular Songs* magazine. "Love life?" Larry replied, "I haven't any." Then he was a confirmed bachelor? "Of course," he said. "Nobody would want *me*."

How strange, how sad! Here is Larry Hart, one of the most succesful writers on Broadway, poised this very moment on the brink of creating the most remarkable run of shows and songs ever produced by a single songwriting team, wealthy, famous, a man with a thousand friends, advertising to the world at large he considers himself totally undesirable. What prompts this duality? Everyone who knew him affirmed that Larry Hart wanted desperately to be loved, perhaps even conventionally married. Yet in answer to the question "Are you a bachelor?" he replies, apparently not bitterly, "Of course. Nobody would want *me*." That he cared enormously about his work is attested to by his partner and by everyone else who worked with him. Yet he shrugs and says it's "simple." This is the defense mechanism of the child, many times disappointed, who pretends not to want what he really wants so that he can't be disappointed again—as he knows he will, because he is undeserving.

Strange, too, that just as Rodgers and Hart began to be really successful, their personal paths began more noticeably to diverge. As Dick's star ascended, he grew more confident, braver, ever more musically and harmonically daring. Slowly, slowly, he also became the dominant partner, as Larry, as indifferent to success as he had been to failure, set his feet on a downhill slide. Rodgers became increasingly disciplined and organized in his work and his private life. Never patient with Larry's intemperate behavior, he gradually began to see it as a threat to the success and the good life their partnership had created.

It need hardly be said again that Larry Hart detested routine. He would go to enormous lengths to avoid it. As for actually working, the trouble with Larry—from Dick's point of view—was that it was hell to get him started. The trouble with Dick—from Larry's point of view—was that he was always trying to. The trouble with Larry—from Dick's point of view—was that he always wanted to be going off someplace to have fun. The trouble with Dick—from Larry's point of view—was that he *never* wanted to go off any-place to have fun.

The very stability of his partner's life pointed up the emptiness of Larry's own. As if in response, the parties, the pranks, the wisecracks, and most of all the drinking began to come thicker and faster than ever, and in their wake came increasing irresponsibility. His daily routine rarely varied. Sam, the barber from the Dawn Patrol, a twenty-four-hour barbershop next door to the Stage Delicatessen on Sixth Avenue, would arrive to give Larry his shave and maybe a facial and a hair trim—Larry was already going bald, and visited scalp specialists with the same unquenchable optimism he did everything else—then stay for a game of pinochle. Larry also liked to play records very loudly, joining in with great delight if little voice.

Work began at noon, ended at four or five. Larry wrote only when the mood struck him, jotting down a random thought, a pleasing phrase, an ingenious idea. Once it was written down, however, he was loath to change a syllable of it. The job, for him, was done. This was hard for Rodgers to take: he was a perfectionist who thrived on schedules and work and discipline.

Larry was a night owl: he went to bed late and slept late. He ate when the mood moved him, spending most mornings in a haze of cigar smoke, still wearing pajamas and a bathrobe, surrounded by a chaos of books and news-papers and magazines. In the West 119th Street days, when he was always much involved with his family, he lived what might have seemed to more conventional souls a noisy and undisciplined lifestyle, but to Larry himself that was normal. He had grown up that way, just as he had grown up accus-tomed to servants, to big spending, to all-night parties. Someone once described life in the Hart household as "happy pandemonium." Larry liked it that way.

He loved doing whatever he was doing, whether it was painting the town red with his cronies, or talking a blue streak with a crowd of gypsies in a West Side bar, enthusing about plays he had seen, songs he had heard, books he had read, ideas he was working on. He thought nothing of inviting a hun-dred people to an impromptu party: one night, he even went so far as to invite the entire Paul Whiteman Orchestra.

He liked baseball, and once insisted on Henry Myers being his guest on a trip to the Polo Grounds to see the Giants play. "He was keenly, genuinely interested during the game," Myers recalled, "and at the end asked me: 'Didn't you find that interesting?' Interesting that he used the word interest-ing." And even more interesting that Myers noticed it.

He and I and another show business friend once had a little three-handed poker game—which I'd never imagined Larry could play—in a smoking room of the old Hotel Astor. We played table stakes, that is, no limit, and Larry was the only win-

ner; we other two lost heavily. We played until very late; at last I suggested we
stop and settle up. Thereat, Larry tossed all his chips onto the table, exclaimed
"Fun game!" and refused to accept a penny.

Irresponsible and raffish though he might often have appeared to be,
Larry loved his creature comforts: all the furniture in the apartment was
comfortable and solidly made, and the walls of his sizable den were lined
with fine mahogany bookshelves. His suits and overcoats were custom made;
his monogrammed shirts, pajamas, and underwear came from Ruby's;
Dorothy Hart said he bought Irish linen handkerchiefs by the dozen and lost
them as fast as he took them out of the box. But because of his constant
activity—this Hart never stood still—he always managed to look rumpled,
and far from glamorous: "dressed from head to toe in faultless bad taste," as
author Michael Arlen is said to have once described him. Larry also avoided
looking in mirrors: the reverse of vanity. If he caught a glimpse of himself,
he would make a moue of distaste.

"He never seemed to have any place to go," one of his cronies remembered. "He'd hang around the bars and saloons on West 45th Street where
actors and show people gathered after the show, and stay there till closing
time, buying everybody drinks, paying no attention to what it cost. He'd
haul out a big wad of bills and spend it without a second thought. All he
wanted to do was talk."

He spent a great deal of his time—far too much, his friends said—with
Doc Bender, who over the years had become a vituperative, self-important,
obnoxious creep, universally despised. He kept a coterie of homosexuals
around him and led a life which, for some inexplicable reason, seemed to
appeal greatly to Larry. The nicest thing any of Larry's friends said about
Bender was that he was a leech. Most of them called him what he seems to
have become: a pimp and a procurer.

As Larry grew more successful, Bender—perceiving Larry as "big-time"—
dug in his claws all the more deeply. It was a merry ride, all expenses paid.
Nanette Guilford recalled Larry giving Doc a hundred-dollar bill to go out
and buy sandwiches for some unexpected guests. When Bender came back
with the sandwiches Larry turned to Nanette. "Look," he grinned, "no
change!" He knew he was being taken; he didn't give much of a damn one
way or the other.

A wisecrack, a clever pun, a prank, a carefully orchestrated practical joke—
these were Larry Hart's meat and drink. In restaurants he hopped from table
to table. On opening nights, he never sat in his seat but walked up and down,
frantically rubbing his hands together (that favorite gesture). He never
seemed to know or care what anything cost. He grabbed every check, bought
every drink, paid every taxi fare. Heaven was a houseful of friends summoned
at ten minutes' notice by telephone, an impromptu, noisy, disorderly all-

night thrash with plenty of food, plenty of friends, plenty of liquor, plenty of jokes, plenty of noise. He needed the excitement, the frenetic activity, as another man needs peace and quiet—a party was a bust if it broke up before dawn. And at the center of it all would be Larry—clowning, talking a blue streak, volatile and magnetic, a fountain of enthusiasm and gaiety.

"But he was lonely in the crowds that he demanded, sought, and collected around him," the singer Mabel Mercer said. "He was the saddest man I ever knew."

28

Twice in a Lifetime

Surprising, original, captivating, fresh, tasteful, smart, delightful, genial, buoyant, gay, sprightly, youthful, informal, bright, likable, tuneful, ingenious, fertile, engaging, joyous, swell—the critics threw bouquets of superlatives at *Babes in Arms*. A major hit, it would stay on Broadway for the rest of the year, earning its seventy-thousand-dollar investment several times over. Its success, hard on the heels of *On Your Toes*—which had closed two months earlier and was now on a hugely successful tour—enabled Dick Rodgers to buy a weekend home at Port Washington on Long Island.

He and Larry Hart were already at work on several more projects; the Janos Vaszary play they'd worked on at MGM about the man who married an angel, which at Rodgers's urging Dwight Wiman had bought from the studio, and a movie offer from Warner Bros. A third proposition they were mulling over was a collaboration with playwright Clare Booth Luce, a musical satire on scientific research; a fourth had come from George Kaufman and Moss Hart, whose *You Can't Take It with You* had been one of the smash hits of the preceding year.

Kaufman and Hart had decided on a new show early in February, soon after Kaufman moved into his new town house at 14 East 94th Street. They enjoyed working together: a case, Kaufman said, of gelt by association. Their first idea was a musical revue to feature Clifton Webb and Ina Claire, with Groucho Marx as a sort of master of ceremonies. Groucho declared himself unavailable, so they moved on to another idea, *Curtain Going Up!* This would be a musical about the writing of a musical, songs by George and Ira Gershwin—currently writing the score for a movie, *The Goldwyn Follies*—in which the real-life playwrights and composers would appear onstage. The Gershwins turned the proposition down not just because they were already working on the Goldwyn movie but because George was unwell.

At the beginning of April, Kaufman joined Moss Hart in Hollywood to work on the movie of *You Can't Take It with You* and to develop yet another idea that had sprung from the latter's fertile mind, a musical about psycho-analysis to feature Marlene Dietrich (her star was dimming a little at the time, and they thought they might persuade her to return to Broadway), the first act of which they had already completed. This, too, failed to come together (although Moss Hart and Ira Gershwin would come back to the idea later in *Lady in the Dark*).

Somehow or other the Kaufman and Hart show-within-a-show idea devel-oped into a musical that would affectionately lampoon Franklin D. Roo-sevelt, called *Hold On to Your Hats, Boys!* (which happened to be the punch line of a current smutty joke). At around the same time, Moss Hart asked Dick Rodgers and Larry Hart, who were preparing to come out to Holly-wood on April 25 to do the Warner Bros. movie, if they would be interested in providing a score for the show. A collaboration with Kaufman and Hart, who had just won the Pulitzer Prize (Kaufman's second) for *You Can't Take It with You*, Larry and Dick enthusiastically agreed.

Rodgers and Hart expected to be in California about three months; an interview they gave as they left New York indicates that plans for the new show were already well advanced. When they came back, Larry said, they would be starting serious work on the Kaufman-Hart idea, which already had a new title—supplied by Dorothy Rodgers—*I'd Rather Be Right*. (It came from a speech by Henry Clay, who in 1850 told the Senate, "I would rather be right than be President.") After they got through with *I'd Rather Be Right*, it would be time to whip into shape *I Married an Angel*; following that, they would do the Clare Booth Luce show.

⌐═══◉═══⌐

The movie they were going out to make would reunite them with their old partner Herb Fields, who had written the screenplay with his brother Joe, using as its basis an unproduced play by revue producer Nancy Hamilton (*New Faces*), James Shute, and Rosemary Casey. It was called *Food for Scandal* or—depending on the source used—*Return Engagement*.

Director Mervyn LeRoy, Larry said, taking a few digs at current musical clichés *en passant*, was convinced that "even the standard musical picture, in which the showgirl takes the leading lady's place, has been done once too often."

> And so he sent for two bad boys of Broadway, Richard Rodgers and me, to lose some money for Warner Brothers by writing another kind of film operetta. LeRoy has a good cast under contract: Fernand Gravet, Ethel Merman, and the ebullient

Carole Lombard. The picture, *Fools For Scandal*, will be done entirely without chorus girls in Ziegfeld sequins; overhead shots of coryphees forming a star or the keys of a typewriter; and crooners singing to an open fireplace, shimmering in which is the face of the girl you can't forget. And so, twice in a lifetime, we are trying to make a musical picture that is not a photograph of a stage operetta . . . by using all the songs as dialogue, by writing all the lyrics in the vernacular of the characters, and by balancing the different musical items one with the other so that we should have a musical score instead of a collection of songs.

Sawed-off, narrow eyed, cigar waving LeRoy—former boy actor, boy tenor, vaudevillian, cameraman, bit player, and gag-writer—had turned director in 1927 and made his name, and launched the gangster movie genre, with *Little Caesar* in 1931. Another LeRoy success was *Gold Diggers of 1933*, in which Ginger Rogers had introduced the Pig Latin version of "We're In The Money" to the screen—a trick Rodgers and Hart had tried unsuccessfully years earlier in *Heads Up!* with "Ongsay And Anceday."

With Carole Lombard and Fernand Gravet starring, *Fools for Scandal* was originally to have featured Fanny Brice, Kenny Baker, and Ethel Merman. Brice was dropped. Ralph Bellamy replaced Kenny Baker, whereupon "Love Knows Best," a "gondolier song" tailored to Baker's soaring tenor, was cut from the picture. Ethel Merman bowed out when she got an offer from Fox to appear with ice-skating star Sonja Henie and Don Ameche in a Roy del Ruth extravaganza that began life as *Bread, Butter and Rhythm* and reached the screen as *Happy Landing*. All-American singer Merman was replaced by ultra-British non-singing Isabel Jeans. With changes as radical as these, clearly what eventually emerged had to be significantly different than the picture originally conceived.

Fernand Gravet had come to Hollywood by way of French movies in which he played the charming, elegant Frenchman (or Austrian, or Hungarian, or whatever). He had played the lead opposite Anna Neagle in *Bitter Sweet*, and his most recent appearance had been opposite Joan Blondell in *The King and the Chorus Girl*, the plot of which was pretty much a reverse twist of the new one (it even had Kenny Baker in it).

Lombard, born plain Alice Jane Peters, was at the pinnacle of a career that had begun in Buck Jones westerns and Mack Sennett two-reelers before she graduated to playing opposite Bing Crosby in *We're Not Dressing*, John Barrymore in *Twentieth Century*, William Powell in *My Man Godfrey*, and Fredric March in *Nothing Sacred*. Quite what she was doing in a musical picture is not clear: maybe director LeRoy felt her wacky brand of comedy would outweigh any deficiencies she might have as a singer. "Anybody that's got heart I can make act, anybody," he was wont to boast. Unfortunately, he reckoned without the leaden script supplied by the brothers Fields.

The story had to do with René, an impoverished marquis (Gravet), meeting movie star Kay Winters (Lombard) in Paris and falling in love with her.

When his suit is rejected, he installs himself as her butler in order to compromise her into marrying him. Philip Chester, her insurance broker swain (Bellamy), completes the triangle. Of course, whether the version that ended up on the screen was quite what Herb and Joe Fields originally had in mind is another matter; probably not.

In addition to the title number, nearer to their rhythmic dialogue than song, Rodgers and Hart produced at least five others for the movie, one of them another "sepia specialty" to be sung in a Paris nightclub scene (by the aptly named Jeni le Gon). It was almost as if Hollywood encouraged them to mine the poorest seam of their inspiration, Dick producing at-best-only-adequate melodies, Larry indulging himself in too-clever internal rhyming that doesn't work in "How Can You Forget?" or, worse, is done at the expense of sense in "Once I Was Young."

Their three-month vacation with pay—Rodgers and Hart no more needed three months to write five songs than Jolson needed to rehearse "Mammy"—turned out to be eventful. Hollywood was at its apogee: in the fall of 1936, movie box office receipts had soared 25 percent over the previous year. During 1937, the industry would produce 778 feature films, more than in any year since 1928. In spite of the fact that the number of jobless still hovered around three million, huge crowds stood on line to see blockbuster hits like *Lost Horizon* with Ronald Colman, Humphrey Bogart and Leslie Howard in *The Petrified Forest*, Judy Garland (singing "You Made Me Love You" to Clark Gable's photograph) in *The Broadway Melody of 1938*, Fredric March and Janet Gaynor in *A Star Is Born*, and Jeanette MacDonald and Nelson Eddy in *Maytime*. And now that he had signed Clark Gable to play Rhett Butler, all the world was waiting to learn who David Selznick would choose to play Scarlett O'Hara in *Gone With the Wind*.

During their stay in Hollywood, a place not unfamiliar with flim-flam, Dick Rodgers and Larry Hart seem to have been indulging in a little flim-flammery of their own. It had to do with one of the most unlikely radio stars of all time, a ventriloquist's dummy, and it came about as follows. In the spring of 1937, while working on *You Can't Have Everything*, the first of a six-film partnership with Alice Faye, CBS-Lux Radio Theatre actor-crooner Don Ameche signed a contract to host NBC's new big-budget Chase and Sanborn–sponsored Sunday night celebrity radio show. His regular guests would be ventriloquist Edgar Bergen and his dummy, Charlie McCarthy. Although the illogic of having a ventriloquist on a radio show seems to have escaped most people, the formula was a hit.

An incidental by-product of Ameche's signing was that his former sponsors, Campana, set out to rebuild their own "First Nighter" radio program, and signed Tyrone Power, a young man to whom Larry Hart had been very close, to star opposite their regular female lead, Barbara Luddy. Shortly after this, Power won the lead in 20th Century–Fox's *In Old Chicago*—co-starring with Ameche and Faye—and went on to become that studio's golden boy.

Back to flim-flam. On June 20, as well as Dorothy Lamour and the regular Bergen-McCarthy duo, Don Ameche's guests on "The Chase and Sanborn Hour" were none other than Dick and Larry. Rodgers and Hart, Ameche announced, had written some songs especially for the show, the first of which would be sung by Dorothy Lamour. Music director Werner Jansen struck up the band, and Lamour launched into—guess what?—"Please Make Me Be Good."

Apart from the addition of a new verse, it was exactly the same song Ken Murray had butchered two years earlier in "Let's Have Fun," although this time, at least, Lamour (and Ameche in the second chorus) sang it properly. Then, following some pretty arch badinage about songwriting with Charlie McCarthy, Rodgers and Hart revealed they had also "just finished a song for Don Ameche." This, surprise, surprise, was the other "Let's Have Fun" song, "A Little Of You On Toast."

Following a chorus played by Dick on the piano, "helping Don to launch it," three more followed. Ameche sang it "straight," then a female quartette praised his charms, following which came a Charlie McCarthy and Dorothy Lamour duet featuring further variations. Despite this crafty piece of Rodgers and Hart gamesmanship, however, the songs fared no better in their second incarnation than they had in their first.

<hr />

At 10:35 in the morning of Sunday, July 11, following five hours of surgery to remove a malignant brain tumor, George Gershwin died at Cedars of Lebanon Hospital. His death was a particularly sad one for the large New York theatrical community then resident in Hollywood; no matter how ruthlessly rival songwriters might compete for lucrative assignments, they were all part of the same family. Gershwin had known everyone, everyone had known George. He was bigger than life, a legend. At every studio there was someone who had worked with him, swapped stories about him, admired him, emulated him: no one could believe he was gone. His score for *The Goldwyn Follies* was still incomplete; it was said—sadly—that one of the last remarks he made was, "That I should have lived to hear Sam Goldwyn say, 'Why don't you write hits like Irving Berlin?'"

George Gershwin was buried in New York; an elaborate funeral service was held at Temple Emanu-El on Thursday, July 15, with a simultaneous service taking place at the B'nai B'rith Temple in Hollywood, which was packed with many of Hollywood's biggest names. At 10:00 a.m., California time, every studio observed a moment's silence synchronous with the commencement of the service. In New York some thirty-five hundred people from the worlds of theatre, music, art, and politics filled the Temple to overflowing, with more than a thousand waiting outside in the drizzling rain. The service was conducted by Rabbi Dr. Stephen Wise; a eulogy was spoken by Oscar Hammerstein II. The honorary pallbearers included Mayor Fiorello La Guardia, George M. Cohan, ex-Mayor Jimmy Walker, Walter Damrosch, and Sam H. Harris. And John O'Hara put into words what most of them were thinking: "George Gershwin died on Sunday, July 11, 1937; but I don't have to believe it if I don't want to."

29

"*The Biggest Opening Since the Grand Canyon*"

*D*ick Rodgers and Larry Hart returned to New York at the end of July, ready to begin work on *I'd Rather Be Right*. They hardly knew whether to be flattered or flustered, they told reporters, "since receiving word that [Mervyn] LeRoy intends using every one of the tunes they submitted for the Gravet film." Incidentally, the same *Herald Tribune* piece of August 1, 1937, stated that "I've Got Five Dollars" had originally been written for *Present Arms*, a claim made nowhere else in the annals of Rodgers and Hart.

Right away, they found there was a problem with *I'd Rather Be Right*. When Larry and Dick had agreed to write the score, Kaufman and Hart had been talking about having Charlie Winninger portray Franklin D. Roosevelt in the show. However, when the writers approached him, Winninger, who had parlayed his his *Show Boat* Captain Andy role into a lucrative career, turned them down; he had all the radio and movie work he could use. Early in June, while working on a Fourth of July speech for "FDR" in the play, Kaufman and Hart realized the ideal actor for the part was George M. Cohan. The problem—and it was a real problem—was they knew Cohan didn't get along with Rodgers and Hart, nor they with him.

What seems to have happened is that they persuaded Sam Harris, Cohan's closest friend and longtime partner, to cable Cohan in Europe, inviting him to appear in a Kaufman and Hart musical without telling him who was writing the score, and Cohan accepted in principle. An apocryphal story that he signed for the show sight unseen is just that: he was far too savvy a campaigner to have done anything so unguarded.

Cohan returned to New York in July, read the script, and—despite some misgivings about the part—signed the contract. On August 1, Rodgers and Hart returned to a fait accompli. They were horrified: after their experience with Cohan on *The Phantom President*, they wanted nothing more to do with him. Moss Hart implored them to go along with the arrangement. Cohan, he told them, had nothing but the highest regard for them; it was the treatment he had received at Paramount that caused the problems. This would be different: it was Cohan's first Broadway musical for a decade, and he was not going to make trouble. On top of that they had Harris as producer. Sam was known to be one of the sweetest men on the Great White Way; if anyone could keep Cohan in line, it was he.

Rodgers and Hart considered their options. It was an either-or proposition: put up with Cohan or withdraw. Since they had already invested time and effort in the project, they decided to go ahead and finish the score, "written variously aboard the Super Chief, at Rodgers's home in Port Washington, in Hart's den high above Manhattan's Central Park, in several nightclubs, on the sands at Atlantic Beach, and in a booth at Sardi's"; they claimed writing the whole score took them less than three weeks. "I don't know if it's true," Rodgers said many years later, "but I do know it's possible."

As they attended casting sessions, a reporter from the *New York Mirror* asked Larry how he and Dick had managed to stay together for so long—an innocuous question or a straw in the wind? "Perhaps we've stayed together because of our personal regard for each other," Larry replied. "Then again, we've worked together for so many years that perhaps we've become indispensable to each other. Or maybe it's our dispositions. I've seen many a team split up because of arguments and petty bickering."

Soon the principals were assembled. Joy Hodges, a Carole Lombard look-alike who'd been in a couple of minor Hollywood musicals, was set to make her Broadway debut as Peggy Jones opposite another newcomer, Austin Marshall, as Phil Barker; Taylor Holmes, who had scored a major success in *Tobacco Road*, was set as Henry Morgenthau, Secretary of the Treasury. Rehearsals were scheduled for September; at the beginning of the month, the collaborators met to run through the songs. It was a remarkable group, perhaps even unique, featuring three of the most successful teams ever to hit Broadway: Harris and Cohan, Kaufman and Hart, and Rodgers and Hart. Apprehensive about Cohan's response, Dick Rodgers persuaded his friend Jules Glaenzer to let them use his elegant Fifth Avenue apartment, which was large enough to hold two grand pianos.

Rodgers played one of these, and rehearsal pianist Margot Hopkins the other. It had been decided that Moss Hart, would do whatever singing was necessary. After some publicity photos were taken, they began. There were some twenty numbers in all, including three ballads for the young lovers: "Have You Met Miss Jones?," "Sweet Sixty-Five," and "Ev'rybody Loves You."

Throughout the performance, Cohan sat in a far corner of the room slumped in his chair. In the hour or so it took them to go through the score, nothing—love songs, charm songs, ensembles, recitatives—seemed to make an impression. Cohan sat unmoving, his face a frozen mask of indifference. At the conclusion, he remained seated silently for a few moments without a change in either his posture or his expression. Then he got to his feet, walked across the room, tipped his hat over his right eye and patted Rodgers patronizingly on the shoulder and mumbled, "Don't take any wooden nickels, kid." Then without a single word about the music he walked out of the room.

They were all stunned; Rodgers tried to reassure the others by saying maybe it would work out the way it had with Chevalier on *Love Me Tonight*, when Maurice had come back the next day raving about the songs. But they knew in their hearts it wouldn't. And it didn't. At which juncture it must be said in his defense that Cohan had some justification for any reservations he might have felt; while Larry Hart's lyrics were never less than consistently brilliant, in every aspect equal or superior to Ira Gershwin's in *Of Thee I Sing*, the quality of Rodgers's score was considerably below his impressive best. And during rehearsals Cohan made his feelings on the subject abundantly clear.

Keeping his distance from Rodgers and Hart during rehearsals, Cohan once again made it cuttingly clear that he liked them not at all. "Tell Gilbert and Sullivan to run over to the hotel and write a better song," he'd say. In this he was to some degree abetted by George Kaufman, who was far more interested in the comedy in musical comedy than the music; he had already alienated Larry Hart by telling Larry that if Hart couldn't supply the kind of lyrics he wanted, he would write them himself.

"Our association with Cohan was at all times disagreeable," Rodgers said later. "He was never happy about the success of the show. I don't know what it was—I never knew at the time. Perhaps he never liked Larry or myself or our tunes. We made every effort to be friendly and cooperative, but we never even had a meal with him. Never, in fact, any conversation whatsoever."

Well before the Boston opening, word had spread that the play was something very special; whether the divided principals liked it or not, their show was news. The very fact that they could depict the highest official of their country onstage and make fun of their own government, the division between the President and his own Supreme Court, and the continuing Depression, was seen as something special in a world where dictatorship was becoming ever more evident. Advance sales in New York were shaping up to be bigger than anything since *Show Boat*. There was already talk of Walter Huston starring in a movie version.

The tryout opening in Boston's Colonial Theatre on October 11 took on the dimensions of a "world premiere." The Cole Porters chartered a plane to fly up, bringing with them a contingent of guests. Also present were Noël Coward, John Mason Brown, Clifton Webb, and Dennis O'Brien. The foyer was a symphony of furs, notable among them Dorothy Rodgers in a full-length mink cape.

During the performance, perhaps because he had such a glittering audience, Cohan's festering grievances surfaced in the number "Off The Record." Instead of performing it as written, he interpolated some lyrics of his own, which put Roosevelt—whom he disliked intensely both personally and politically—in a particularly bad light. Having committed this unpardonable sin, he compounded it by jovially telling the audience he would probably get fired for having done so.

All hell broke loose backstage. Practically tearing what was left of his hair out, Larry Hart demanded that either Cohan be taken to task for this insultingly unprofessional behavior or he would quit the show and take his lyrics with him. Rodgers vehemently backed him. After the performance Sam Harris and George Kaufman exacted a promise from Cohan that he would never again do such a thing. Cohan agreed, confiding later to a friend he would never have made such a promise to anyone except Kaufman. That might have been that, but the following morning the New York *Herald Tribune* front-paged the fracas:

COHAN REFUSES
TO SING LYRIC
ABOUT AL SMITH

IN BOSTON TRYOUT HE CUT
LINES, PUT IN HIS OWN
AND WRITERS OBJECT TO IT
BUT ALL'S HAPPY NOW

"GILBERT AND SULLIVAN OF
U.S." REWRITING THE PART.

More recriminations ensued: Cohan accused Rodgers of planting the item; Rodgers and Hart protested about the "Gilbert and Sullivan" crack—being compared to the British duo was one of their pet hates. Cohan was airy about his transgressions. "It was just something about [Governor] Al Smith and the Liberty League," he told reporters. "Al's been an awful good friend of mine. It's all straightened out now, as far as I'm concerned. I've reached the point in life where I refuse to fight any more."

Rodgers fumed, knowing the Al Smith lyrics had been dropped weeks earlier. Hart was lofty: "You understand that a writer must take a firm stand that no one collaborate with him," he told the papers. Kaufman smoothed things over by throwing a large party at the Ritz to celebrate having three of his plays on in Boston at the same time. Not to be outdone, Cohan threw another party the following night, and before the company left town for another session of tryouts at Ford's Theatre in Baltimore beginning October 24, Sam Harris gave yet another to cement good feeling.

In Baltimore, as work continued, the unrepentant Cohan grew ever more intransigent; he wanted more "character-building" numbers. "The World Is My Oyster" disappeared; "Tune Up, Bluebird" became "We're Going To Balance The Budget." The lyrics of "Off The Record" were toned down, making "FDR's" direct allusions to Hearst and Du Pont slightly less libelous. Cohan's judgment also resulted in the original title ballad being dropped and replaced by an upbeat song with the same title. When the charming but so-so "Ev'rybody Loves You" was also dropped during rehearsals, Rodgers and Hart were less than delighted. This left them with only one moneymaker— after all, it was from love songs, not patter, they derived their income—and they were already irked that their royalty from the show was less than that of the authors. The usual deal was librettist and composers got an equal deal; on *I'd Rather Be Right* Kaufman and Hart took 8 percent to Rodgers and Hart's 5.

A final explosion of temperament took place when "Have You Met Miss Jones?"—the replacement song for "A Treaty, My Sweetie, With You," requested by Kaufman or Cohan or both—stopped the show cold. "At last," Dick Rodgers said pointedly, "this is our chance. Now we can have an encore and maybe somebody will remember a couple of bars." Kaufman lost his temper, "but only by proxy," Rodgers recalled. "He wouldn't come up to me and tell me so. He sent Moss Hart to do the dirty work. We had a great big argument on the street in front of Ford's Theatre. In the end, Kaufman won. He was bigger, richer, and older."

Demand for tickets to see the show when it opened in New York was intense. Cohan's coming back in a musical was enough in itself to ensure a good advance sale: according to *Variety* nearly $250,000 came into the box office before the opening, which was good news for Sam Harris, whose bill for mounting the show was now up to around $150,000. The *Los Angeles Herald Express* reported that Walter Chrysler was offering a 1938 coupe for four good seats (top ticket was $4.40), and that scalpers were getting up to $250 a ticket. On the day the show opened—the same day Fiorello La Guardia was re-elected Mayor—the *New York Sun* printed the lyrics of "Off The Record" on its front page.

The story was—for all its satire—musical comedy simple. On the Fourth of July, Peggy and Phil go to the park to hear a concert. They are troubled because they can't get married until President Roosevelt balances the budget. Phil falls asleep, and in his dream they meet the President, who takes pity on their plight and summons his Cabinet to see what can be done. Suggestions for balancing the budget include hundred-dollar postage stamps, a tax on government property, agents employed as pickpockets so people won't know their money is being taken, and getting rid of Baltimore. All good ideas: but every time FDR tries to make one of them law, the Supreme Court dashes out of the bushes and tells him it's "unconstitutional."

In a "Fireside Chat"—the fireside wheeled onto the one-set stage—FDR appeals to the women of America to stop wearing makeup and donate the three billion dollars they save to the government. A Federal Theatre unit, which has to give a performance whenever three or more people are seen together, does an ersatz Strauss number, "Spring In Vienna" (the locale was changed to Milwaukee subsequent to Hitler's annexation of Austria the following March). FDR asks to see the Wagner Act: two acrobats spring across the stage. He goes on the radio with a variety turn to convince the country to elect him to a third term (prompting his cabinet to predict "His Chances Are Not Worth A Penny"—he isn't as good as Jack Benny). So it goes; and it ends when, after a touchingly patriotic speech, FDR advises Phil and Peggy to get married anyway. Phil wakes up from his dream, and they decide that's what they'll do.

On opening night, November 2, *I'd Rather Be Right* stopped the traffic on 52nd Street: the show had the biggest opening, said critic George Jean Nathan, since the Grand Canyon. Two squadrons of police did what they could to contain the heaving crowds—"the most insufferable crush, confusion, and amiable uproar Fifty-second Street has ever known"—who had turned out to see a star-studded celebrity audience that included Paul Muni, Loretta Young, Gloria Swanson, Edna Ferber, Fredric March, Florence Eldridge, Samuel Goldwyn, Elsa Maxwell, David Selznick, Kitty Carlisle, Franchot Tone, Joan Crawford, Beatrice Lillie, Irving Berlin, Ina Claire, Libby Holman, Fannie Hurst, Grace Moore, Jack Warner, Gertrude Lawrence . . .

Pausing only for his opening-night ritual of kissing his wife, Richard Rodgers led the orchestra into the overture of a memorable evening. Only two small misadventures marred the opening; Cohan tripped over a cable backstage and had to play the first performance with his leg in a special rubber cast, and Beatrice Kaufman, who wore a new fur coat to the premiere, had it stolen from her seat at intermission.

At a huge post-premiere party at the Hotel Carlyle hosted by Kaufman and Rodgers—"he season's most illustrious gathering of professional celebrities," according to Lucius Beebe—Dick Rodgers and Larry Hart announced plans for a European trip. Larry would leave on Saturday, November 13, on board the *Rex* to travel in Italy and do some work on the libretto of their new show *I Married an Angel*, already about half complete. Dick would follow ten days later on the *Normandie*; they would rendezvous in London. Yes, he confirmed, the score was also about half done. In addition, they were considering a three-month Hollywood assignment, an offer to write a show for the Opéra Comique in Paris, and possibly writing another non-theatrical piece along the lines of "All Points West."

The morning's reviews confirmed what everyone already knew: *I'd Rather Be Right* was a smash hit. *Vogue* magazine published a full-page collage of all the famous faces seen at the premiere. Takings for the first full week at the 1,334-seat Alvin were $35,600, and business was further boosted by the fact that, for the first three and a half weeks of its nine-month run, there was only one other musical competing with it—*Babes in Arms*. In two short years Rodgers and Hart had become the undisputed kings of Broadway.

30

A Special, After-dark Existence

rom the beginning of 1938, when he took off with Doc Bender for all points east, Larry Hart's personal life began its metamorphosis: from an erratic but fairly ordered existence into a frenetic round of work and parties and vacations and late nights and booze and parties and talk and booze and . . . other things. He was making a lot of money, more than he knew what to do with: Rodgers and Hart were earning twenty thousand dollars a year from ASCAP royalties, 6 percent of the receipts from their shows—say two thousand a week from *I'd Rather Be Right* alone—and as much as fifty thousand dollars for a six-week movie assignment, not to mention outright sales of film rights (such as the twenty-one thousand MGM paid for *Babes in Arms* around this time). Larry's share of this bounty produced an annual income in excess of sixty thousand dollars. "To put it into context, this was in the middle of the Depression," producer Leonard Spigelgass said, "and nobody had any money. But Larry was *loaded*."

There were, however, more than enough hangers-on ready and willing to help him spend every penny of it; there is no limit, the adage goes, to the number of people who will drink free champagne. Once Larry Hart began making big money, his fate was sealed. He was a free ride, a soft touch, a money tree: all you had to do was shake him, and down it came. "He was the greatest party-giver of all time," Milton Pascal said. "Nobody could *ever* pick up a check with Larry around. He used to have all these parties, six hundred people, where you met everybody in the business."

Frieda Hart enjoyed her son's parties enormously, but when she got tired she would go off to bed and leave the youngsters to their fun. One night, Dorothy Rodgers went into her room and found her reading.

"Doesn't the noise bother you, Mrs. Hart?" she asked.

"No, it hass neffer boddert me," Frieda replied, her German accent untouched by five decades in America. "Except maybe dot night ven der whole Paul Viteman bend came in."

Frieda determinedly maintained her own, self-contained lifestyle. Fridays she had her ladies' day penny poker session. Sunday the roast chicken was served exactly at one, in spite of the fact that Larry was more often than not still in bed. Even when there would only be three at the table, Frieda always prepared enough food for a large family. Although Teddy had married in 1938, and he and his wife Dorothy had their own apartment on West 57th Street, they were still expected to come to dinner every night. When Larry threw a party, Frieda brought out her best china and silver. "God, those nine-course dinners in his house with all the porcelain and miles of silverware on either side," Leonard Spigelgass recalled. "And all those servants— those elderly servants!"

Naive to the point of blindness, Frieda saw only what she wanted to see: lots of people having a wonderful time, and her beloved, talented, successful son the sun around which they all orbited.

One time, [Spigelgass continued,] a woman singer got on to a table and did a striptease, all the way. And all of a sudden Frieda comes in! Oh, she says, you poor child, you must be freezing, and she goes and gets her fur coat and drapes it over her. Another time, a Sunday night, Eddie Robinson calls up from the lobby. Now the penthouse had only one door, and Big Mary opened it to see Eddie standing there. And she said, "Aaaarrghh, Little Caesar, Little Caesar!" and hid in her bedroom for the rest of the night!

Spigelgass met Larry through Davy Morris, who played the part of Sidney Cohn in *On Your Toes*.

Davy and I went to Downey's steak house on Eighth Avenue. Larry and Bender were there, drunk. I knew all his lyrics, and could provide intelligent conversation. We got along splendidly. He was spending all his time with Bender then, and a lot of stupid people. Handsome men, but stupid. Bender was kind of a comic. Mrs. Hart hated him. But if you wanted Larry, you took Bender. Lovers? Those two? Oh, God, no. He was a procurer, a pimp, an arranger, men or women.

Larry loved actors. There were often lots of blacks at his parties, which was very unusual in those days. Of course, what was an orgy *then* wouldn't be an orgy *now*. I mean, kissing with the lights on! Shocking!

You have to understand, homosexuality in that period was on two levels. To the world at large you were beneath contempt, but *inside*, inside you were a member of the most exclusive club in the world. No ordinary CPA could get into that circle, Larry Hart, Cole Porter, George Cukor. That was *the* world. Those houses on 55th Street with the butlers—you were king of the golden river!

So what was to be the pattern for the rest of Larry Hart's life was set on that 1938 European trip, which Larry claimed was his and Dick Rodgers's first vacation in six years. From this time forward, right after New Year's, he and Bender would take off for somewhere warm—sometimes a cruise, sometimes Palm Beach in Florida, or Palm Springs in California. Although it doesn't appear that Larry's dentist friend actually gave birth to the phrase "going on a bender," their trips together were usually precisely that. Then back to New York to recover, to work on the spring show, which would be followed by another long summer vacation: the Caribbean, perhaps, or Mexico.

What sexual adventures they got up to, no one knows; but since Doc made no attempt to conceal his proclivities, and it was widely rumored that he was Hart's pimp, it was assumed these were as predatory as the ones indulged in by Cole Porter and his chum Monty Woolley, who spent many a happy evening cruising sleazy bars along the New York waterfront in Porter's Rolls, picking up sailors for what they called "fucking parties."

"Whatever sex Larry indulged in, and nobody is ever sure of what that was," Joshua Logan said, "it was assumed Doc Bender found it for him. Larry never talked about his own private, sexual life. He had a special, after-dark existence." Perhaps with good reason; the homosexual scene was still, of necessity, underground: sex between consenting male adults was as taboo in the late 1930s as it had been forty years earlier for Oscar Wilde.

Illegal or not, there were plenty of people ready and able to provide whatever the paying homosexual might prefer. The better-known New York pimps like Harry Glynn, Matty Costello, and Jack Alexander could send around a sailor or truck driver or dancer posing as a messenger with a "special delivery" package to be delivered in person to hotel room or apartment. The going rate during the Depression was ten dollars for a white man, five for a black, perhaps twice or three times that much at places like Clint Moore's all-black male brothel in Harlem.

How happy Larry was about his situation is another matter. His deep affection for his partner—they had already spent nearly two decades together—was undiminished; but, as George Kaufman had said, a collaboration was like a marriage without sex. And toward sex Larry was still, to use Edith Meiser's exact word, ambivalent. While he professed to like girls, there was no question he liked boys, too, and there were plenty of temptations of both kinds when he worked and when he played. Too, the vacations he took with Bender were the only complete freedom he ever got. Living at home with Frieda and semi-resident aunts or uncles on extended visits, he had no place of his own to take a lover, even had he found one.

There was another problem. "When it came to sex," a young man who claimed to have been his lover said, "he left an awful lot to be desired. Please, believe me. He had a fetish. A fetish about sleeping with anybody. Even if he had sex. He'd get up and go into the closet. You'd wake up and you'd find him in the closet." But why? "Sex frightened him. He didn't know what to do or how to do it. But he wanted to, desperately."

On November 23, 1937, with Larry Hart somewhere between Naples and Budapest, Dick Rodgers sailed for Southampton, pausing only to give the *Telegram* a thumbnail sketch of his partner that owed a great deal to artistic license. Larry had grown up on the streets of Harlem, playing stickball with Milton Berle, Morrie Ryskind, Billy Rose, and Georgie Price (he said, although it was demonstrably untrue).

> He is not athletic, has yet to play a game of golf, is a poor tennis player. His favorite sport is to watch a prizefight. He cares little for poker, but plays pinochle with George S. Kaufman, Marc Connelly, and Groucho Marx. He worries too much, works too hard, gets excited too easily. I love the guy.

During their month's stay in London, Dick and Dorothy Rodgers were guests at a party attended by Lady Aberconway, the novelist G. B. Stern, and David Horner celebrating the birth of W. Somerset Maugham's grandson, Nicholas Somerset Paravicini, born the preceding October; another was given in their honor by Lady Sybil Colefax. Dick also talked to C. B. Cochran about doing a new musical with him. He told reporters about the new show he and Larry were preparing to star Vera Zorina in New York, and repeated the old story of how "My Heart Stood Still" was written, this time placing the near-crash at Versailles.

Contrast his vacation with Larry's, as related to a reporter of the *New York Post* at Dinty Moore's (steaks $2.50) subsequent to his return with Dick and Dorothy on the *Normandie* immediately after New Year's. "I went to Italy," he said. "Naples, Rome, Florence, Venice. Then to Vienna. From Vienna to Budapest. I didn't like Europe. They've got no women over there like the American women." Four major Italian cities and nothing about Mussolini and the rising tide of fascism? A week in Vienna just three months before the *Anschluss* and nothing about war clouds? That wasn't what Larry Hart wanted to talk about nor what the newspaperman wanted to hear. Instead he printed the story of how Larry had run across Jimmy Rogers (onetime band-leader at the Lombardy Hotel where Dick and Dorothy lived) playing for dancing at Chez Moi in Florence. And? "We all got tight as hoot owls!"

The rumors of war had not yet touched America seriously; the country was big-band swing mad. And Manhattan! You could hardly walk a block without seeing one of the famous names: Benny Goodman, the man who started it all, playing his own brand of hard-driving swing in the Madhattan Room of the Hotel Pennsylvania on Seventh Avenue at 33rd; Jimmy Dorsey at the Terrace Room of the Hotel New Yorker, Eighth and 34th; Artie Shaw at the

Blue Room of the Hotel Lincoln, Eighth and 44th; Red Norvo at the Palm Room of the Commodore, Lexington and 42nd; saccharin-smooth Guy Lombardo at the Roosevelt Grill, Madison and 45th; Les Brown and his Band of Renown at the Edison's Green Room on West 47th; Bob Crosby and his Bobcats, with their forceful Dixieland, at the Lexington on 48th.

And the ballrooms: Woody Herman at Roseland, 51st and Broadway; Chick Webb stompin' at the Savoy, Lenox and 140th; Duke Ellington at the Cotton Club, which had moved south from Harlem into Times Square. And movie show-houses like the Paramount, standing on the Seventh Avenue site of what had once been Shanley's Restaurant between 43rd and 44th streets, where you could hear Tommy Dorsey, or Loew's State at Broadway and 45th, where Jimmie Lunceford was playing his relaxed brand of swing, or the Strand two blocks up for Latin American music by Xavier Cugat.

Radio did more than its share of spreading the new gospel. Every night the airwaves were filled with the music of the big bands. Magazines like *Metronome* and *Down Beat* kept the fans informed: they knew the name of band personnel and vocalists the way kids knew baseball players—Harry "Sweets" Edison was third trumpet in the Basie band, Bunny Berigan had Buddy Rich on drums, Gordon "Tex" Beneke played second sax with Glenn Miller, Martha Tilton sang with Goodman, Billie Holiday with Artie Shaw, Mildred Bailey with Red Norvo.

In March 1938, the big record sellers were no longer show tunes or even movie songs (unless you included those from Disney's new, full-length cartoon feature *Snow White and the Seven Dwarfs*) but band numbers: on Victor—Benny Goodman's band playing "Don't Be That Way"; on Brunswick—Duke Ellington, with "I Let A Song Go Out Of My Heart" and Red Norvo with "Please Be Kind," the most-played song on the air of the moment (famous for the parody which goes, "This is my first affair, so what goes where?"). The only "theatrical" song in the "Most Played on Air" listings—and that from their last movie score—was the Gershwins' "Love Walked In"—in thirty-fifth place.

Rodgers and Hart were not impressed by swing. If he could be convinced it was "something more than the old ad-libbing that has been going on for the past twenty years, I'd be more interested in it," said Dick. Agreed, said Larry: "Benny Goodman only does better what Ted Lewis did years ago." And anyway, he didn't like their songs "swung"—he felt it destroyed the subtlety of his lyrics.

Whether they liked it or not, swing was here to stay; and with the paying customers preferring to listen and dance to the big bands, putting on a musical comedy was tougher than it had ever been. Since *I'd Rather Be Right* only a couple of producers had risked it. The Shuberts brought in three contenders: *Hooray for What!* opened on December 1; it had begun life as an anti-war satire and became a madcap solo vehicle for Ed Wynn, the Perfect

Fool, that lasted until the end of February. They had less luck with the Evelyn Laye–Jack Buchanan vehicle, *Between the Devil*, scored by Arthur Schwartz and Howard Dietz; even "I See Your Face Before Me" and "By Myself" failed to please enough customers to keep it going more than twelve weeks. Their third offering, *Three Waltzes*, yet another retread of the lives of the Strauss family, was old hat before it opened. Two revues, *Right This Way* and Leonard Sillman's *Who's Who*, closed after, respectively, fifteen and twenty-three performances.

This was the unpromising climate in which Dwight Wiman readied *I Married an Angel*. The show had really come alive during the preceding summer, when Wiman returned to New York after having seen the London production of *On Your Toes*, which starred Jack Whiting as Junior. It also featured the debut of twenty-year-old Berlin-born ballerina Brigitta Hartwig, who had made her professional debut at age eleven and changed her name to Vera Zorina because it sounded more Russian. Wiman mentioned her as a possibility to Dick for the show; when Rodgers met her at a party in Hollywood—Zorina was appearing in *The Goldwyn Follies*, he making *Fools for Scandal*—he wired Wiman enthusiastically HAVE JUST MET ZORINA THATS OUR ANGEL.

By March, Larry and Dick had most of the score finished; they had their angel; they also now had a brilliant new director, Joshua Logan; they even had the Shubert Theatre booked for a tentative May opening. What they still didn't have was a book. Dick decided, as he had done before, to get Larry away from the clutches of Doc Bender and the temptations of Manhattan with a trip down to Atlantic City, where they could work uninterrupted on the libretto together.

On the four-hour train journey down to Atlantic City they found themselves, for no particular reason, talking about Shakespeare, and Larry observed how remarkable it was that no one had ever adapted one of his plays for the musical theatre. The more they talked about it, the more they realized they were onto a great idea. But which play? For the rest of Dick's stay on the boardwalk, *I Married an Angel* took a backseat as they tried to find what Oscar Hammerstein II described as "the door in and the door out." Once you had that, he always said, all the rest was only the middle. Finally Larry came up with the door in. Teddy Hart was always being mistaken for another well-known comedian, Jimmy Savo. People said they looked like twins. How about casting them in a musical version of that well-known spoof on twinnery, *The Comedy of Errors*?

With that, Larry was on fire; within a couple of days they had the whole idea blocked out. Exciting though the new idea was, however, it still didn't solve their problems vis-à-vis the more urgent task of completing *I Married an Angel*. Larry elected to remain in Atlantic City to finish working on the first act while Dick returned to New York. When there was no sign of him after a few days, he called the Traymore Hotel and talked to his partner. Larry admitted he was hopelessly stuck.

31

Musical Comedy Meets
Its Masters

oshua Logan, a former movie dialogue director for Selznick (*The Garden of Allah*) and co-director for Walter Wanger (*History Is Made at Night*), had just made his directorial debut on Broadway with a smash-hit Wiman production, *On Borrowed Time*, starring Dudley Digges and Dorothy Stickney. He was just thirty years old:

I was sent down to Atlantic City, [he recalled,] where Larry was staying in one of the big hotels finishing the book. . . . It was a four-hour train trip down to Atlantic City, and quite difficult to get to from New York City, and I was met by Larry in a limousine. I had never met Larry before, and he must have been nervous to meet me, because he was very over-talkative, he was smoking and puffing away at his eternal cigar, laughing too much, rubbing his hands together in a typical gesture he did whenever he was either excited or trying to get another idea, whatever it was.

I cannot use anything but almost embarrassingly warm terms about him. He was the most lovable, cuddly, honey bear. He was very small, slightly over five feet—less than that, really, but he wore elevator shoes, you know—and he was not a beauty, he had a rather gnomelike face and body, and a very explosive personality.

He drove us up to the hotel, and I immediately expected to sit down and talk about the script. But not at all; he got out a card table and we went into a game which he called "Cocksucker's Rummy." And I said, "Why is it called that?" And he said, "Because you can do anything. You can cheat, bluff, lie, you can hide cards in the middle of the deck, steal from your partner, anything goes. The only thing that's wrong is if you get caught." Well, I found that very easy to learn. And he would howl with laughter, you know, at the whole idea of playing a game in which you could cheat. He loved anything to do with larceny—it delighted his soul. I think it was because of the chicanery his father practiced, because the old man was some sort of crook. He thought anything his father had done that was under the table, or wrong, was hilariously funny.

All that day they played Larry Hart's particular brand of rummy. Then it was dinner at some great lobster place Larry knew. Next day, more cards. And the next. And each evening a new place for dinner, a new place to have drinks.

"But when are we going to get to work?" Logan kept asking.

"Plenty of time for that, plenty of time," Larry assured him.

Well, pretty soon it was Sunday, or whatever day it was we were due to go back on. We had two hours to wait for the train, and Larry decided, without any conference with me at all, to write the second act, right then. I said, "What'll we do, talk about it on the train?" And he said, "No, I'll have it finished by the time we get on the train, and we can tell stories and have some fun." He put a huge pad of paper on the little card table and just sat and scribbled, we call it chicken-scratch, maybe four or five lines on each page, and threw the pages over his shoulder.

I kept thinking, He's going to run out of steam pretty soon, but not at all, he just kept going and going and going until he had about a hundred pages. I spent my time just trying to keep them in some kind of order; it was difficult to tell what followed what, because you couldn't read anything—there wasn't one word on any one of them that could be read.

This they brought back to the "principal," as Larry had taken to calling Dick Rodgers. As Larry handed the pages over, Logan said, Dick's expression was "very skeptical. And even more so when he looked at it."

"This isn't any good, Larry," Dick said.

"Oh, why, it'll be wonderful!" Larry argued. "We can put that on the stage tomorrow."

"Larry," Dick said, "you know this isn't any good. I don't think even you could read it."

They argued. Larry protested. Dick refused to budge. Larry sulked. "He always did this," Logan said. "It was his way of protecting what he had written, his lyrics. It was always great, even when it was terrible. He was like a small boy, always playing hooky, just to see what he could get away with."

With the cast set—English-born Dennis King, who had come via "serious" acting to the musical stage with *Rose-Marie* and *The Vagabond King*, would play the lead opposite Zorina, with Walter Slezak and Vivienne Segal providing the light relief—Larry was finally coerced into doing some serious work on the libretto. Even so, Logan continued:

We found we were going into rehearsal with no second act. . . . People were asking me about the second act, and there wasn't any! Oh, the plan was there in the original, it had just never been put down on paper. So Dick said, "We'll meet at Larry's apartment tonight and write the second act." Well, I wasn't used to writing a second act in a night! But when we met at Larry's place, I found that *I* was writing the second act, putting it down in plain old pen to paper. I couldn't do shorthand, we had no dictating machines or anything, but in a sense the three of us dictated what should go down, and we stayed up all night long, and got someone to type part of it for us next morning, and by the end of the day I had gone into the second act. And I sort of learned, as I talked to people, that that's the way Dick and Larry always did their shows, and they'd been writing books for years!

The finished story line was light-fantastical: disillusioned Budapest banker Count Willie Palaffy vows never to marry unless he can find an angel. Whereupon, an angel appears and he marries her; unfortunately, her heavenly honesty almost ruins his life. When she learns the wisdom of a little lie, she can no longer remain an angel. All ends happily, of course. Now, with the story worked out, Logan set to work to stage it—only to discover there was no score for the second act, either:

I knew we had a sequence where they had to sing something at an all-night party in Budapest, while everybody in the cast was waiting up to go to Willy's—to Dennis King's—bank and pull out all their money because they were angry with him for some reason. I knew there had to be some big number there, and I figured they'd write a song that everybody could be part of. Larry had said, "You must give everybody a chance to be funny and step out of character for a while." He had a great feeling for how a show should go, that I guess he had developed from knowing so much about the early German musical comedies he had translated for the Shuberts. Any rate that night I, fortunately, slept; but they sat up all night and finished the score of the second act.

Next day Larry and Dick proudly presented Logan with the song they'd written. Logan couldn't believe his eyes or ears. It was called "At The Roxy Music Hall."

Now the Roxy Music Hall was as far from Budapest as I could possibly think of, and I said, "How on earth are you going to fit this in?" and Larry said, "We'll do a takeoff on the music hall and everybody in it will take off their own particular part that they think is funny, for instance Dennis King, because he sang "And To Hell With Burgundy" [in *The Vagabond King*], will sing a song against wine, "To Hell With Burgundy," and Audrey Christie and Vivienne Segal will be the Rockettes but instead of the whole chorus line they'll do it with just two of them, and then Walter Slezak will do a kind of takeoff on [vaudevillian] Ted Shawn, and Zorina will do a ballet that takes off the ballet."
And I said, "But how do you get into it?" And Larry said, "Oh, we'll just bring up the subject of New York, and then she'll start to sing it." And I said, "Oh, no, no, no, you've got to write it in *lyrics*," and Larry *immediately* ad-libbed an introductory phrase—it later became the verse—which Dick Rodgers equally quickly set to music.

Logan started thinking how to stage the number, but he just couldn't see how it could be made to work. "How can you do this?" he said to Larry. "It has nothing to do with anything."

"You can't sing about Budapest all night long," Larry said. "We've got to do something different, that's all, and give people a little fun. They'll forgive anything if it's good!"

Logan capitulated and staged the number, but he still didn't have his heart in it. "I still couldn't imagine how it would work," he said.

The song Larry had envisaged Dennis King guying himself in was called "Men Of Old Milwaukee" and was more or less a retread of the skit song "And To Hell With Mexico!" Rodgers and Hart had written twenty years earlier for the second *Garrick Gaieties*. Having consigned Port or Rhine wine ("let us call it swine wine") as well as claret, Moselle, champagne, and sherry to damnation, it did indeed end, as promised, on the line "And to Hell with Burgundy!" Slezak's Ted Shawn parody is harder to identify: it may have been "Women Are Women," the lyric of which is still missing.

It was next decided that Zorina would perform her specialty, the "ballet that takes off the ballet" choreographed by Balanchine, in front of a Salvador Dali-esque backdrop; surrealism was currently very "in," and all New York was breathlessly awaiting the artist's first visit. In despair, Logan asked Larry Hart to please explain exactly what surrealism had to do with a banker who marries an angel in Budapest, or, for that matter, the ballet.

"Absolutely nothing!" Larry told him.

"Then why is it in the show?" Logan inquired.

"So Zorina can dance," Larry snapped, impatiently, "and so Balanchine can fry Dali's ass. What the hell are you trying to do, make this Ibsen?"

———

Among the other songs especially created for *I Married an Angel* was one whose title Larry (again utilizing his impeccable ear for the phrase of the day) lifted from a best-selling phenomenon of the time, a self-help book by Dale Carnegie called *How to Win Friends and Influence People*. "I'll Tell The Man In The Street" sprang from from yet another phrase used by FDR in one of his Fireside Chats. "A Twinkle In Your Eye" was written specially for one of Larry's favorite performers, forty-year-old Vivienne Segal, who could "sing scabrous lyrics like a lady."

The daughter of a Philadelphia physician and "angel," opera-trained lyric soprano Vivienne Sonia Segal, had made a stunning Broadway debut at age sixteen in a Shubert production, the Romberg operetta *The Blue Paradise*, and gone on to star in Jerome Kern's *Miss 1917*, two years before Rodgers and Hart even met. She appeared in Ziegfeld's 1924 *Follies* and played the lead in *The Desert Song* and *The Three Musketeers* before being enticed to Hollywood to appear with Jeanette MacDonald and Ramon Novarro in MGM's film version of Kern's *The Cat and the Fiddle*. To her enormous chagrin, when the film was finished, Louis B. Mayer cut her part to ribbons to keep the focus on MacDonald; whereupon the feisty Miss Segal point-blank refused to make another film, ever. She didn't, either.

When Dick and Larry were in Hollywood making *Love Me Tonight*, Vivienne was starring in a stage production of *Music in the Air*. She recalled that after seeing the show,

> they came backstage to say hello to me, and of course, the first thing they said was, "You're a comedienne!" I said, "I've been trying to convince people of that for years, but no one would listen." So Larry Hart then said to me, "I'm going to write a part for you, I'm going to write a comedy part for you." And I thought, hmm, they all say that. Well, I bumped into him several times in New York after that . . . and said, "Well, where's that part you were going to write for me?" He said, "When the time comes, when I get the right thing, I'll let you know." Well, one of the truly marvelous things about Larry was that friendship had nothing to do with business. I could have been his sweetheart and if he didn't think I was right for something he was doing—no soap. So finally, I bumped into him as we occasionally did, at Sardi's or on the street somewhere, and he said, "You're in my next play." I said, "That's nice (ha ha) I'm glad to hear it. And what is this play?" And he said, "When you're called for you'll find out. And you'll be good in it."

While the show was in rehearsal, Rodgers got a call from a friend who ran the Warner office in New York to advise him that *Fools for Scandal* would open at the Radio City Music Hall on March 24.

"Great!" Rodgers said, "we'll all go and make a party of it."

"Not exactly, Dick," said the friend. "In fact, I'm calling to ask you to promise that you won't go to to see it."

It turned out that most of the score Rodgers and Hart had written for the movie had been discarded; what was left, apart from some rhythmic dialogue, had been reduced to mere incidental music. As the *Hollywood Reporter* noted on March 15, the remaining songs were "all right, but one wonders why they are in the picture at all." No doubt Rodgers and Hart felt exactly the same way.

During rehearsals, Larry got especially friendly with a young dancer, a friend of Balanchine's named André Eglevsky, whom Joshua Logan recalled as

> "a very talented man, because not only could he do all these leaps and dances—he was a ballet dancer—but he could also draw the dirtiest pictures you have ever

seen. They were quite remarkably well done, very, very erotic, and Larry thought
these were just hilarious, he loved anything in that way. He was constantly a
naughty boy, yet he was innocently naughty boy, he never did anyone any harm in
his life. He was a constant friend of the underdog, and anyone particularly who
was looked down on by Mrs. Grundy. Respectable and proper was not only repul-
sive, but stupid and dull.

Larry flouted the rules because he thought they were stupid. If people
were shocked by the idea of a man of his age having a close relationship with
a young dancer, the hell with them. That was one of the reasons he was so
fond of George Balanchine, who had the same irreverent attitude toward his
"art" that Larry did. Balanchine claimed to have based the Zorina ballet on
Othello, although Logan confessed he could never figure out how. One day,
watching rehearsals, the Russian fumed at the half-hearted way the dancers
were performing.

"I going down on thot stage end geeving dose leetle ballet girls haaaal!" he
vowed.

Logan watched him go, curious to see the kind of hell Balanchine would
dish out. His face set and angry, Balanchine marched up the steps on to the
stage and stood in front of the dancers.

"Leetle ballet girls," he pleaded sweetly, "please leeft legs *leetle* bit higher."

Rehearsals over, tryouts of *I Married an Angel* were scheduled for three
days in New Haven followed by twelve at the Shubert in Boston commencing April 19. At the dress rehearsal in Boston, Joshua Logan got his first
experience of the other side of Larry Hart. "I didn't realize before this that
Larry had been drinking quite a lot," he said,

> but it had gone fairly well, the usual mistakes, lighting and things, and Audrey
> Christie started to sing "At the Roxy Music Hall." She finished the verse, then
> sang "Now come with me, and you won't believe a thing you see," and all of a sud-
> den I heard the most fascinating, terrifying, nervous-sounding sound and I looked
> up, and there to my right, like a little windmill, was Larry Hart screaming, yelling,
> waving his arms, doing something with his cigar, and talking, almost as if he was
> speaking in a foreign language, repeating a phrase "Nonowsiggasnmyshow,
> nonowsiggas, nonowsiggas!"
>
> I said, "Larry, please, what are you trying to say?" And he said, "No now singers
> in my show! No now singers in this show!" And I said, "No what kind of singers?"
> "No *now* singers!" he yelled. I said, "What are 'Now' singers?" and someone
> explained to me, "He means that Audrey added the word 'now' instead of just
> singing 'Come with me.'" Larry was so upset by this that he went up on to the
> stage and sat on the proscenium to protect his lyrics. And I remember while he was
> sitting there, we went on with the show, he just sat there, little tiny Larry Hart,
> and Walter Slezak leaned over to me and said, "May I make a suggestion for the
> decor? I think you ought to have another little Larry Hart sitting on the other
> side."

One of the love songs in *I Married an Angel* is the poignant "Spring Is Here"—was this the "Spring Song" they originally planned to write for *Jumbo*? Its lyric is often considered one of Larry Hart's most autobiographical. Was it? "If you look through all his lyrics," Josh Logan said, "you'll find that the most touching are those about unrequited love. And 'Spring Is Here' is one of the greatest examples of it. I think, and my theory is borne out by many other things, that Larry was really in love with Vivienne Segal, for whom he wrote it—for the character she played."

Teddy Hart's wife Dorothy said Larry had asked Vivienne Segal to marry him more than once; she always turned him down, she said, because her first husband had been a drunk, and she didn't want the same problem all over again. "I loved Larry in one way," Vivienne Segal said, "but not that way."

"I guess even Vivienne knew that you can't marry Larry Hart," Joshua Logan said. "It would be impossible. Like putting salt on a bird's tail. He just wouldn't sit still that long, much as he loved Vivienne—and he *adored* Vivienne. His heart was always available to be caught, to be had, to be hurt, to be wrung out. But it was very seldom that he was able to say that clearly. I think it's one of the most touching things about him as a human being, that his life was that."

I Married an Angel—"With 50 Lovely Dancing Angels"—opened at the Shubert Theatre in New York on Wednesday, May 11, 1938. To Rodgers, sitting with his wife in his usual aisle seat in the back row of the theatre, it seemed the show was dying on its feet: the Edgar Bergen–Charlie McCarthy skit didn't work; Vivienne Segal's soaring soprano was muted by a heavy cold; the laughs were few and far between, the applause perfunctory. After a couple of stiff drinks in the Astor Bar at intermission, it seemed to him the second act seemed to play a little better, but he left the theatre convinced the show had flopped. He could not have been more wrong. The critics raved, notably Brooks Atkinson, who declared, "Musical comedy has met its masters, and they have reared back and passed a Forty-fourth Street miracle." Logan recalled how they all rushed about gleefully quoting that one at each other: all, that is, except Larry Hart, who refused to believe Atkinson meant what he had said. It was an object lesson to Logan that every one of the critics singled out the Roxy Music Hall sequence, confirming Larry's assertion that the public would forgive anything if it was good. Good? It was terrific!

To understand why Dick's rinky-tink tune and Larry's affectionate send-up of the Radio City Music Hall stopped the show, it must be borne in mind that the theatre, gigantic even by New York standards—three mezzanines,

sixty-two hundred seats, sixty-foot-high proscenium, the Radio City Rockettes ("the world's finest precision dancers") filling its four-hundred-thou-
sand-dollar stage—was managed by none other than Samuel L. "Roxy"
Rothafel, the man who had given his name to that other, now-passé wonder
of earlier times, the Roxy Cinema. The song, however, was performed by
just *two* "showgirls"—Audrey Christie and Vivienne Segal—and in one, at
that!

Angel settled in for what would be a nine-month run. With *I'd Rather Be
Right* still drawing crowds, *Babes in Arms* and *On Your Toes* sold to the
movies, and a new show based on Shakespeare's *Comedy of Errors* already in
preparation, Rodgers and Hart were at the peak of their careers. And hard
upon Larry Hart's return from a Caribbean vacation to celebrate his forty-
third birthday, they received the ultimate American accolades: a two-part
profile in *The New Yorker* and an appearance on the cover of *Time* magazine.

32

━━━◈━◈━◈━━━

"If It's Good Enough for
Shakespeare . . ."

*M*argaret Case Harriman's anecdotal two-part *New Yorker* profile, with its William Auerbach Levy caricature, was charming. The first part appeared May 28, the second a week later. It repeated many well-worn Rodgers and Hart tales, including the "dummy" lyric for "There's A Small Hotel" and the story about their writing "It's Easy To Remember" for Bing Crosby after they got back to New York. Miss Harriman later consolidated her pieces in a book called *Take Them Up Tenderly*.

Time, as always, went straight for the jugular, depicting Larry Hart as a "tiny, swarthy, cigar-chewing bachelor who lives with his mother, whom he describes as 'a sweet, menacing old lady,' scowls at white ties, gives manners-be-damned, whiskey-by-the-case all-night free-for-alls, gets bored with people and keeps on picking up new ones."

"Rodgers takes the world in his stride; Hart is tempted to protest, fume, explain, deprecate," *Time* continued, citing an occasion in Khartoum where Larry remonstrated with a desk clerk because the hotel didn't carry *Variety*. "On the surface" Rodgers, living by the clock, managing his own affairs and holding Hart in check, appeared to be the businessman of the pair. But Larry Hart, who "employs a business manager, who 'runs a temperature' when he does not feel like working, who has to be yanked out of bed late in the day by a determined Negro servant girl, and who prefers to meet a question with a wisecrack rather than an answer, very likely knows to a fourth decimal place the dollars and cents value of his 'temperament.'"

In estimating Rodgers and Hart's 1938 income at a hundred thousand dollars, a considerable sum even for those post-Depression days, *Time* put its finger on one of Larry's major problems: he was increasingly irresponsible about money. Concerned by his partner's profligacy, Dick Rodgers persuaded Larry to let his own business manager, William Kron, take care of his finances. Kron was an astute accountant who was already handling money matters for Jerome Kern, Edna Ferber, and others. This in turn led to a new game for Larry called Running Away from Kron.

The stagestruck Kron would frequently turn up at Larry's apartment early in the evening, ostensibly to discuss business. If caught, Larry would distract him with dirty jokes, which the straitlaced Kron rarely got, and invite him to join whatever party was planned for the evening (which was Kron's real reason for coming, anyway). More often, a spy in the lobby would call the penthouse, and Larry would yell out, "Kron!" Whatever festivities were going on were abandoned as everyone tried to get out of the building before Kron got in. It was like dodging the truant officer; Larry loved it.

As fall drew in, work began on the new show, which now had the title *The Boys from Syracuse* (on Broadway, where there are few coincidences, it was no secret that Jake and Lee Shubert, born Szemanski in Lithuania, had immigrated to Syracuse, New York, and there got their start in the theatre). Larry had contacted George Abbott soon after his return from Atlantic City and asked him to work up a tentative script, intending to collaborate on it with him. When Rodgers and Hart saw Abbott's treatment—borrowing only the setting of Ancient Greece and the principal characters of Antipholus and the Dromio twins—it was so completely in line with their thinking, and so totally in keeping with the bawdy Shakespearian tradition, that they never felt the need to change a word.

Some indication of Abbott's approach may be had from the opening announcement, made before the show itself begins. "This," the speaker tells the audience, "is a drama of ancient Greece. It is a story of mistaken identity. If it's good enough for Shakespeare, it's good enough for us!" Abbott used only one line of the Bard's work: "The venom clamors of a jealous woman poison more deadly than a mad dog's tooth." As it was spoken by the Seeress, Abbott had Jimmy Savo stick his head out from the wings and proudly announce to the audience: "Shakespeare!"

Perhaps the fact that George Abbott's book was so right explains how, while *Boys* was in rehearsal, Rodgers and Hart found time to meet with John Krimsky, Entertainments Director, to discuss a musical to be staged during the following year's New York World's Fair, due to open April 30. Dick and Larry decided to pass; Krimsky was looking for something nearer Radio City than Broadway. The assignment, *American Jubilee*, eventually went to Arthur Schwartz and Oscar Hammerstein II. It was not a success, although Arthur Schwartz was always fond of the ballad "How Can I Ever Be

Alone?" and Hammerstein aficionados sometimes have fun with "Tennessee Fish Fry."

The score of *The Boys from Syracuse* is a delight; "Falling In Love With Love"—with a whirling waltz melody that is said to have been Hart's favorite—is one of Rodgers and Hart's most elegantly bittersweet creations, while "This Can't Be Love" demonstrates yet again Larry Hart's impeccable gift for the throwaway love song. The specialty numbers "What Can You Do With A Man?" (introduced by Teddy Hart and outsize Wynn Murray— "acres and acres of beauty going to waste")—and "He and She" (Murray and Jimmy Savo), the story of a married couple so perfectly awful that when they died and went to Heaven all the angels moved to Hell, were witty and apposite. The ensembles, so casually erudite they sometimes take the breath away, included "Come With Me," which sings the praises of going to prison, where you don't have to get up for work or walk the dog, and in your isolation you're never bored by politics and can in good conscience "miss a row/ of tragedies by Sophocles and diatribes by Cicero." Only one number, "Cinderella," was discarded; it would seem probable this was the song "My Prince," also known as "What A Prince," which appeared in their next show, *Too Many Girls*.

The cast, too, was full of talent. Teddy Hart and Jimmy Savo played the twin Dromios. Ronald Graham (who had shone in *Virginia*, a Laurence Stallings–Owen Davis failure of the preceding year with songs by Arthur Schwartz and a new collaborator, Albert Stillman), and Eddie Albert (real name Edward Albert Heimberger, brought over from the cast of *Room Service*) played the Antipholus twins. In addition to Wynn "Johnny One Note" Murray, the leading ladies were Marcy Westcott and Muriel Angelus, a newcomer to Broadway from England. "Larry was a demon," she remembered; "he was like Peck's Bad Boy. I could never get to know him, he was here one minute and around the world in forty seconds."

No apter description of Larry Hart exists: here one minute and around the world in forty seconds. He had always been unpunctual; now he simply didn't turn up for meetings at all. Unlike before, however, he could no longer be readily found at Ralph's or Dinty Moore's or any of his regular hangouts and dragged back to work. He just disappeared; and no one knew where he was. "Larry was always a sweet and lovely guy," Irving Pincus said. "But by the time he was doing *The Boys from Syracuse*, he was sweet, lovely, and screwed-up."

"I was in Miami, at the Roney Plaza Hotel, when they were writing *The*

Boys from Syracuse," Gene Rodman said. "Larry . . . would come down in shorts, a commander's cap, lifts on his sneakers. A Greek god he wasn't . . . He would sometimes look around and say, 'I love it down here, it's the only place I can relax. New York is crazy—all that night life. Here, it's calm and relaxing.' The funny thing is that while he was saying it, he was jumping around, rubbing his hands together, changing from one chair to another, between each sentence. He had a handsome young masseur with him on the beach and they went everywhere together all the time he was in Miami."

George Abbott recalled that Larry was drinking heavily and would be absent for two or three days at a time during the preparation of the show:

> This didn't bother me because he was as quick as lightning when he was there. If we needed a new verse, he'd pick up a pencil and paper, fidget himself into the next room for a few minutes and then come back with what we needed. I remember that this was how he wrote the verse for "Falling In Love With Love"; he scratched it on the back of an old piece of paper while Dick and I talked about something else.
>
> Nevertheless, Dick was very concerned about Larry's growing addiction. For one thing, he saw his collaborator gradually deteriorating; secondly, he knew from experience that when a show got on the road, it needed a lyric writer ready for emergencies. Dick's fears were realized; when we went to Boston [on November 7, 1938], there was no Larry. But everything in the show fell into place so easily we didn't need him.

It was just as well; on November 17, with the show still trying out, Larry was admitted to New York's Mount Sinai Hospital "for observation." He had convalesced at home for several weeks and was too busy worrying about his health to miss being at the rehearsals. He was then sent to the hospital only when tests revealed a persistent spot on one lung. In fact, Larry was in Mount Sinai for a month, not a week. He was suffering from pneumonia, contracted during the New Haven tryouts. For the first time, Larry Hart missed his own opening: when the curtain went up on *The Boys from Syracuse* on November 23, he was still confined to bed.

※

Once again, Rodgers and Hart had a smash hit. "If you have been wondering all these years just what was wrong with *The Comedy of Errors*," said critic Richard Watts of the *Herald Tribune*, "it is now possible to tell you. It has been waiting for a score by Rodgers and Hart and direction by George Abbott." Once again, a triumph. Once again a long-running show—seven months this time—and a film sale, to Universal, hard on the heels of the opening. Two unused songs, "The Greeks Have No Word For It" and the lilting "Who Are You?," were recycled for the movie, which starred Allan Jones as the Antipholus twins (trick photography) and Joe Penner—who'd changed his catchphrase from "You naaaahsty man" to "Wanna buy a duck?"—as the Dromios. Such success was all the more remarkable in a theatrical climate where practically nothing was working: between April 1936 and November 1938, thirty musicals had opened. Only four of these had run for more than 200 performances, set against 315 for *On Your Toes*, 289 for *Babes in Arms*, and 290 for *I'd Rather Be Right*.

Of the current crop, Cole Porter's *You Never Know* had lasted less than ten weeks, in spite of the presence of Clifton Webb, Libby Holman, Lupe Velez, and "At Long Last Love." Kaufman and Hart had managed only 105 performances with *Sing Out the News*. *Knickerbocker Holiday* featured Walter Huston as Peter Stuyvesant (and a chorus line that might have been lifted bodily from *Dearest Enemy*), but the show, even with Huston singing Kurt Weill's "September Song," added few laurels to director Joshua Logan's crown, and never recouped its cost during its twenty-one-week run. Not until a second Cole Porter show, *Leave It to Me*, which came in two weeks before *Syracuse*— featuring Mary Martin, making her Broadway debut, singing "My Heart Belongs To Daddy" (with Gene Kelly as one of her chorus boys)—did anyone offer Rodgers and Hart serious Broadway competition.

On November 29, a bulletin announced Larry was expected to leave the hospital in about ten days. Fifteen days later, he was "out for an airing." A letter to Mel Shauer dated December 14 confirms he had been ill with pneumonia the past couple of months with a small spot on his lung, was now convalescing, and was planning to have a month in Miami with Frieda commencing the following week. The show—which he would see for the first time at the Saturday, December 17, matinee—was a big success, he said, characteristically adding that this was great for Teddy.

A week later, he attended an important event in his partner's life—one which, for the very first time, excluded him completely.

On Christmas Day 1938, Paul Whiteman and his orchestra presented Richard Rodgers's *Nursery Ballet* at Carnegie Hall. Later broadcast over NBC's New York station WEAF, it remains the only Rodgers composition that has no poetic text or was not founded in some play or movie; written especially for his daughter, Mary, the three-movement suite, orchestrated by Roy Bargy, began with "The March Of The Clowns," followed by "A Doll Gets Broken" and finally "Little Girls Don't Fight." It is charming, if lightweight; someone described it as musical comedy without singing, which is perhaps why there is no mention of it in Rodgers's autobiography.

Right after Christmas, Larry Hart headed for the sunshine of Florida, telling the newspapers (and no doubt anyone else who'd listen) that he and Rodgers were planning a musical based on *The Arabian Nights*. He came back briefly to New York long enough to confer with Dick and George Abbott about a new idea Abbott had come up with shortly after *Boys* opened. It had originally been written as a screenplay by George Marion Jr., who had worked with Larry and Dick on *Love Me Tonight*. Although Marion, son of a famous actor and brother of one of the most successful Hollywood screenwriters, had never written a Broadway libretto, Abbott thought this one, a rah-rah college football musical, had promise.

On February 11, Larry and George Abbott left for Miami; according to report, Larry had already done some research for the project at the New York Public Library. Four days later, Dick Rodgers joined them to work on what had now become "a trio of ideas." One of these was a Shubert production, *On the Line*, which the Shuberts were hoping they could persuade Al Jolson to star in. The second was a vehicle for Vera Zorina that they had discussed with Dwight Wiman and Joshua Logan. And the third was, of course, the college football musical that became *Too Many Girls*. Larry was already doing intensive research for it with Doc Bender, not in the New York Public Library, but in the Cuban-flavored nightclubs of Miami's Carter Causeway.

By the time they got back to New York on February 22, Rodgers and Hart were able to state categorically that the *Arabian Nights* idea was a nonstarter. It was also "definite" that they were "not doing a show with Robert L. Ripley and Harry Kaufman." The Shuberts had gone off the boil; the Guy Bolton–Matt Brooks–Eddie Davis idea about a radio cowboy who goes out west and meets some real bandits eventually passed to producer Vinton Freedley, who retitled it *Hold On to Your Hats* (that must have been some joke), signed Al Jolson to star, and enlisted Burton Lane and E. Y. "Yip" Harburg to write the score. It would not see the lights of Broadway until September 1940. As for the Zorina idea, so far that was all it was, an idea.

Until the two Georges were available—Abbott was producing *See My Lawyer*, a madcap comedy that introduced Milton Berle to Broadway (in a role originally intended for Danny Kaye) supported by Millard Mitchell, Gary Merrill, and Uncle Miltie's childhood playmate Teddy Hart; George Marion Jr., was hard at work finishing the college football book—Rodgers and Hart were at something of a loose end. They went to Chicago, where *I Married an Angel* was playing the Grand Opera House, to appear—as "America's Gilbert and Sullivan"—on "Tune-up Time," a CBS radio program sponsored by Ethyl Gas, the "guaranteed no-knock fuel."

"I'm Dick, I'm 152 pounds, five-feet-eight tall, thirty-six years old—and I must have been a beautiful baby," said Rodgers.

"I'm Larry. I'm five feet tall, I'm fourteen years old, and I'm *still* a beautiful baby," piped Hart. "I may be little, but I'm *cute.*"

Singer-emcee Walter O'Keefe chided him for plugging someone else's songs: "I thought you were going to say I'm five feet tall and I must have been a small hotel."

Then singers Kay Thompson (what Larry would have called a "well" singer), Kenny Baker–clone Ray Heatherton from *Babes in Arms*, O'Keefe ("who never knocks anyone because this is a no-knock show"), and the André Kostelanetz orchestra—with piano introductions by Rodgers—variously performed "You Took Advantage Of Me" (two choruses in just over a minute), "My Heart Stood Still," "The Lady Is A Tramp," "With A Song In My Heart," and "Manhattan."

On May 26, it was reported that Larry had lunched with Mae West, who wanted him to write a play based on an idea of hers. Nothing happened, of course. Writers are constantly approached by people with terrific ideas offering what seems to them an irresistible deal: the writer takes the idea and writes a book, and then they share the proceeds fifty-fifty. The recommended response is, "Wait, I've got an idea, too. Why don't *you* write the book, and we'll share the proceeds of *that* fifty-fifty." It usually does the trick. Maybe Larry was too polite, or maybe he just wanted to have lunch with Mae West.

In August, Larry and Frieda made their long-deferred move ten blocks uptown from the Beresford at 81st Street to the twin-towered Ardsley at 320 Central Park West, where realty agents Slawson & Hobbs had on offer a large terraced duplex on the twenty-first floor. This was an ideal arrangement now that Teddy was no longer at home. Frieda could have her own place on the top floor; Larry would take the lower floor for his bachelor apartment. Clearly it never became home the way the Beresford had been.

Dorothy Hart described Larry's den as an enormous room with an even larger terrace. Unlike the one at Beresford, however, it was not lined with bookshelves; for a long time, hundreds of books remained stacked in shipping cartons on the landing. Larry's bed, an uncomfortable sleeping couch in one corner, somehow conveyed a bleakness at odds with the fact that Dorothy Rodgers had furnished and decorated the interior.

Immediately after moving in, Larry had a huge, heavy, soundproofed door installed between the two floors, so that Frieda should not be disturbed by any late-night revelries below. And, one imagines, to ensure the personal privacy that Larry Hart had forgone for so many years. Here, at last, he could do what he liked. Alas, it turned out to be a fatal self-indulgence, for all his cronies moved in, too, and the place became a sort of show business Grand Central. Dorothy Hart recalls seeing William Kron and Larry's chauffeur placing bets on the races over the telephone—doubtless with Larry's money—while Larry himself lay semi-conscious in bed. She does not say he was too drunk to care; but it seems unlikely it can have been otherwise.

In September, casting began for *Too Many Girls*. If his autobiography is read between the lines—as all such works should be—it can be inferred that, among his many other talents, George Abbott was a dancing fool who liked nothing better than to spend his evenings in Havana tango bars; so the fact that the show would have a Cuban atmosphere made him "particularly excited." One of the characters in George Marion's libretto—hewing to the Good Neighbor policy—was Manuelito, supposed to be the best football player ever to come out of South America. But most of Hollywood's "Latin" types were either of the George Raft variety or comedy players like Leo Carrillo. Where were they going to find a Latin American who could handle comedy, dance, and sing? Where else but in a Cuban nightclub?

New York socialite Brenda Frazier had found a new "in" place where the smart set could let their hair down: Mario Torzatti's La Conga, just off Broadway. There were two bands at La Conga: the one that introduced the dance that was all the rage, led by a handsome young Cuban whose full name was Desiderio Alberto Arnaz y de Acha, but who called himself Desi Arnaz; and a second, Pancho's, famous for tangos, and also for a vocalist named Diosa Costello, who, according to Desi, "could shake her ass better and faster than anybody I had ever seen."

Larry Hart and Doc Bender had seen Desi Arnaz at La Conga in Miami; Doc, it was said, was mad about the boy. Just a few years before his death, claiming he wanted to keep his stories for an autobiography he was then writing, Desi Arnaz declined to talk about Doc Bender and Larry Hart. What we are left with is the recollections of some others, and what Arnaz himself wrote. At least one intimate friend said Larry was crazy about Desi, too. Was Desi responsive? "No," he said. "Larry never went that far. But Desi—he tried everything. He was another mad character; and a drinker, a real drinker."

Whatever their personal relationship, Larry knew Arnaz was ideal for the part of Manuelito. Night after night Hart and Bender ringsided at the New York club, bringing one after the other Dick Rodgers, George Marion Jr., and eventually George Abbott, who—predictably—not only watched the show but joined in the conga line. At the end of the evening Abbott invited Desi Arnaz to audition for the part.

After Abbott left La Conga that night, Arnaz said, Larry Hart and Doc Bender took him to one side and asked him if he had ever read a script before. Desi didn't even know what a script was.

"That's what I was afraid of," Larry said. "I'll tell you what I'll do. I'm going to bring you a script, and we'll work with you before you read it for him. At least you'll have an idea of what it's supposed to be, but for God's sake, when you read for Abbott . . . pretend it's the first time you've seen it. He would raise holy hell with us if he found out we had given you a script before you went over there. He likes people to read cold for him."

Two days later, after a particularly long night on the town, Desi turned up for his audition. First, he sang. "Well, he's loud enough," Abbott said to Dick Rodgers. He gave Arnaz a script and asked him to read a scene he had marked, where Manuelito is trying to decide which college he wants to go to in America. He makes only one condition: it has to have lots of girls.

"Oh," says the scout, "you mean co-educational."

"Tha's it," says Manuelito. "Co-operational."

Desi Arnaz was good and ready—all too good and all too ready. As he thrashed through the familiar lines, Abbott looked at Dick Rodgers; they turned to see Larry heading for the exit.

"Larry!" Abbott yelled, pinning him to the spot."You gave him the script, didn't you?"

Larry did some acting of his own. "Who? Me? How? Why?"

"Because he hasn't looked at one goddamned word of that scene. He did the whole thing like a big ham, emoting and waving all over the place. You taught him, didn't you? And, I may add, you did a bad job of it."

Finally they all confessed, and "Dizzy," as Abbott insisted on calling his new star, got the part. As always with a George Abbott show, the cast was full of new talents, many making their Broadway debuts. In short order, Diosa Costello was signed up, too. A young actor named Eddie Bracken was brought into the cast from the road company of *What a Life*, the play (produced and directed by Abbott) that launched the character Henry Aldrich ("Coming, Mother!") into a seemingly interminable radio and movie series. Perennial ingenue Marcy Westcott from *The Boys from Syracuse* (Larry Hart used to say, "I bet she pisses ice water") was given the romantic lead, and Mary Jane Walsh, who had been in *I'd Rather Be Right*, was her college chum.

Another Abbott discovery, Richard Kollmar, was set as the hunky lead, Clint Kelley. Among those cast as students were Leila Ernst and a handsome blond dancer named Van Johnson. Like Arnaz, Van took Doc Bender as his agent. And like Arnaz, he became an enthusiastic participant in Larry's late-night parties. Six months later, when Kollmar quit the show, Johnson replaced him. Kollmar went on to woo and wed syndicated columnist and bad-mouther Dorothy Kilgallen—he was the other half of her radio show "Dorothy and Dick"—and still later became a producer.

The story, although contrived and artificial, was pretty harmless. To keep her out of trouble, rich daddy sends daughter Consuelo (Westcott) to Pottawattomie College ("an institution so backward it plays football on Fridays"), where virginal co-eds wear yellow beanies to show they're still virginal. Daddy also hires four all-American football players as (undercover) guardians of Consuelo's virtue: Clint (Kollmar), Jojo (Bracken), Manuelito (Arnaz), and Al (Hal LeRoy). Consuelo, of course, falls in love with Clint; when she finds out he's working for Papa, complications ensue.

<hr />

Book ready, actors cast, score—more or less—complete, *Too Many Girls* went into rehearsal. Diosa Costello and Desi Arnaz had a number called "She Could Shake The Maracas," which, taken in conjunction with Desi Arnaz's description of her dancing style, needs no further explanation. Kollmar and Westcott had two love songs, "Love Never Went To College" and the evergreen "I Didn't Know What Time It Was," a further example of Larry Hart's ability to translate a street phrase into a hit song. In late thirties slang, anyone who didn't know what time it was wasn't with-it, wasn't hep, dig? For their finale, and for Mary Jane Walsh, Rodgers and Hart stood their love song to Manhattan on its head with "Give It Back To The Indians," a diatribe on New York's lack of charm almost as appropriate today as it was half a century ago.

While they were rehearsing in New York, all the kids from the show—not to mention George Abbott and George Marion Jr.—went down to La Conga, where Desi Arnaz and Diosa Costello were still performing. Abbott and Marion "both loved to dance," Arnaz recalled. "Abbott [was] really a great rumba dancer. He would dance with Diosa all night between shows." You can see where he would. One night, he told Desi Arnaz they had to find a way to get his drum-beating conga-chain act and Diosa Costello's dancing into the show; the best place for that would be the first-act finale.

As it stood, Dick and Larry's first-act finale was "Look Out!," a cheerleaders' number built on a "shave-and-a-haircut" rhythm which worked its way through the names of all the real-life college football teams. But it was a march, and Arnaz simply couldn't adapt it to the conga rhythm. He told the rehearsal pianist and the conductor, who looked at him sadly. This eighteen-year-old Cuban kid was going to tell Rodgers and Hart to change one of their songs? Just then Dick came in. The problem was explained.

"What's the matter with you guys?" he said to the musicians. "We'll have the first chorus straight, and then when Desi starts the conga beat, we'll change it to fit his thing. To tell you the truth, I like it better his way."

And so, "while I would bang the drum and lead the shouts," Arnaz wrote, "[Diosa Costello] would be shaking her nice Puerto Rican ass all over the place. All the chorus girls and boys, the whole cast, covering the entire stage at different levels, would get into this big conga line, answering the shouts, kicking their legs, throwing their arms up in the air while the drums kept getting louder and louder. It was really frantic."

In rehearsal, however, Larry Hart disappeared again. "He was really needed this time," Abbott said, "for we had to juggle things around during the tryout and to do quite a lot of patchwork. The second act opened with eleven girls singing about the woes of being deserted by their athlete lovers during football season. Rodgers thought it would be a good idea to have the first act open with eleven men singing about the fact that the football season was the only time when they were important; during the rest of the year, they were lifeguards, telegraph boys, or camp counselors. Since Larry could not be found, Dick turned to and wrote the lyric himself." What is significant about "Heroes in the Fall" is not that Rodgers pinch-hit for his colleague (he'd done that before) but that when he did turn up, Larry neither protested nor essayed a rewrite.

Although Desi Arnaz placed it emphatically in New Haven on September 28, the show tried out only in Boston, beginning October 2, just a month after the storm that had been gathering in Europe finally broke and Britain declared war on Germany. A new song had been added to the scene prior to the first-act finale, a small riposte to the juggernaut of swing dominating the airwaves and the record industry. It was sung by Marcy Westcott, Eddie Bracken, Mary Jane Walsh, Richard Kollmar, and Hal LeRoy, and it put into words Dick Rodgers's pet hate: the way the big bands obliterated his melodies. Its title was self-explanatory: "I Like To Recognize The Tune."

Desi Arnaz recalled that "after the opening night performance, Mr. Abbott had quite a few notes for everyone in the cast."

When he finished with us, he turned to Larry Hart and said, "Larry, we need at least two more choruses for 'I Like To Recognize the Tune.'" The number had stopped the show cold that night, and even though the cast were prepared with two extra choruses, which they sang, the audience still wanted more. . . .

Larry looked in his coat pocket and took out an envelope, put it on top of the rehearsal piano onstage, borrowed a pencil and started to write. That's how we left him as we went to a restaurant across the street from the theatre for the party which Rodgers and Hart and George Abbott were giving for the cast and friends who had come up from New York. In about a half-hour Larry came in, went to Mr. Abbott and Dick Rodgers and said "What do you think of this?" as he handed them the envelope. He had written three more choruses as good as, if not better than, the ones he had written before.

On Wednesday, October 18, 1939, as German armies pushed the French back to the Maginot Line, *Too Many Girls* opened at the Imperial in New York. "All the important critics were there," Arnaz recorded, "Walter Winchell, Dorothy Kilgallen, George Jean Nathan, Brooks Atkinson, Ed Sullivan, Louis Sobel, Danton Walker, Leonard Lyons, Earl Wilson, and many others."

As Arnaz remembered it, the show was great. "All the jokes got big laughs. Mary Jane Walsh stopped the show. Hal LeRoy stopped the show; so did Marcy Westcott and Richard Kollmar with 'I Didn't Know What Time It Was.' Diosa Costello stopped the show. Bracken stopped the show. I got all my laughs, thank God, and the first-act finale really got a standing ovation. We all thought we had a big hit. But as Bracken said, 'You are never sure in New York until those reviews come out.'"

After the show everyone went to La Conga. "The joint was packed," Arnaz said. "Everybody connected with the show had decided to come and wait for [the reviews] at La Conga instead of Sardi's. When we finished the last show, around 4:00 a.m., Diosa and I sat at the table with the Rodgerses, Larry Hart and his mother, George Abbott and Marcy [Westcott], the George Marion Jr.'s, Brenda [Frazier] and Peter [Arno, then her boyfriend] and Dick Kollmar and Dorothy Kilgallen [already involved romantically]."

About half an hour later Arnaz saw Polly Adler—Desi was a regular customer at the house of New York's most famous brothelkeeper—heading toward their table with all the newspapers. "As she approached, she hollered in that big, deep voice of hers, 'Cuban, you are the biggest fucking hit in town!'

"She shook up that table pretty good," Arnaz recorded, "including me, and I don't 'shook up' too easy." After Polly went back to her table, Dorothy Rodgers asked innocently who she was. Peter Arno took the plunge, although one wonders why Larry Hart didn't: there is an enduring tradition that, at one time in his speckled career, his O.M. had partly financed—and doubtless frequented—one of Polly's uptown establishments. No matter: the reviews were great, three and a half stars in the *Daily News*, more than any show since *Babes in Arms*. The adjectives were of the type apparently reserved solely for Rodgers and Hart: fresh, exhilarating, tuneful, youthful, sprightly, colorful.

It was beginning to look as if Rodgers and Hart could do no wrong. The secret, they said, was simple: don't have a formula. And don't repeat it. So what next? They were already talking to Alex Yokel about musicalizing *Three Men on a Horse*, and Max Gordon was talking about commissioning them to adapt Donn Byrne's novel *Messer Marco Polo* (a project Kern and Hammerstein had already worked on and abandoned). Neither idea jelled. Instead, they got involved with a show so bad a trained seal walked away with the notices.

33

———※—※—※———

The French Have a Word for It

Higher and Higher, as the new show was to be called, had been in development since the preceding winter. The idea had originally come from Irving Pincus, a onetime Brant Lake campmate of Larry's, whose family owned the Alvin Theatre. During the run there of *The Boys from Syracuse*, Pincus bumped into Larry often. "There was an annual affair in New York called the Butler's Ball," he recalled:

It got space in the newspapers because the maids and butlers who worked for the top families dressed up to go to it at the Waldorf Astoria Hotel. One night . . . Larry and I were sitting around Ruby Foo's Restaurant, and I told him I thought the Butler's Ball would make a good musical and I had some idea of a story about it. He had been drinking brandy and he was so far gone that when he spoke he was incoherent. It must have been three thirty in the morning.

Next day was a Saturday, a matinee day at the theatre. I was upstairs in the office when the doorman told me Dick Rodgers wanted to see me. Me? I thought he must have meant my older brother, Norman, who was the important one, but the doorman said no, he wants to see you. I went down, and Dick said, "We'd like to buy the musical you told Larry last night." That floored me. I couldn't believe that, drunk as he was, Larry could have remembered even seeing me. But he did.

Then I had to remember what I told him. I went back upstairs and typed out an outline of my story as best I could. All three of us had Howard Reinheimer as our attorney, and he handled the negotiations.

Although Pincus did not know it, he had hit upon what exactly what Rodgers and Hart had been searching for—a vehicle for Vera Zorina. With this in hand, they went to Dwight Wiman, and he enthusiastically added his support to the project. Meanwhile, *Too Many Girls* intervened, and so the story line was still in embryo as 1939 came to an end, and Larry Hart made noises about finally settling down to writing the "Broadway" book he had been talking about for years. Quite what this might have been, we will never know; if he ever put pen to paper, then, like the autobiography "Always on a Bender" that Doc Bender is supposed to have been writing at the time of his death in 1964, it has disappeared.

The 1939–40 theatrical season, in which *Too Many Girls* was the opening big gun, promised much but delivered little. The *George White Scandals* lasted only 120 performances, although Ann Miller's "Mexiconga" number won her the Diosa Costello role in the movie of *Too Many Girls*. In spite of the presence of Vincente Minnelli as director, and a score that included "All The Things You Are," the new Kern and Hammerstein show, *Very Warm for May*, was an unparalleled disaster for producer Max Gordon; it also marked the end of Jerome Kern's Broadway career.

Taking a leaf from Rodgers and Hart's book, the Rockefellers tried an adaptation of Shakespeare's *A Midsummer Night's Dream*, transposed to 1890s Louisiana and featuring Louis "Satchmo" Armstrong as Bottom. Even with a melodic score by Jimmy van Heusen that included "Darn That Dream," *Swingin' the Dream* was an expensive thirteen-performance flop.

On December 6, songwriter-turned-producer Buddy DeSylva brought in the last musical of the thirties, Cole Porter's *Du Barry Was a Lady*. With a book by DeSylva and Herbert Fields, sumptuously mounted (the designer was Raoul Pène du Bois, who had also done *Too Many Girls*), it starred Ethel Merman and Bert Lahr. Its highlights included two memorable duets: Lahr and Merman offering each other the ineffable, unique "If you're ever up a tree, send for me . . . down a well, ring my bell" promises of "Friendship"; and Betty Grable (making her Broadway debut at long, long last) and Rodgers and Hart graduate Charles Walters (later a talented movie director) comparing notes on a swell party in "Well, Did You Evah?" Originally developed as a movie vehicle for Mae West—was this the idea she discussed with Larry Hart that summer?—*Du Barry* was one of the very few shows produced in the latter half of the decade to outrun the Rodgers and Hart hits.

Throughout the fall, Dick Rodgers was again working on a project that did not include his partner. Although Larry had no time for the upper stratum of New York society ("a lot of crumbs held together by their own dough," he called them), Rodgers was vastly attracted by the perceived glamor of the Four Hundred, and clearly yearned to be one of them, as Gershwin had so yearned before him. Thus he was persuaded by wealthy art and ballet patron and social butterfly Gerald Murphy—who figures largely in the story of the disintegration of Scott and Zelda Fitzgerald, and whose life motto was "Living well is the best revenge"—that the Paris-based Ballets Russes de Monte Carlo should expand its repertoire with ballets of American origin. A young dancer named Marc Platt provided the idea for a libretto, and Rodgers wrote music for what became *Ghost Town*. Given a sumptuous production by impresario Sol Hurok, it was premiered at the Metropolitan Opera House on Sunday evening, November 12, 1939, with Rodgers conducting the orchestra.

One of Larry's friends, Gene Rodman, was at the premiere: "I saw Larry coming up the aisle and was going to talk to him when I saw he was so ossified, plastered, he could hardly navigate. Much as I loved him, I didn't speak to him. He could be impossible when drunk." Irving Eisman confirmed that Larry's growing addiction encouraged in him the too-ready combativeness of the chronic lush, ready to take offense at any slight, real or imagined. "When he would drink a lot he would become belligerent and egotistical about his work," Eisman recalled. "It was so untypical of him, boasting how good he was. But only when he was drunk. Not when he was sober, never."

Dorothy Hart was with Larry one night when they bumped into Humphrey Bogart and his then-wife Mayo Methot at Reuben's at about 2:30 a.m. The Bogarts had what might be called a hard-hitting relationship—his pet name for her was "Slugger"—and had just been thrown out of the Algonquin Hotel for bad behavior. They had no baggage and no place to go, and were considerably the worse for wear. To his sister-in-law's surprise, Larry wanted nothing to do with the battling Bogarts. They were sloppy and drunk. When Larry wasn't drinking he had no tolerance for anyone else drinking, she said. She felt this was typical of the ambiguity of his personality.

Another by-product of the drinking was forgetfulness. Larry Hart's cavalier attitude to his hats and coats was almost a byword; he constantly left both in some bar and had to be hurried to the childrens' department at Wanamaker's or Macy's or Saks or Bloomingdale's to get another, which he would lose with equal indifference two nights later. Milton Pascal records an occasion when he was at the Hart apartment, and Marc Connelly and Tallulah Bankhead phoned up from the foyer: they were having dinner with Larry. "He asked me to go down and talk to them," Pascal said, "and tell them he would be ready in a moment. About five minutes later he came down. When he saw me he was really surprised. 'Hey, glad to see you!' he said. 'Look, why don't you come and have dinner with us?' He'd completely forgotten I had been up there with him!"

"We were in Ruby Foo's restaurant," Irving Eisman said, recalling another occasion. "and Larry was falling down drunk. He wouldn't leave, and I asked him what was wrong. He said, 'I'm waiting for my change.' He was so far gone, he didn't know he hadn't even asked for the check."

Legend to the contrary, however, Larry wasn't always on the town. At home, Dorothy Hart said, he was never the noisy, boisterous Larry Hart of the Broadway stories, but quiet, very quiet. That didn't mean he couldn't be infuriatingly impulsive—he was. Sometimes just as the family would sit down to dinner, everybody ready to eat, he'd say, "I'm going in to take a shower" But other times, he enjoyed sitting around the table after dinner with whoever was there and talking. It was an aspect of his character a lot of people just didn't know about or didn't want to, she said, recalling many nights sitting with Larry playing cards, hearts or rummy. And how he'd laugh and sing a gleeful little song when he stuck her with the joker. His favorite game was pinochle.

Sunday was the family dinner, the delicatessen dinner, she recalled. Often there would be two or three young cousins, and ninety-two-year-old Uncle Willie and seventy-seven-year-old Aunt Rachen, and, of course, Frieda. There was always another uncle and aunt somewhere around, and Larry loved it, just loved them all. It was as if he was another person, she said, someone not part of the Broadway scene at all. He wouldn't even talk show business.

Christmas saw the biggest parade of hype mounted by the movie industry in years as *Gone With the Wind* opened in New York. Lines also formed to see Bette Davis in *Dark Victory* and James Stewart in *Mr. Smith Goes to Washington*. Jukeboxes were playing "Annie Doesn't Live Here Any More" and the Bonnie Baker hit "Oh, Johnny!" In London, like children sticking out their tongues at Hitler, music hall comedians were boasting "We're Gonna Hang Out The Washing On The Siegfried Line." Isolationist Senator William Borah referred to the conflict in Europe as "the phony war," and the epithet stuck; pollster Elmo Roper found that 67.4 percent of all Americans wanted no part of it. The country was much more interested in knowing whether Tennessee could outplay Southern California in the Rose Bowl; USC answered the question decisively, winning 14–0.

Right after the holidays, rehearsals were scheduled for the new Rodgers, Hart, Logan, and Wiman show. *"Higher and Higher* was an attempt by Dwight Wiman to repeat the success of *I Married an Angel*," Logan said.

Well, it turns out that Dick and Larry didn't find it very easy to write that story. Dick had gone to Clare Booth Luce, who suggested the whole story be told through the eyes of the servants, which made it almost impossible. So in the end we brought in a girl that I happened to know named Gladys Hurlbut, and she worked on it for a while, then decided that I should be her collaborator, so we wrote it—God help me!

Whatever Logan's motive was for bringing in actress-writer Gladys Hurl-but, it could hardly have been her track record, which consisted of a couple of movie stories, two flop plays, and a so-so stage career. According to Irving Eisman, Larry was extremely unhappy about Logan's bringing in another writer at all—"fuming" was the word Eisman used—but bowed to the inevitable. In the story Logan and Hurlbut came up with, the servants of a suddenly bankrupt millionaire deck out Minnie, one of the maids, as a debu-tante, hoping to marry her off to a wealthy playboy; complications ensue when things go wrong, and the customary happy ending follows.

By the time the show was ready to cast, Zorina had long since opted for what turned out to be a pretty uninspiring Hollywood career (which would include being replaced by Ingrid Bergman halfway through filming the part of Maria in *For Whom the Bell Tolls*). She appeared opposite Eddie Albert (with Donald O'Connor playing the young Junior) in the Warner Bros. movie version of *On Your Toes* (released October 21, 1939, with such Rodgers and Hart songs as were used at all reduced to background music, and only the two dance numbers retained) and was currently filming *I Was an Adventuress*, a romantic comedy that turned out to be neither.

It was decided that Budapest-born Marta Eggerth—the wife of opera star Jan Kiepura, and star of many filmed operettas in Germany and Austria—would play the role of Minnie, the maid. The book now had to be revamped to suit the completely different talents and personality of the new star. "And Larry, although he wrote some marvelous lyrics, really I think deep in his heart didn't like it very much," Joshua Logan said, "and began to drink."

Logan had some problems of his own. After a spur-of-the-moment trip to South America, he had returned to Broadway to direct a new play by Paul Osborn, *Morning's at Seven* (a title Larry would steal for one of the songs in the new show). In great form for the first five days of rehearsal, he suddenly collapsed with a fever of 104° and was rushed to the hospital. Dwight Wiman pinch-hit for him until he was able to take over during the tryouts, but Logan felt the only-moderate reception the play got when it opened on November 30 was due to his absence. He was still "down" when they started work on *Higher and Higher*. "I think by the time I did it, I just . . . I didn't like it, there was something wrong with it. And I think Larry must have felt it, too."

With a cast that included Jack Haley, the beloved Tin Man from *The Wiz-*

ard of Oz, Shirley Ross, Lee Dixon, and Robert Rounseville—and a chorus that included June Allyson and Vera-Ellen—ready to go to work, Rodgers looked around for his partner to start writing. There were a total of thirteen songs in the finished score, although exactly which Larry wrote and which he did not we can only guess. Most of them were comedy numbers like "A Barking Baby Never Bites" and "I'm Afraid"; only one or two were at all memorable, most notably the haunting "It Never Entered My Mind." Marta Eggerth's big number was "Nothing But You," which has a lyric insipid enough to suggest Hart didn't write it; her other number, "From Another World," isn't much better. "Disgustingly Rich" was the only one retained when the show was filmed a year or so later.

Apparently, Larry Hart began doing his disappearing act while the show was in rehearsal. Whether he was in New Haven for the tryout that began March 7 is not recorded. His absence meant Dick Rodgers had to write some of the songs, or extra choruses for existing ones, without Larry even being there. "They weren't very good," Logan recalled, "but at least they kept the curtain open." On top of that, "Gladys Hurlbut had seen this performing seal at a county fair in Woodstock, and we had to work that into the story. She was sure it would steal the show." That everyone agreed to this nonsense is some indication of how bad things were. As a result, Logan "got into a terrible depression, and did the whole thing in this depression. I don't know why they listened to me. I didn't like it, I didn't like anything, I didn't like life."

The second and much longer tryout was at the Shubert in Boston, from March 12 through 30. Some of these, at least, Larry attended. Elinor Hughes of the *Boston Herald* interviewed Rodgers, Hart, Logan, and Wiman just prior to a midnight conference. "We've been having trouble with a number in the last act," Dick Rodgers told her, "a number called 'Life, Liberty, And The Pursuit Of You.' It was all wrong from the start: Jack Haley and Shirley Ross shouldn't have sung it, it didn't tie up with the situation, and it never went well. The only excuse that I have to offer is that we had to write the show in a terrific hurry. So we did the song and it was unsatisfactory from the start. It's out of the show now and we have another one in ['I'm Afraid'] that we hope is much better."

"Larry told me," interposed Dwight Wiman, "that apart from the fact that the song was bad, the lyrics worse, that the people who were doing it were baffled by its inappropriateness and couldn't put it across, and that it didn't belong in the show at all, it was a good song." At this, Elinor Hughes reported, Hart grinned and said nothing; perhaps because he knew it wasn't one of his songs?

"There are so many things to think about," Rodgers went on. "Take, just for instance, the song called 'Ev'ry Sunday Afternoon.' In writing the music I had to remember to make it simple, fitting the characters of the people in the story who were to sing it; Larry, in writing the lyrics, had to think what these people, who worked hard and had only a short time each week to talk to and be with each other, would think about and say. In short, songs must suit the characters and situations in the story first and foremost, since you'll never get far writing for personalities of individual players alone."

Untypically, Larry seems to have had little to say; he had always dominated such interviews in the past. He told the reporter the story about how Max Dreyfus (although he was careful not to name him) had turned them down when they tried to sell him the songs they later used in *Garrick Gaieties*, and how, after they became successful, he was angry with them for not letting him publish them. "Think this over and write a story," he told her. "Where are the good new musical comedy composers? Whom can you think of that's come along in the past thirteen years, except Arthur Schwartz?"

"The show did very well in Boston," Joshua Logan remembered, "and everyone kept saying, 'What are you worrying about, it's going wonderfully well!' I said, 'Because it's awful, it just doesn't feel right.' The problem was, although the songs themselves were perfectly good, they didn't come out of anything that interested you. So when Shirley Ross sang 'It Never Entered My Mind' about a man you just didn't care about, it just didn't have any impact."

They opened *Higher and Higher* at the Shubert in New York on Thursday, April 4, 1940. Backstage, Josh Logan sensed the audience's reaction was less than enthusiastic, although everyone seemed to love the performing seal. After the show, as the audience crowded the center aisle to leave, there was a small but perfect silence. Then into that silence, Logan said, an enthusiastic voice rang out. "Darleeng, zat was zee muss vonderful fuck in zee world!"

> Many people turned to see who was talking. It was the young French wife of an American lawyer. She was using the French word for 'seal' which is 'phoque.' Her husband tried to keep her quiet, but she protested, 'I deed not say zee *show* was good—I shuzz say zee *fuck* was good—zee *fuck*, only zee *fuck*, and oh, how I loved zat *fuck*.' By this time the whole aisle was laughing.
>
> I stood in my box saying to myself, 'At last, a funny scene—but it's not onstage.'
>
> Next morning the notices were as gruesome as I knew they would be. The show ran as long as the advance sale lasted, a few months.

Five days later, Germany invaded Denmark and Norway. The "phony war" was over. As George Kaufman ominously prophesied when he heard the news, now they were shooting without a script.

34

Bothered and Bewildered

*J*ohn O'Hara was a man of enormous contradictions. *Appointment in Samarra*, *BUtterfield-8*, and brilliant stories for *The New Yorker* had made him one of America's best-known writers. He was, in his own words, "America's shyest novelist," a claim he qualified somewhat by adding that the second shyest was Ernest Hemingway. Thirty-five years of age, a warm, witty, and delightful companion when sober, he was given to black rages and brutish behavior when drunk, which was not infrequently.

One of O'Hara's most successful fictional creations was a character who in many ways personified his own cynical outlook. He appeared in a series of stories which took the form of letters written to a successful bandleader friend by a cheap, amoral nightclub master of ceremonies in the Middle West who signed himself "Yr Pal Joey." The first of these had been written in 1938 shortly after O'Hara finished his novel *Hope of Heaven*. He told his wife Belle he was going to take a room at the Ben Franklin Hotel in Philadelphia and write; instead he went on a three-day drunk.

When he sobered up, he purged his soul by writing a piece about the nightclub heel, Joey, someone who was an even lower low-life than he felt himself to be. *The New Yorker* bought it immediately and asked for more. He quickly wrote two more, and although he soon tired of the device of the letters through which Joey spoke, he found "the more I wrote about the slob the more I got to like him." Joey had developed into a rounded character, an aggressive two-bit hustler who wants to get into the big time but never will; in 1940, the stories were collected and published in book form as *Pal Joey*.

O'Hara had gone to Hollywood and was working for 20th Century–Fox doing "polish jobs" on screenplays at a thousand dollars a week, one of which was coincidentally the Vera Zorina movie *I Was an Adventuress*, the so-called romantic comedy about jewel thieves that featured a dance sequence choreographed by George Balanchine. One night at a dinner party George Oppenheimer, who'd written *Nana*—the coincidences multiply—made a suggestion that changed O'Hara's life. "You have a play in that Joey character of yours," he said. The more he thought about it, the more O'Hara came to believe that Oppenheimer was right, and that if anyone ever did write a play about Joey it would be an enormous success. In October 1939, he sat down and wrote a note to Dick Rodgers, then trying out *Too Many Girls* in Boston.

> Dear Dick,
>
> I don't know whether you happened to see any of a series of pieces I've been doing for *The New Yorker* for the past year or so. They're about a guy who is master of ceremonies in cheap nightclubs, and the pieces are in the form of letters from him to a successful band leader. Anyway, I got the idea that the pieces, or at least the character and the life in general, could be made into a book show, and I wonder if you and Larry would be interested in working on it with me. I read that you two have a commitment with Dwight Wiman for a show this spring, but if and when you get through with that I do hope you like my idea.
>
> All the best to you always. Please remember me to the beautiful Dorothy and say hello to Larry for me. Say more than hello, too.

Dick was sold on the idea before he finished reading. He showed the letter to his partner. Larry, who knew the seedy ambience of Pal Joey's world from the inside out, was as enthusiastic about the project as he. Both of them could see from the start that here was a chance to do something totally different from anything they—or anyone else—had ever done before. That challenge greatly appealed to them.

Work on *Pal Joey* began immediately after the opening of *Higher and Higher*. On April 12, Larry and Doc Bender took off on vacation to Taxco in Mexico, where, Larry told a reporter, he was to work on a new show "understood to be for George Abbott." Just two days later, Moss Hart announced he was planning to musicalize Robert Sherwood's 1926 play *The Road to Rome* with a score by Rodgers and Hart, but quite clearly Larry and Dick were so eager to do *Joey* that nothing else interested them.

Writing started in early May. Rodgers had already talked to George Abbott about directing the play; by this time "O'Hara had a script, but it was a disorganized set of scenes without a good story line and required work before we would be ready for rehearsal," Abbott wrote. There was a lot of rewriting; O'Hara, a seasoned scenarist, took this in his stride. "Getting a musical together is one long process of backing and filling," he wrote, "hedging and trimming."

In July, Abbott signed to produce and direct, and casting of the principal actors began. After he'd taken a look at a young actor Larry Hart had seen in a William Saroyan play, *The Time of Your Life*, Rodgers invited him to audition for the part of Joey. His name was Gene Kelly, and he recalled:

I do know when I auditioned, Larry wasn't there. Richard Rodgers and John O'Hara were, though. If George Abbott was there, I do not remember. However, during *Time of Your Life* I had gotten to know Larry Hart in local saloons, not in Sardi's or the Stork Club or "21", but the cheap saloons around Eighth Avenue and 45th Street, and Larry Hart would come in there. He loved the actors, and he loved to hang around with them, and I got to know him, not closely, but in a fun kind of way. He'd come in and we'd be around the bar, and he'd tell stories, usually chomping on a cigar. He was a marvelous little fellow, and of course we all admired him for his great talent, and so he had a ready audience. We all looked up to him.

Gene Kelly did a "stupid thing" during his audition; he sang a Rodgers and Hart song, "I Didn't Know What Time It Was," not realizing in his naivete that this might be construed as trying to curry favor. It made no difference, anyway. "I guess it was just a day or two and they said I was hired for the part," he said.

O'Hara was thinking in terms of Marlene Dietrich for the role of Joey's rich patroness, but Rodgers and Hart—especially Larry—preferred Vivienne Segal, who would sing the kind of lyrics he had in mind "like a lady." Auditions continued throughout August. June Havoc, the original "Baby June" Hovick, sister of Gypsy Rose Lee, the dancer-stripper whose intellectual pretensions would be guyed in one of the show's songs, was cast in a small part. Jack Durant, an acrobatic dancing comedian who had appeared in a couple of Alice Faye movies, was tapped for another. Leila Ernst, Jean Casto, Van Johnson, and a young dancer named Stanley Donen were also signed.

O'Hara's book dealt with cheap nightclub emcee Joey Evans—what they used to call "a heel"—who ditches his naive girlfriend Linda to romance a rich older woman, Vera Simpson—"Mrs. Chicago Society." Vera (O'Hara had deliberately given her Vivienne Segal's initials) sets him up in his own nightclub, and thus exposes herself to a cheap blackmail attempt by one of Joey's former colleagues, dancer Gladys Bumps, and her boyfriend "heavy" Ludlow Lowell, who's muscled his way into being Joey's "agent." This fails when Linda warns Vera what they are planning and Vera calls her friend the Police Commissioner; but she decides to kick Joey out anyway. In a final scene Linda offers a happy ending, but Joey tells her he's got these big-time plans, maybe he'll shoot her a wire from New York. Is he downhearted about what happened? The hell he is: he's not that smart.

The fifteen songs were written in a three-week blaze of activity. Following what had now become their normal practice, Dick wrote the melodies for the love songs and Larry fitted lyrics to them; for the "situation numbers" Hart blocked out the lyrics first and gave them to Rodgers to set. Joshua

Logan recalled Larry phoning, often, to read him something he had just written. "The thought of putting something over on the censors was delicious to him. 'Listen to this, he'd say, listen to this. "Lost my heart, but what of it? My mistake, I agree. He's a laugh, but I love it, When the laugh's on me." You get it, Josh, you get it? The laugh's *on* me, you know, on top of her.' He thought that was terribly funny."

Writing the kind of cynical, callous, suggestive lyrics needed for characters like Joey and his older benefactress was a paid vacation for Larry Hart. He was writing for Gene Kelly, whom he idolized, for Vivienne Segal, whom he adored: all he had to do was let rip. Frivolously mock the intellectual pretensions of a stripper not unlike Gypsy Rose Lee? Bring on a reporter who's seen it all, and "Zip"! A couple of sleazy nightclub numbers? Throw in the corny colors of "That Terrific Rainbow" and the cheesy sentiments of "Flower Garden Of My Heart." A love song to be sung—completely insincerely—by a heel who's trying to "make" a not-too-bright "mouse"? "I Could Write A Book." The sensuous soliloquy of an older woman musing on the debatable charms of her kept lover? "Bewitched, Bothered And Bewildered." And every step of the way Rodgers matched his partner, producing melodies ranging from the torrid to the tawdry, from the virulent to the voluptuous. In fact, when he first played "Bewitched," Abbott took him to one side and said, "Dick, don't you think that melody is too sweet for the kind of lyrics Larry has written?"

"If it is, it'll be the first time," Rodgers retorted.

When they were halfway through the score, an ominous report appeared in the New York newspapers that they were splitting up. On September 10, they took the trouble to issue a formal denial: there was no truth in the rumors. "We've been parting for twenty-two years and we still are," Rodgers told reporters. "It's ridiculous," Larry added.

Asked whether he recalled what their relationship was during the making of *Pal Joey*, George Abbott replied tersely, "Dick [was] angry at Larry's drinking, but [there was] no talk of [a] split-up." Nevertheless, the first visible crack in the public facade of Rodgers and Hart had appeared. And during the next year and a half it would become ever wider. No matter how loyally Dick protected his partner, he had to admit to himself that Larry was becoming increasingly difficult to work with, and to face the probability that one day not too far down the road he might no longer be able to work at all. Meanwhile, talk of a revue for the Shuberts to star Bea Lillie and Carmen Miranda early in October came to nothing; Rodgers and Hart had enough on their minds.

Chorus calls for *Pal Joey* began late in October; choreographer Robert Alton needed seventeen chorus girls and eight male dancers. Each call drew as many as five hundred hopefuls, although not many could meet Alton's requirements—ballet as well as tap. On November 11, rehearsals began at the Biltmore and Longacre theatres. George Abbott described the working atmosphere:

We started rehearsals unsure of what the total effect was going to be. After a week Bob Alton thought the show was hopeless and wanted to quit, but Rodgers persuaded him to stick it out. Jo Mielziner made a major contribution to the production by suggesting that the curtain of Act One be a scene in which Joey envisions his future in the magnificent club which his new girl friend is going to buy him. It cost ten thousand dollars to build the set, a good deal of money in those days when a musical had a budget of one hundred thousand dollars, but I accepted the suggestion unhesitatingly. This is a perfect illustration of how many collaborators there really are in a musical comedy.

June Havoc's recollections reflect her performer's perspective—and ambitions:

We rehearsed in two theatres. On the stage of one the actors worked with Mr. A., as he was carefully called. The other stage was noisy and exciting. That was where Robert Alton worked with the dancers. At the first reading, I had sat in the circle with Mr. A. and the actors. My part was very small. I didn't appear until the second act was well under way.

Havoc made something of a crusade out of getting Gladys into the first act, badgering production stage manager Jerome Whyte, pushing herself into choreographer Alton's sight line. Finally Alton installed her into a comic routine up early in the show. "Go tell Mr. A. O'Hara has to find a way to get you on for this," he told her, and a short scene was written to establish Gladys so she could do the number. But Havoc wanted more, more:

A few days later, after rehearsal was over and the company had departed for the evening, I watched from my hiding spot in an upper box. Rodgers had brought in a new song. He played and sang. Everyone laughed. "Tomorrow," he said, "find a funny voice who can really move, okay?" Auditions were planned, the golden group departed. I emerged from my secret place, pouncing on Jerry Whyte. If he hadn't snagged a copy of "Terrific Rainbow" for me so I could work on it all that night, I could never have beat out my competitors at the auditions the following day.

Next, Rodgers came in with "Flower Garden Of My Heart." Everyone loved it. What they needed now was "a tatty soprano" to sing it. Auditions were set. Once again the shamelessly voracious Havoc decided to go after the spot: "That night, after blessing Jerry again, I blessed the vocal coach who had given me a few five-dollar lessons. That investment had liberated a high B-flat I hadn't known what to do with until now. As extra insurance, I sang in my high voice while on point[e]."

One gets the feeling she would have painted herself blue as well, if she thought it would have fattened her part; but she got the number.

> Each time a new number fell my way, John O'Hara wrote something to get me
> onstage, so that by the time we opened in Philadelphia, I had five songs and a nice fat
> part. I sang high, I sang low, I danced fast, slow, and on point[e]. And, for the ending
> of my first number with Jack Durant, I carried 180 pounds of blackmailer offstage
> while we sang the last sixteen bars of "Plant You Now, Dig You Later" in harmony.

While all this was going on, Dick Rodgers was experiencing havoc of a different kind, and this time not with his partner but with his producer and his librettist. The deeper they got into it, the more Abbott was convinced the show would be a failure. A play about a disagreeable subject with disagreeable people could never be escapist theatre. "Perhaps because of the daring nature of our show, he thought he was sticking his neck out a bit too far," Rodgers guessed. "Apparently people must have told him that it wasn't commercial to do a show with such a disreputable character for a hero, and he was apprehensive about losing money on so risky a venture."

Next came problems with the writer. Imagining when he handed in the book his work on the show was over, John O'Hara, who later admitted he was suffering from depression at the time, decided to stay away from the theatre in order not to be "that perennial Broadway nuisance, The Author." But as June Havoc pointed out, they needed him to write in bits of business to get people on and offstage, to link new production numbers, to add or elide dialogue scenes. George Abbott would attempt to fix a scene, but it was difficult. "I know what's wrong," he would say, "but I can't do it."

"O'Hara came to rehearsals very little," Abbott confirmed. "When I needed rewriting I would do it on the set, and later he would drop into the theatre, look over what I had done, go to an empty dressing room, rewrite the new material and depart. I am sure that *Pal Joey* must have been important to him; but I can never remember his demonstrating any approbation or enthusiasm—nor, on the other hand, any criticism. He seemed disinterested, but I am sure that this was just his manner."

Whatever it was, it bothered the hell out of Dick Rodgers. "There were periods during which I didn't hear from him for several weeks, and I couldn't get him on the telephone, " he said. "Finally, in desperation, I sent him a wire: SPEAK TO ME, JOHN, SPEAK TO ME." When even this appeal failed to elicit a response, Larry Hart went to O'Hara's apartment. A maid let him in, and he went to the bedroom door. "Get up, Baby," he said. "Come on, come on, you're hurting George's feelings." The idea that anyone could hurt George Abbott's feelings was sufficiently bizarre to get O'Hara back to the theatre; he and Larry became drinking buddies thereafter.

Late on the evening of Sunday, December 8, as news came through that during the preceding month German air raids on Britain had killed nearly five thousand people and maimed another six thousand, the *Pal Joey* company took the train down to Philadelphia for final rehearsals and tryouts at the Forrest Theatre. In addition to the cast and musicians, there were three freight cars of scenery, no part of which had any of the cast or the production staff seen assembled in one place. Lighting had to be planned and fixed; the chorus and principals had to rehearse with the orchestra none of them had yet even heard. Rehearsals were called for eleven; for most of the succeeding week they continued until three or four the following morning, often at the nearby Hotel Erlanger.

During rehearsals, Gene Kelly said, Vivienne Segal and George Abbott had "some kind of a run-in. Larry was completely upset because he adored Vivienne, and Abbott, when he had trouble, he'd just say, 'Fire them,' no hesitation. And one day he was talking about getting rid of Vivienne, I just overheard this, it had nothing to do with me at all, and I was amazed at Abbott's complete confidence that he could supplant this very able woman. But Larry Hart, he was . . . he almost broke down."

The storm passed; Vivienne stayed. The dress rehearsal was a disaster, but that was nothing new. Opening night on December 16 was a success, and after that, everyone relaxed. Larry Hart, Gene Kelly, John O'Hara, and O'Hara's pal novelist Budd Schulberg, who had come down to hold O'Hara's hand during the tryouts, became an inseparable drinking quartet. "Let me buy you a stimulant," Larry would say each night after the show, and off they would go. On more than one occasion Kelly, O'Hara, and Schulberg had to carry Larry from a bar to a taxi and back to his hotel.

After the Philadelphia opening, they went to the Adelphi Hotel to wait for the reviews. O'Hara had a pile of telegrams in front of him. "I bet he sent them to himself," someone cracked. Later they watched the floor show at the Adelphi, which was so bad they realized that in *Pal Joey* "instead of a violent caricature we had on our hands only a somewhat underexposed photograph."

"I remember Larry as being a lot of fun, very cheerful," Kelly said, "although nervous, but a very cheerful guy to have around, keeping everybody 'up.' I never knew him in any bad or neurotic moods. Bender? Sure I knew him. He was always there. Although, you know, I don't remember ever seeing him smile. He came down once or twice to Philadelphia. They were fairly constantly together. Wherever you saw Larry running around, he was being shadowed by Bender."

The Philadelphia reviews for *Pal Joey* were very good, but there was some tightening to do. One of Joey's songs, "I'm Talking To My Pal," sung with a muted trumpet response just before the big first-act dream-sequence finale, was cut; with five songs, three dances, and fifty-one pages of dialogue, Gene Kelly had enough to do. "It was a charming lyric and a charming song," Kelly recalled, "but it was just too long to do that and do the ballet at the end of the first act."

They also needed more choruses for "Zip," which had stopped the show cold. It was in a scene where Jean Casto, playing the reporter Melba Snyder, interviews Joey about his new nightclub. She is onto his lies from the start: been there, done that, heard it all before.

Larry's lyrics were not just topical; in each twenty-three-line chorus he was having wicked, witty fun with all the fads and "names" of the day, some of them still famous, like Pablo Picasso, others—like the socialite movie star Countess di Frasso, formerly Dorothy Taylor, sister of the president of the New York Stock Exchange—long forgotten. He works in a mention of the Yankee Clipper, the latest "in" place to eat; goes on to give Walter Lippmann, America's most influential syndicated columnist, a dig in the ribs; slyly teases playwright William Saroyan, who earlier in the year had been awarded (but refused) the Pulitzer Prize for *The Time of Your Life*, the play in which Gene Kelly had been spotted; pairs Zorina, a major star of her day, with Cobina, actually Cobina Wright, one of the two famous debutantes of that season (the other was Brenda Frazier); and aims the final punch line at Margie Hart, one of the many strippers competing for Gypsy's fame. Show-stopping stuff, but as Gene Kelly said, before the curtain went up that night, they needed more:

> So next day, Larry was up very late and got to the theatre late as was his wont—I know because he and John O'Hara and I used to go out and drink every night and we had a great time, great fun. I was up for ten o'clock rehearsal because I was a kid and I had all the energy in the world. And I remember them saying, "Where's Larry, where's Larry?" And he came in and said, "Oh, yes, let me see," and in four-letter words he wrote down the rhyming scheme to bring it back, like Lincoln doing the Gettysburg Address, and literally, it was a piece of old wrapping paper. And that night, it was all done, he had three, four, five choruses, whatever was wanted. And as fast as the girl could learn them they went into the show.

Hart's night's work was another whirlwind high-culture tour that if anything was superior to his first effort, with his intellectual stripper pronouncing Dali's paintings (currently all the rage) passé, musing on whether the Metropolitan Opera could ever be made to pay, noting that English people pronounce the word "clerk" as "clark," and deciding that anyone who said "clark" was "a jark." Nor did she care for Whistler's mother, Charlie's Aunt (a big hit on Broadway), or Shubert's brother (the universally detested impresarios who also hated each other). The close was again a joke on one of Gypsy's competitors, this time Sally Rand. A third verse featuring Toscanini, Jergen's hand lotion, Tyrone Power, Rip van Winkle, and the fact that either Mickey—Mouse or Rooney—made her "sicky" took a last-line poke at Lili St. Cyr, yet another ecdysiast. It was a pyrotechnical display of lyric-writing virtuosity, and audiences loved every word of it, even the five-syllable ones like "intellectual" and "heterosexual," "mystic" and "mysogynistic."

"What was so amazing," Gene Kelly continued,

> was that everything that he put in was so right currently, so topically funny and such high class humor that if it were published in *The New Yorker* it would have made you laugh without the music. Every line got belly laughs. Just to underline the point, every girl who played that—we had Jean Casto, she'd never sung before. She stopped the show every time. If she'd be ill, an understudy would do it, or later, a replacement, or a chorus girl would do it, it made no difference. Nobody failed to stop the show, so you have to deduce it was Rodgers and Hart's song.

Still the Jonahs pursued them: the story editor of a movie company, whose judgment was always respected, did absolutely nothing for George Abbott's confidence by telling him to let the show die out of town rather than tempt disaster with it in New York. His advice was ignored, and *Pal Joey* opened at the Ethel Barrymore Theatre in New York on Christmas Day 1940, before one of the most distinguished first-night audiences of the season.

Abbott's misgivings are perfectly understandable; even today, over half a century away, some of the dialogue gives a hint of how different a musical this really was. In one scene, Joey puts on an act to impress a kid who works at Mike's club. Gladys overhears and challenges him.

GLADYS: Now you're an aviator?

JOEY: What's it to you?

GLADYS: *(mimicking him)* Tonight I might tell you some of my experiences. The big aviator! Were you ever up in an elevator, for Godsakes?

JOEY: You bore me.

GLADYS: What was that one you used to tell? How you were a rodeo champion?

JOEY: You bore me. Anyway, you never heard of me.

GLADYS: I heard about you. remember that tab show you used to be in? My sister was in that show. I heard all about you.

JOEY: Yeah? Which was was your sister?

GLADYS: The one you didn't score with.

JOEY: That must have been the ugly one.

In another scene, Vera has come back to the club, just as Joey has bet Mike she will. Just as she knows he told Mike she would.

VERA: So he fired you, but you said, "She'll be back, I know her kind." Right?

JOEY: I said, go ahead.

VERA: You thought it over. "How can I get her to come back?" By the way, how'd you get my number?

JOEY: Easy. The press agent of this joint has a 1919 Social Register.

Vera goes ahead, anyway: she's "Bewitched, Bothered and Bewildered"—but she's wise to herself. After Joey is used as a pawn in a pathetic little blackmail scam, Vera decides he's more trouble than he's worth, and tells him so. Joey reacts badly.

JOEY: Come on, say it. This is the brush-off. Those punks gave you a scare, and you're walking out.

VERA: A slightly brutal, though accurate way of putting it. You can keep the club . . .

JOEY: Are you trying to kid me? You got some other guy, that's why I'm getting the brusheroo. I get it now—"Take him"—you meant me. All right—go on back to him.

VERA: I have a temper, Beauty, and I want to say a few things before I lose it.

JOEY: Lose it. It's all you got left to lose.

Vera and Linda—the "mouse" Joey has been stringing along—compare notes on Joey. Linda tells Vera if she wants him she can "Take Him." Vera decides against it; Joey just isn't smart—in fact, she knows a movie executive who's twice as bright, she tells Linda. Finally, both agree they'll be better off without him. Joey gets the gate; but of course, being Joey, he makes out like it was his idea, and we last see him shooting his shopworn line at nice sweet Linda, too arrogant to realize she's probably his last best hope.

Hardly surprising, then, that as the curtain came down the audience divided sharply into two camps: those who thought the show was awful, and those who thought it superb. The critics were similarly sundered; nothing about *Pal Joey* encouraged a tepid response.

After the show, Larry Hart hosted one of his no-holds-barred parties while they waited for the reviews. He and John O'Hara were particularly eager, Gene Kelly said, for the critics to recognize the new seriousness the show was bringing to the musical stage. At around midnight, Larry "had a friend at the *Times* read him the tear sheets of the Brooks Atkinson review," Kelly said. "When he heard it, Larry broke down. I was standing right by him, he said, 'What, what?' He broke down, cried, sobbed, because he wanted Atkinson, for whatever reasons, to say this show was a milestone. He locked himself in his bedroom and wouldn't come out. All the other reviews, comparatively, didn't matter."

Brooks Atkinson's reaction mattered most because Larry believed him to be the critic who best understood what he and Rodgers were trying to do. It had been Atkinson who only three years earlier had declared that in Rodgers and Hart musical comedy had met its masters, so it is instructive to read just what he actually wrote.

"If it is possible to make an entertaining musical comedy out of an odious story," he began, "'Pal Joey' is it. Taking as his hero the frowsy night club punk familiar to readers of a series of sketches in The New Yorker, John O'Hara has written a joyless book about a sulky assignation. Under George Abbott's direction some of the best workmen on Broadway have fitted it out with smart embellishments."

Tough, but so far not hostile. "Rodgers and Hart," Atkinson continued, "have written the score with wit and skill." He praised Robert Alton's dances, Jo Mielziner's "high-class" scenery, and the talented performers. Gene Kelly's performance was "triumphant." Vivienne Segal sang some scabrous lyrics in a singularly sweet voice. June Havoc, Jean Casto, and Jack Durant's acrobatic dancing—which also stopped the show opening night—were singled out. It was in his final lines that Atkinson cut deepest and hurt most. The story, he complained, kept harking back "to the drab and mirthless world of punk's progress. Although 'Pal Joey' is expertly done, can you draw sweet water from a foul well?"

He always insisted his criticism was well intentioned rather than destructive; either way, it devastated Larry Hart. He believed *Pal Joey* to be the best work he had ever done. If this was the reaction of the critic he most admired, what the hell was the point of any of it?

Despite Atkinson's strictures, largely favorable critical reactions ensured *Pal Joey* settled in for a healthy run. True, it puzzled a lot of the people who came to see it. "The ladies would come in from Westchester for the Wednesday matinees," Gene Kelly said:

> They were frigidly cold, they were sub-zero. On some songs they'd just sit there grimly and stare at us and there was hardly a patter of applause. There was one song, "Our Little Den of Iniquity," which I did with Vivienne Segal, sitting on pillows out 'in one' by the footlights. We dreaded doing that on Wednesday matinees. But it was a pleasure to do the Friday and Saturday night shows, because then we got the swinging group. Then we got the people who came back twenty, twenty five times to see it.

———❀———

This is not too difficult to understand. In a year when songs like "Blueberry Hill," "Polka Dots And Moonbeams," "Dance With A Dolly With A Hole In Her Stocking," and "When You Wish Upon A Star" were big hits, songs like Rodgers and Hart's "Den Of Iniquity" asked a lot from audiences unused to such not-so-subtle innuendoes as "sep'rate bedrooms . . . one for play and one for show."

———❀———

Dick Rodgers and Larry Hart had always believed that the only thing worth doing was the thing you believed in most; and both of them believed *Pal Joey* was different, special, and important. "It seemed to us," Rodgers said, "that musical comedy had to get out of its cradle and start standing on its own two feet, looking at the facts of life." In spite of many highly articulate supporters, however, audiences were never completely won over, and although *Pal Joey* had a very respectable 374-performance run, it was never perceived as a smash hit.

Bad luck continued to dog the show. Just before it was premiered, the radio networks and the songwriters' union, ASCAP, became involved in a life-or-death struggle over performance royalties. ASCAP raised the performance fees they charged broadcasters to such heights that the networks responded by forming their own performing rights society, Broadcast Music, Inc., which in turn resulted in the banning from broadcasts of all songs by ASCAP members. Not until late in February 1941 did ASCAP accept a Department of Justice consent decree which resolved the deadlock. Thus the score of *Pal Joey* lost its most powerful medium of dissemination, and surefire hit songs like "I Could Write A Book" and "Bewitched" were condemned to temporary obscurity.

A decade later, in a completely changed moral climate, *Pal Joey* would become the most successful musical revival in Broadway history. "It is true," Brooks Atkinson would write, "that *Pal Joey* was a pioneer in the moving back of musical frontiers, for it tells an integrated story with a knowing point of view. Brimming over with good music and fast on its toes, it renews confidence in the professionalism of the theatre."

Unfortunately, by the time Atkinson revised his opinion, Larry Hart was not around to see his faith in the play vindicated, to hear "Bewitched" become the number one song in the Hit Parade, to watch "I Could Write A Book" become a standard. Filmed in 1957 with Frank Sinatra as Joey, Kim Novak as Linda, and Rita Hayworth as Vera—even sanitized out of all recognition, it still had bite—*Pal Joey* can now be seen as a milestone in musical theatre, what one critic called "the nearest Broadway has come to producing its own *Beggar's Opera*." Unquestionably the finest achievement of the Rodgers and Hart partnership, *Pal Joey* is notable for one other reason: it marked the beginning of the end for Larry Hart.

35

I Could Have Been a Genius

Early in 1941, shortly after Larry Hart and Doc Bender got back from their annual trip to Miami, RKO Pictures, encouraged by the success of *Too Many Girls*, gave producer Lou *(Flying Down to Rio)* Brock approval to go ahead with another Good Neighbor movie based on his own story about a Texas oil millionaire who goes to Argentina to buy a famous race horse. With James Ellison, Maureen O'Hara, Diosa Costello, and Buddy Ebsen cast to further the Good Neighbor policy, Brock commissioned Rodgers and Hart to write a score for the movie, which was to be called *They Met in Argentina*.

Rodgers and Hart composed a dozen songs; the producers in Hollywood showed how impressed they were by dropping five of them from the finished print. Maybe they were smart, at that; certainly those that remained, including "Simpatica," "Lolita," "Never Go To Argentina," and "Amarillo," were no better than they had to be. But neither was the picture, directed in black and white by Leslie Goodwins, who later specialized in low-budget "Mexican Spitfire" nonsenses featuring Lupe Velez, and RKO all-purpose director Jack Hively. Robert Dana of the *Herald Tribune* not unfairly described it as "an American musical at its worst." It was the sole Rodgers and Hart offering for the spring of 1941, or indeed for over a year. "The reason, I'm afraid, was Larry," said Rodgers.

This was very unfair; in fact, not a single musical opened on Broadway between *Lady in the Dark* in January and *Best Foot Forward* in October. And since Dick Rodgers was co-producer of the latter—a fact he kept secret because "things were getting sensitive in the Larry Hart area"—he knew better than anyone that the climate for anything that wasn't frothy and escapist was not altogether encouraging. And, not to put too fine a point upon it, Rodgers and Hart had nothing to work on anyway.

In May, around the time *They Met in Argentina* was released to an indifferent public, Dick and Dorothy Rodgers sold their Port Washington home and moved to a beautiful fifteen-room colonial house on the Black Rock Turnpike in Fairfield, Connecticut. While this was going on, adding fuel to the speculation the partnership was breaking up, Larry Hart was enthusiastically working on a project of his own.

On January 10, 1941, a play written by former music teacher and actor Joseph Kesselring had begun what was to be a record-breaking three-year run at the Fulton Theatre. A hilariously gruesome comedy starring dumpy, delightful Josephine Hull and fey, nervy Jean Adair as two pixilated poisoners, John Alexander as the engagingly barmy brother who thinks he is Teddy Roosevelt, and Boris Karloff as their dangerously insane nephew Jonathan, *Arsenic and Old Lace* was like nothing ever seen on the Broadway stage, before or since; it was an overnight and overwhelming hit.

Karloff, incidentally, was making his Broadway debut. It had come about when, during rewrites, someone suggested a what-if: what if Jonathan had undergone plastic surgery to make him look like Karloff, who had become famous for playing movie monsters? The next logical step was obvious, and Karloff was offered the part. During the run of the play, producers Lindsay and Crouse, or "the Beamish Ones," as Karloff dubbed them, delighted in teasing Karloff—well known for his parsimony—paying his two-thousand-dollar weekly salary in nickels, redrawing his contract so he got only twenty-five dollars a week plus whatever was tossed onto the stage. Karloff retaliated by demanding a raise to cover the cost of makeup powder. A few nights later he got a huge gift-wrapped box that looked like something from F.A.O. Schwarz at Christmastime. When he opened it, he found it packed with tooth powder, foot powder, baking powder, roach powder, Seidlitz powders, powdered eggs, and even gunpowder.

Almost back to back with *Arsenic and Old Lace*, another play on a not dissimilar theme by Owen Davis opened at the Belasco. *Mr. and Mrs. North*, which was based on a 1940 Francis and Richard Lockridge novel, *The Norths Meet Murder*, concerned itself with a young couple who return to their apartment in Greenwich Village to find a corpse in the closet and another murder in the offing; all is saved by the scatterbrained Pam, who solves the crime with the aid of a cooking recipe. All this comedic mayhem gave producer Alex Yokel an idea.

He had found a novel by Richard Shattuck called *The Snark Was a Boojum*. It had to do with three expectant mothers stranded in a New England snowstorm, an eccentric's will, an escaped maniac, and a great deal of Lewis Carrolliana. Yokel was convinced this jabberwocky could be made into a hit play and suggested Larry Hart adapt it for Broadway. Larry, who was very fond of Yokel, figured he owed him one: after all, hadn't Alex given Teddy Hart a chance in *Three Men on a Horse?* He set to work with his usual no-holds-barred enthusiasm, meanwhile telling the papers he "had all sorts of plans for fall." Fortunately, perhaps, no one asked him to be specific.

He completed at least seventy pages in his chicken-scratch but could not whip anybody into being enthusiastic about what he'd written: not only did it not work, it just wasn't funny. He abandoned it abruptly. Then, according to Dorothy Hart, Alex Yokel came over to the Hart apartment intending to spirit Larry away someplace where he could complete the adaptation. Yokel, who had not had a hit since *Three Men on a Horse*, was desperate. So desperate that, according to Dorothy Hart, he attempted to physically remove Larry from the Hart apartment, and if his brother Teddy had not been present would probably have succeeded, for Larry was in no condition to know what was happening. What ensued was a disgraceful tug-of-war with the producer and Teddy fighting over Larry's supine—and by inference drunken—body.

In the end, Yokel gave up on Larry Hart and later persuaded Owen Davis to do an adaptation, but by September 1943, when it opened on Broadway, the moment had passed; *The Snark Was a Boojum* lasted precisely five performances. It would be interesting to know if Larry Hart ever saw it.

Although Rodgers and Hart were not seeing a lot of each other, Rodgers was still very actively seeking a new project for them to work on. He was congenitally unable not to work: he once told Alan Jay Lerner that if he didn't compose regularly, he actually became physically constipated. Right now, he was excited about something that had come his way from his friend and former neighbor Edna Ferber, who had asked him if he would be interested in musicalizing her novel *Saratoga Trunk*, then being serialized prior to its publication in book form.

Rodgers said yes, but with reservations. Unsure that Hart was in anything like the kind of shape he would need to be in to write a libretto, he suggested they talk to Oscar Hammerstein II. On June 29, Edna Ferber wired Hammerstein, who was in Hollywood, saying how happy they would be if he would write the book. The same day, another wire from Rodgers arrived: LARRY AND I SIT WITH EVERYTHING CROSSED HOPING THAT YOU WILL DO SARATOGA TRUNK WITH US.

Ferber, of course, had been a great fan of Oscar Hammerstein since *Show Boat;* but Rodgers knew only too well that it had been a long, long time since Hammerstein had written a hit. True, his lyric to Jerome Kern's song "All

The Things You Are," from the failed *Very Warm for May*, was being heard constantly on the radio, but other than that, the word on Hammerstein was, "Oscar can't write his hat"—which makes Rodgers's choice the more puzzling. If it was just a book, why not Herbert Fields, why not Moss Hart, George Abbott, John Cecil Holm? The answer has to be that none of these men wrote lyrics.

As things transpired, difficulties with the author, who was "acting up," according to Rodgers, and further complications with Warner Bros., who already owned movie rights, caused them to abandon the project the following month. But another door had been opened, as is clear from Rodgers's reply to Hammerstein:

> I can say this, however, I was delighted and warmed by several things in your letter. Even if nothing further comes of this difficult matter it will at least have allowed us to approach each other professionally. Specifically, you feel that I should have a book of "substance" to write to. Will you think seriously about doing such a book?

There was nothing casual about this: it was a direct overture toward partnership. Since Rodgers was well aware that Oscar Hammerstein had rarely written a book for which he did not also provide lyrics, any such partnership would of necessity exclude Larry Hart. Surely not even Rodgers, demonstrably insensitive to the craft of lyric writing, could have envisaged a collaboration between the two lyricists.

On August 17, hard on the heels of the abandonment of the Edna Ferber project, it was announced that Rodgers and Hart would write a score for a show based on a series of pieces by Ludwig Bemelmans about the fictional Hotel Splendide which had appeared in *The New Yorker*. The book would be written by Donald Ogden Stewart, a sometime member of the Algonquin Round Table group, friend of Hemingway and Dorothy Parker, and hugely successful screenwriter. (One of his most recent screenplays, *The Philadelphia Story*, starring Katharine Hepburn, Cary Grant, and James Stewart, had just won him an Oscar.)

According to Irving Eisman, Hart didn't want to do the show. But then, Larry didn't want to do anything. He was caught up in the classic alcoholic dilemma: the worse his problems got, the more he drank; the more he drank, the worse his problems got. From the summer of 1940, his disintegration had accelerated so alarmingly that even his closest friends, only too familiar with his alcoholic excesses, his erratic behavior, and his irresponsible way of life, were shocked. Word on the street was that Larry was drunk by noon, and that by night he was indulging in back-alley sex with anyone he could pick up.

Hart had already been admitted to Doctors' Hospital on a couple of occasions as a voluntary patient. According to Dorothy Hart, he never admitted in any way that he was an alcoholic. He insisted he could stop drinking whenever he wanted to. But these short stays—just a week or so—were not long enough to do him any good, though at least he then received proper food and care. But too often, he left before he was discharged.

Now the problem was getting worse. Dick Rodgers talked seriously to the family about having Larry committed to an institution. When Teddy Hart and his wife refused to countenance such a proposition, Dick shook his head despairingly. "He'll have to be committed," he repeated. Meaning, one supposes, he'll never go any other way.

Rodgers knew as well as the family what Larry's feelings about psychoanalysis were. He had even put them into a lyric: "Waking up to find that you're a girl is too good for the average man." But they could no longer ignore the drinking, the frequent mental blackouts, the disappearances. Dorothy Hart recorded that she and her husband tried on a number of occasions to get him to consider therapy, but Larry would have nothing to do with it.

On one occasion while Larry was in Doctors' Hospital drying out, Rodgers talked with psychiatrist Dr. Richard Hoffman about his partner's problems. Hoffman offered to drop in on Larry from time to time, passing himself off as just another member of the medical staff. That night, Larry phoned Rodgers. "Your witch-doctor was in to see me," he said, and that was the end of that.

Was there then no way to help him? "Whether Larry would ever submit to having help was always the question," Rodgers said. "He never would, he never did. I tried, I tried very hard to get him to have help, which he always refused. He would become panic stricken at the idea. He was conscious of the fact that he was destroying himself. But it was compulsive. He couldn't help it."

Probably the greatest torment of all to Larry Hart was his conviction that no woman ever could or ever would want him. There is no question that many women—beautiful women—were attracted by his wit and charm. But, as John O'Hara so perceptively noted, Hart was a lonely man who "knew better than anyone else that he was a disappointment to the lady admirers who had counted on swooning."

Another factor in his disintegration was the inescapable passage of time. By the end of the 1930s, his brother and nearly all his friends were married. Their domesticity sharpened his own sense of the emptiness of his life, the pointlessness of having fun when there was no one to have fun with. He grew careless about his personal appearance; sometimes he was so disheveled that acquaintances who saw him coming would cross the street to avoid him. And he drank, and drank, and drank; beset by his own feeling of worthlessness, he picked up shabby characters even more worthless, people even his racier friends shrank from. An anecdote of Dick Rodgers's illustrates the point— not to mention the teller's almost wilful blindness to its real content.

I had an appointment to meet Larry for lunch one day and he called it off, had his maid phone to say he couldn't make it. And I couldn't establish contact with him for a week. And then one day he showed up, he met me at Tony's on 52nd Street for lunch. He looked terrible, he needed a shave, his face was bruised, and one side of his jaw was very badly swollen, it looked as though he had a golf ball in there. And he said, "After lunch I'll tell you what happened to me." And he ordered scrambled eggs and coffee, things that were easy to eat, ice cream, and I said, "So tell me what happened."

He told me this man had arrived at his apartment late one afternoon with a letter of introduction from a mutual friend on the coast. Larry saw him, and invited him to stay for dinner. And the man had some cocktails and some wine with the meal, and about ten o'clock Larry thought the man was "showing signs" and ought to go home. And he wouldn't go. So Larry, wanting to get him out of the apartment, because his mother was there, said, "Well let's go downtown, we'll go someplace and have a drink."

They got down to the sidewalk in front of Larry's apartment house and Larry said to him, "Now don't you think you'd better go home and sleep this off?" This fellow called Larry a very dirty name and took a swing at him and hit him on the chin, and Larry went down, and the taxidrivers, who were standing around waiting for fares, formed a circle and watched. You know, Larry was a little bit of a fellow and he didn't stand a chance, and finally somebody broke it up and the man disappeared and a friend of Larry's, who happened to be there, took him to his house, afraid to take Larry back to his own apartment because they didn't want his mother to see all the blood. The friend put him to bed in his apartment.

Larry told me this story and [I said] I hoped that somebody had called the police and they put this character in jail. And Larry said, "Well, as a matter of fact I feel sorry for him." And I said, "Why? This horrible guy, why do you feel sorry for him?" And Larry said, "Well, I broke his wrist, knocked out three teeth, and kicked him where it hurt the most, and he won't be out of the hospital for a week."

In the summer of 1941, George Abbott sent Rodgers the script of a musical he was going to produce and invited Dick to join him in the venture. Rodgers agreed on condition that his partnership be entirely silent. "I wouldn't put my name on the show because things were so sensitive," he said. "I didn't want to call attention to the fact that I was working away from Larry, because I thought that would be harmful." The book of *Best Foot For-*

ward—another college football story—was by John Cecil Holm. Ralph Blane and Hugh Martin, who had worked as vocal arrangers on a couple of Rodgers and Hart shows, were given their first chance to write a full Broadway score.

During the tryouts in Philadelphia, Rodgers unburdened himself about Larry to George Abbott; Abbott's prognosis was even more pessimistic than Dick's own. Without hesitation, Rodgers picked up the telephone and called Oscar Hammerstein, who had moved to a beautiful farm near Doylestown, a few miles outside the city, about the same time the Rodgers family bought their house in Fairfield. They were old friends; if anyone could understand the agonies he was going through, it would be Oscar. And he was right.

He told Hammerstein about the Hotel Splendide project, and how impossible it was to get Larry to work. He confided his feeling that, whether he liked it or not, their collaboration was nearly over, and wondered whether Oscar would like to give some thought to becoming his lyricist. They talked—there is no documentation for this, but what had gone before makes it unimaginable that they did not—about the possibility of their collaborating on something more substantial than just another musical comedy. In the meantime, Oscar said, he was reluctant to step in and displace Larry. "But I'll tell you what I will do," he offered, with typical generosity. "You get Larry to work on the Bemelmans thing, and if he falls by the wayside, I'll gladly step in and finish the job for him without anyone being the wiser."

Dick's search didn't stop there. Sam Marx related that, soon after Ira Gershwin returned to the Broadway stage with *Lady in the Dark*, "Ira told me about Dick calling him up to have dinner. Dick talked about the problems he was having with Larry and sort of hinted that he had to make a change. He never asked point blank but sort of talked around it." The day after the meeting, Gershwin received a call from Max Dreyfus, who asked him if he'd enjoyed the evening with Dick, and wondered if Ira would care to think about working with Rodgers on some future musical project. It would have been an interesting alliance, but Ira said no.

Larry, too, was adrift, but in a different way. He spent a lot of his time drinking moodily into the small hours at the Lambs Club, the show business rendezvous on West 44th Street near the Algonquin Hotel. Songwriter and arranger Alec Wilder, who was trying to kick his own hard-liquor habit, remembered going in there one night and asking for a particular brand of beer. When the barman told him they had none, Larry Hart—whom Wilder had never met—came over to the bar and had them send out for a case. "He wasn't showboating," Wilder said. "It was just he desperately needed to be needed, to have someone he could be generous to."

Alan Jay Lerner, recently out of college and with a career of writing some of the best lyrics ever created for the Broadway stage still some years ahead, also met Larry around this time at the Lambs Club. It was Hart, Lerner said, who first encouraged him to believe he might have a future as a lyric writer. "He was kind, endearing, sad, infuriating, and funny," he wrote, "but, at the time that I knew him, in a devastating state of emotional disarray. We became good friends, not because he found me particularly fascinating but because he was so terrifyingly lonely and I worshiped him so that I made myself available to join him at any hour of the day or night, usually for gin rummy which I played badly because I was not interested, and he played badly because he was usually drunk."

On another occasion, Larry turned to him and said, "I've got a lot of talent, kid. I probably could have been a genius. But I just don't care."

Pal Joey, transferred to the Shubert Theatre September 1, was still running, although its cast had changed: Jack Durant and June Havoc had been lured away to Hollywood, and the U.S. Navy had taken Gene Kelly, who was replaced by Georgie Tapps, who had had a small part in *I'd Rather Be Right*. He was not Larry Hart's favorite performer. Jean Casto recalled seeing Larry standing toe to toe with George Abbott, shaking his finger somewhere in the region of Abbott's navel, screaming, "How could you do this to the show with this terrible man, this Georgie Tapps? How could you *do* this to me?" And Abbott shrugged and said, "We could afford him."

This apparently determined Larry Hart to conduct his own guerrilla warfare. Larry Adler recalled going to the Shubert to buy a ticket and meeting Hart outside: "He said, 'You're not going in, are you?' "And I said, I was, and he said, 'Please don't,' and I said, 'What's the matter, this is my only chance to see the show, Larry, and I want to,' and he said, 'Please don't, I don't want you to see it with that fellow in it,' and he looked as if he were about to cry. He was unshaven, and he was quite drunk. So out of respect for Larry . . . I didn't see it until many years later, as a revival."

The war in Europe still dominated the news. In August, President Roosevelt had met with British Prime Minister Winston Churchill at Placentia Bay in Newfoundland to formulate the Atlantic Charter, which firmly placed the United States alongside Britain and against the forces of the Axis—Germany, Italy, and Japan. With German forces outside the gates of Moscow and Leningrad, Hitler and Mussolini met in conference in East Prussia. The first U.S. Liberty Ship was launched in Baltimore. SS troops massacred thirty-four thousand Jews at Babi Yar. On October 9, the President asked Congress to permit the arming of merchant ships. No one doubted now that war was near.

A couple of weeks after *Best Foot Forward* began what would be a ten-month stay at the Barrymore Theatre in New York, it was announced that Rodgers and Hart would be writing a theme song for a WOR radio show hosted by pundit Clifton Fadiman "to present in terms of radio what we are defending in America and the need for defending it." Whether, as seems likeliest, this was "Keep 'Em Rolling" (later used in a 1942 Universal short with the same title starring Metropolitan Opera star Jan Peerce), or some other lost song, cannot be established.

On Broadway, Herbert Fields and his sister Dorothy brought in another smash hit with Cole Porter songs. *Let's Face It* featured Eve Arden, Mary Jane Walsh, Edith Meiser, and Nanette Fabray. Although hardly anything in the score lasted, the show made a major star out of Danny Kaye, who in turn made a major item out of Porter's "Let's Not Talk About Love." The only other show to appear was something George Jessel had written in collaboration with Harry Ruby and Bert Kalmar called *High Kickers*, a look back to the days of vaudeville which starred Jessel and his old pal Sophie Tucker, playing herself. Jessel was still as sharp-tongued as ever. An actor friend told him about a performance he had given. "I had the audience glued to their seats!" he boasted. "How clever of you to think of it," murmured Georgie.

At the beginning of November, Larry Hart, Dick and Dorothy Rodgers, Jane Froman, and André Kostelanetz agreed to make a short "concert" tour of Army camps near Toronto. Jane Froman would sing, Kostelanetz would play, Dick and Larry would talk and perform some of their songs. In the evening they would feature in a local radio program. On the night of their departure, Larry headed for the club car of the train and spent the whole journey drinking himself stupid.

Next morning, there was a reception in their Toronto hotel. Before he even reached the room in which it was being held, Larry was violently, publicly sick. When they him got back upstairs Dorothy Rodgers confronted Larry and demanded he turn over all his liquor to her. Shocked—it was the only time in all the years they had known each other that Dorothy revealed she knew about his drinking—he handed over the bottles. As she was leaving, he meekly asked her when it would be okay for him to have a drink, and she told him as soon as that night's show was over.

Larry got through the day—although what torments of withdrawal he suffered can only be guessed—and even managed to do the tour of the camps and attend the city hall reception. But by the time they did the radio broadcast that evening he was in bad shape, and during the Rodgers and Hart segment—they read a war savings appeal—he was shaking so badly that he was hardly able to hold the pages of the script. He got through it somehow, but as he came offstage, he grabbed the tumbler of whiskey which, true to her promise, Dorothy Rodgers had waiting for him, and drank it in down like water.

As winter drew in, Rodgers found at last the vehicle he had been looking for. Back in 1932, a gangling, boyish actress named Katharine Hepburn had taken Broadway by storm in a Julian Thompson play called *The Warrior's Husband*. Set in Asia Minor in mythological times, it was a costume romp in which the sexes are reversed, the effeminate men staying home to do the chores while the Amazon women hunt and fight. This unnatural state of affairs is only righted when their island is invaded by a band of Greeks seeking the girdle of Diana, the source of the women's prowess. Hepburn had played Antiope; Irby Marshal was Hippolyta, Queen of the Amazons; and Romney Brent portrayed her effeminate husband Sapiens.

Apart from an athletic entrance that had her leaping down a narrow stairway with a stag over her shoulder and pitching it at the feet of Hippolyta, all Hepburn mentions about the play in her memoirs is that she started at $150 a week; after about six months (she recalled—actually the show ran less than three) her salary was cut to $75 a week, exactly what she was paying her maid.

The play had been filmed in 1933 by 20th Century–Fox, with Elissa Landi as Antiope, David Manners as Theseus, and Marjorie Rambeaux as Hippolyta; on November 24, it was reported that Rodgers and Hart were dickering with the studio for the rights. Rodgers said the play came to him via the author's agent, Audrey Woods. He liked it; Hart, "who had been going through an extended depression, perked up noticeably when I told him about it. In fact we decided to write the adaptation ourselves."

Rodgers also decided that he would use the experience he had gained as co-producer of *Best Foot Forward* to produce the new musical; just to be on the safe side, he teamed up with their old friend Dwight Deere Wiman. To Dick's chagrin he discovered that Wiman was prone—in his own clubby, socializing way—to disappear as often as Larry, leaving Dick to round up the "angels" needed to finance the show. One of these was a wealthy cigar manufacturer named Howard Cullman; another, more surprising, was Richard Kollmar, the former leading man of *Knickerbocker Holiday* and *Too Many Girls*. He offered Rodgers a thirty-five-thousand-dollar investment in return for associate producer status. Done.

Only one man was ever considered as director, in spite of a highly visible nervous breakdown that had put him out of circulation for a year: Joshua Logan. It was an enormous show of faith, and Logan knew it. "The contracts were signed and several meetings were held," he wrote. "Even Larry, who had been stepping up his disappearing acts lately, appeared at one. My enthusiasm for the show was growing when I received a letter from the War Department. 'Greetings,' it said."

36

Nobody's Hart

On December 7, 1941, that "day that will live in infamy," as President Roosevelt termed it, Japanese bombers attacked Pearl Harbor; America immediately declared war. Three days later, the other members of the Axis, Germany and Italy, declared war on the United States, and the following day Japanese forces occupied the island of Guam. On Monday, December 22, the Selective Service Act was introduced, requiring all men between the ages of eighteen and sixty-four to register, and all between twenty and forty-four to hold themselves in readiness for conscription. The same day, Wake Island fell. It was into this un-Christmas-like atmosphere that Eddie Cantor's first Broadway show since *Whoopee* opened on Christmas Day at the Hollywood, the first "wartime" musical.

Banjo Eyes, an adaptation of *Three Men On a Horse*, featured songs by Vernon Duke that were serviceable rather than sensational; the hit of the evening was an interpolation that became one of Cantor's trademark numbers, "We're Having A Baby (My Baby And Me.)" The only other new show that month was a revue, also with Vernon Duke songs, *The Lady Comes Across*, directed by Romney Brent (you have to admire his endurance, if nothing else) and choreographed by George Balanchine, which had originally starred Ray Bolger and Jessie Matthews. Both quit the show before it reached New York, where it lasted only a humiliating three performances.

One door closes, another door opens. All along, Rodgers and Hart had wanted only one person to play Sapiens, the soppy King-hero in their adaptation of *The Warrior's Husband*, and here was Ray Bolger falling into their laps. For Hippolyta they chose Benay Venuta, a lusty-voiced young singer who had understudied and later replaced Ethel Merman in *Anything Goes*; for Antiope, Constance Moore, a sweet-singing graduate of Universal, Fox, and Paramount programmers making her Broadway debut.

By report, the book took Rodgers and Hart only three or four weeks to write; one detects in its cavalier approach to the classics much more of Hart than of Rodgers. In this incarnation, Homer is portrayed as a pompously vain author who gets miffed when a herald can't remember his triple rhymes; Hercules, who "enslaved himself because he misbehaved himself," is a bone-head who got his reputation by accident; Theseus is forever "on the make." As with the Greeks, so with the Amazons: when Antiope catches herself using the word "sweating," she apologizes to the men and changes it to the more genteel "perspiring." Later, she is chided for "swearing like a long-shorewoman." When Hippolyta announces she is off to the war, Sapiens sulks. "I never dreamed I'd be a war groom," he complains. "Now I won't even get to visit Nigerian Falls."

Under the title *All's Fair*, the show was scheduled to begin a month of rehearsals in the first week of April 1942, at the St. James Theatre. Dwight Wiman interceded with the U.S. Army, who deferred the immediate induc-tion of director Joshua Logan and scenic designer Jo Mielziner. Bob Alton was again choreographer, Irene Sharaff the costume designer. Songwriter Johnny ("Body And Soul") Green was signed as musical director. Green was a bit too free with his praise for Larry's taste. "Look," he muttered to Joshua Logan, as Green gushed over someone's performance, "the green is corn."

Everything about the show was looking good—except Larry, who was on another binge. He disappeared again. Vivienne Segal, touring with *Pal Joey*, was astonished when he turned up in Cleveland, "accompanied by this guy, this awful man. You just had to look at him," she said, "to know he was a pimp. And he was going to go on the street and find a tart for Larry. A male tart." She telephoned Doc Bender, who was still her agent. "He came on the next plane," Vivienne said, "and took Larry back to New York." Whether Bender was looking out for Larry or protecting his investment we shall never know; probably a little of both.

Just a few days after this, Larry's doctor, Harold Hyman, went over to the Hart apartment with Dick Rodgers, where they found Larry lying in a semi-coma on his bed. They got him across to Doctors' Hospital at 87th Street and East End Avenue, not inappropriately in the heart of what was then New York's "Little Bohemia." Doctors' Hospital was not your run-of-the-mill medical establishment; contributions from 180 of the city's richest inhabi-tants had financed its construction. Its elegant furnishings were a match for those of any hotel on Park Avenue; there were rooms which could be rented by the day by relatives of patients who wished to remain close to them.

Dr. Hyman signed Larry in for a month's drying out; taking care of business, Dick rented a room adjacent to Larry's and had them send up the piano that had been brought in when Cole Porter was a patient. And there, willy-nilly, he made his partner work on the score; it took about a month to write. In the circumstances, for what they produced to have been any good at all would have been remarkable. In fact, most of the thirteen songs that comprised the original score were considerably better than adequate, and included two gems: "Wait Till You See Her" and the affecting "Nobody's Heart."

There are lines in the latter song between which one might read unutterable sadness. It was as if Larry Hart had realized at long, long last that even his love for Dick had died, and there was nothing left to live for. As soon as he got out of the hospital he dived straight back into the bottle.

"Dick was forced to cover a lot for Larry, who kept on disappearing," Josh Logan wrote. "Once when I asked Dick where Larry was, Dick shrugged. 'God knows. I've asked his most disreputable friends, and even they're worried.'"

Which brings us to the unanswered, perhaps the unanswerable, question. Just where *did* Larry Hart disappear to? Those closest of all to him never so much as hinted at an answer and appear to have preferred not to know. Yet he must have gone somewhere. Turkish baths, of course, sometimes; but not for a week at a time. A year earlier, a perfectly decent room with private bath, shower, radio, and ice water at the Hotel Piccadilly on 45th Street cost only $2.50 for a single, $3.50 for a double, but finding vacant rooms in wartime New York was no longer easy. Which leaves only a few other possibilities. One is that Larry Hart had a secret hideaway somewhere in the city, a place he told no one about and to which he went to live what Josh Logan called his "secret after-dark life," but this seems a pretty far-fetched proposition. What seems likelier is that he drank all the way down the bottle until he wound up in a flophouse someplace and then, when he climbed back out of his drunken stupor, navigated back to his own apartment. A third possibility is that he stayed somewhere where nobody knew who he was.

This last idea is not as outlandish as it might seem. Polly Adler's "house," for example, elegantly furnished with French reproduction furniture, four-poster beds, and antique mirrors, its rooms candlelit and perfumed, catered to the kind of clientele that could pay upwards of a hundred dollars a night for its pleasures—and its privacy. That is not to suggest, of course, that Larry made his home at Polly's; but there were places like it all over New York—high class and low, providing male or female company—where someone could "disappear" for as long as he liked, do whatever he wanted, and know no one could find him.

"Everyone had his own juicy theory of what Larry did on his midnight prowlings," Joshua Logan said, "but does it really matter?" He is right, of course: it doesn't matter at all. Any half-baked student of psychology could make a pretty accurate guess at Hart's sexual preferences—and hang-ups—from the things he wrote and the things he said. What matters is what they

did to him—or what he did to himself—and no one in the world could have
prevented that from happening.

═══◈═══

When Larry finally did turn up, he was in poor shape. Benay Venuta remembered him at rehearsals, still suffering from the aftermath of pneumonia, eyes
watering, dabbing his runny nose with a handkerchief and looking pretty
wretched. Larry loved her brash singing style. "He told me, 'I'll write a big
number for you, babe, you'll stop the show with it.' And he did. He wrote
'Ev'rything I've Got' while we were in rehearsal, and it turned out to be one
of the smash hits of the show." Poorly though he was, Larry still took impish
delight in following around the lovelorn Johnny Green—who was having a
passionate affair with Constance Moore—and teasing him with a scatological
version of the song.

All's Fair moved to the Shubert Theatre in Boston on Sunday, May 10;
they opened the following night. The show was still being reshaped. Several
new songs were added during the three-week tryout, including "For Jupiter
And Greece," "Life With Father," and "Now That I've Got My Strength,"
which replaced "Life Was Monotonous." Two others were jettisoned, and
an important cast change was made when Richard Ainley, playing Theseus,
was replaced by Ronald Graham, another casualty of *The Lady Comes Across*.
One other charming bit of business was added, a dance routine built around
"Life With Father," in which Ray Bolger was partnered by his "mother,"
sixty-year-old Bertha Belmore; she had played the same role in the original
production of *The Warrior's Husband*.

Now called *By Jupiter*, the show opened at the New York Shubert on
Wednesday, June 3, and was an instant smash. It started the season and also,
as someone later remarked, probably ended an era. The only other major
musical to come in during the year would be Irving Berlin's *This Is the Army*.
Reminiscing about *By Jupiter*, Joshua Logan said:

> The thing I remember most about that show was the last beautiful ballad that
> Larry ever wrote in his life, and that was "Nobody's Heart." Outside of that, I'm
> not sure that Larry had his whole heart in the show. Oh, he wrote "Gateway of the
> Temple of Minerva" to get it out, it was usable. And another one, "Careless Rhap-
> sody," which didn't sound like Larry Hart at all. "Life with Father" came in at the
> last minute. And then there was "Wait Till You See Her."

It was one of the prettiest waltzes Rodgers ever wrote, and Larry's lyric,
with its unconventional rhyming scheme, had a lovely, matching tenderness.
They all loved it, Logan especially:

That of course is one of the tragedies to me of my musical career. It was written very early on, and it was a song that we were just mad about, everything about it. But every time we tried to stage it, it didn't work in the story. So finally we tried it with a whole new set of dances and costumes, and Bob Alton worked on it, and we put it in and we opened on Broadway, and we knew it wasn't any good. And we took it out, the night after the opening. It just didn't fit the plot, the timing was wrong. There were some hit songs in [the show] that I don't think were that good, not Larry Hart's top. It wasn't as good a show as *I Married an Angel*, but it was a bigger hit because of the fact that everybody was coming to New York because of the war. And it ran for two years because of that.

Almost as *By Jupiter* settled in, Dick Rodgers was taking on another under-cover co-producing job, again in partnership with George Abbott. *Beat the Band* had a book by Abbott and George Marion Jr. (who, it's said, had helped Rodgers to complete some of the lyrics for extra songs in *By Jupiter*) that was little more than an update of the plot of *Poor Little Ritz Girl*, with songs by the selfsame Johnny Green. Abbott remembered the show as "the poorest job of producing and directing that I ever did."

> The first mistake was in thinking that it was any good; the second was in the cast-ing [Jack Whiting and Susan Miller]; and the third and biggest was in not aban-doning it when it obviously was a failure . . . After a bad opening in Boston, I telephoned Dick Rodgers and asked him if he would come up and give me a hand. He came, found the mess too deep-seated to be able to offer any constructive sug-gestions, but did stay around to keep us company.

One reason Rodgers was not able to offer any assistance was that he had other things on his mind; they might also account for his having stayed in the pleasant company of George Abbott. For at this particular moment, Dick was going through personal hell, as Joshua Logan remembered. "Just before I went overseas, Dick came to me [in June] and he said, 'I'm going to have to try without Larry, I can't go through this any more, I just haven't got it in me.' And I said to him, 'Have you thought of anybody to take his place, who will you get?' And he said, 'What would you think if I got together with Oscar Hammerstein?' and I said, 'Oh, my God, it would be marvelous.'"

What had happened was that around the time *By Jupiter* was trying out in Boston, Oscar Hammerstein was working on a couple of projects, one an adaptation of Georges Bizet's opera *Carmen*, for which he would write a new book and lyrics. Don José would become Joe, a black soldier at a camp in South Carolina; Escamillo, the bullfighter, heavyweight champ "Husky" Miller; and Carmen, a worker in a parachute factory, Carmen Jones.

The second project was a musical adaptation of a Lynn Riggs play, *Green Grow the Lilacs*, which had been produced by the Theatre Guild in 1931 with Franchot Tone in the lead. While he was in Hollywood in May, Hammerstein had talked it over with Jerome Kern. Earlier in the year he and Kern had been discussing an adaptation of the Edna Ferber novel *Saratoga Trunk*, which Billy Rose had grandiosely announced he would produce with their score, but the musicalization had fallen through yet again. Kern was not enthusiastic about the Lynn Riggs play, reminding Hammerstein it had run only sixty-four performances; turning a flop into a hit musical was really working against the odds.

In the face of Kern's lack of interest, the Theatre Guild decided to look elsewhere. Theresa Helburn and Lawrence Langner, neighbors of Dick Rodgers in Connecticut, asked him to read *Green Grow the Lilacs* with a view to doing it as a Rodgers and Hart show. Rodgers read the play and "saw" it at once. He called Larry, and they agreed to meet at the office of Chappell & Co., their music publishers, presided over by the elderly Max Dreyfus.

Rodgers was waiting in the boardroom when Hart arrived. Larry was haggard and pale and obviously had not had a good night's sleep in weeks. Dick knew he could no longer avoid telling Larry what was on his mind. "I began by telling him I wanted to get started on the new show right away but that he was obviously in no condition to work," he said. Larry wearily shook his head. Writing *By Jupiter* had exhausted him. He didn't want to plunge immediately into writing another show; he was going down to Taxco for a long, long holiday. It wasn't just he was too exhausted to work, he really didn't think the Lynn Riggs play could ever become their kind of musical.

When Rodgers couldn't talk him out of it, he knew "things would go from bad, which they were, to worse, which none of us could afford." He decided to make a stand. If Larry would enter a sanitarium, he would have himself admitted along with him, and they could work there, together. He told Larry the only way he was ever going to lick his problem was if he got off the street.

Larry wouldn't listen. He wanted to, he *had* to go to Mexico.

Faced with what he saw as his partner's intransigence, Rodgers got angry. He insisted he wanted to go ahead with *Green Grow the Lilacs* and added that if Larry wouldn't do it, he was thinking of asking Oscar Hammerstein to work with him. That Larry replied without anger is some indication of how low his self-esteem had sunk.

"Sure," he said softly, "you ought to get Hammerstein as your collaborator. I don't know why you've put up with me all these years."

He got up to leave, stopping only to say he thought Dick was making a mistake, that *Green Grow the Lilacs* couldn't be turned into a good musical. Then he was gone. Rodgers claimed that after Larry left he sat down and cried like a baby; he knew a long, unique, and wonderful partnership had just walked out the door.

37

To Keep My Love Alive

odgers's prediction proved entirely correct: when Larry Hart came back from Mexico in the late summer, he had to be carried off the train on a stretcher. Once again he was checked in to Doctors' Hospital. He called Nanette Guilford and asked her to come see him. "We had little in common anymore," she said, "but of course, I went. I think Larry was having his second or third bout with pneumonia. Dick called him while I was there. Larry picked up the phone, and I heard him say, 'No, Dick, no, no, no, no.' And this kept on. 'I don't see it,' he said, 'I just don't see it.' They were talking about *Green Grow the Lilacs*, and I said, 'Why don't you do it if he feels that strongly?' And he said, 'No, Baby, I don't see it, I just don't see it.'"

Clearly Rodgers had still not entirely given up hope that Hart could be persuaded to participate in the new show; but Larry was adamant. Besides, he had lots of plans, lots of plans. Toward the end of the year an item appeared reporting that Rodgers and Hart had dropped a Mexican musical called *Muchacha* they were working on; whose imagination this may have been a figment of, we can only guess. Larry was meanwhile noisily discussing several new projects.

One of these, brought to him by author-playwright Paul Gallico, was a story called *Miss Underground*, for which Gallico proposed they enroll the talents of refugee Emmerich Kalmán, the famous Viennese composer. Doc Bender grandiosely told the world he would produce it at the New York Hippodrome, and Larry liked it enough to set to work; with great energy he produced seventeen or eighteen lyrics. Those which survive (at least eight seem to have been irretrievably lost) easily withstand comparison with his other work. Comedy songs and patter numbers like the shamelessly parodic "Alexander's Blitztime Band" combined German dialect with American slang as cleverly as Larry had woven New World idiom together with Olde Englysshe twenty years earlier in *A Connecticut Yankee*.

A love song, "It Happened In The Dark," appears to be not quite finished; unpolished, anyway. Perhaps nearer Hart's usual standard was the emphatically antiromantic "Do I Love You? (No, I Do Not!)." The interlocutory "Messieurs, Mesdames" has lots of nice little inside jokes and even offers clues to who Larry had in mind to play in the show: Vivienne Segal and Wilbur Evans (who had just appeared in a revival of *The Merry Widow* at Carnegie Hall). "Otto's Patter" (also listed as "The One Who Yells Loudest Is The Captain") was a farcical dialect song in the same vein as a then-current, and hugely popular, anti-Hitler Bronx-cheer comedy song, "Der Führer's Face," from the cartoon *Donald Duck in Nutzi Land*. Trying to imagine these lyrics actually being sung on a Broadway stage in 1942 requires a considerable feat of imagination; to modern eyes they are perilously close to parody of the "Springtime For Hitler" sequences of Mel Brooks's movie *The Producers*. Joshua Logan always said Dick was a superlative editor of Larry's work. Perhaps that was what was missing here.

Whatever one feels about the content, however, it is difficult to reconcile work like this with the stories of Larry Hart being continually, hopelessly plastered, yet Emmerich Kalmán's daughter, Yvonne, indicated that the whole experience was a miserable one for her father because Larry so often came to their meetings falling-down drunk. She thought that might have been the reason the show was never produced.

Not entirely so: "The story was very interesting," George Balanchine wrote, "and I thought the score was beautiful." What it needed, and never seems to have had, was someone who could go out and successfully raise the money to put it on; clearly the almost universally detested Doc Bender was not that somebody. It was always difficult to find backers for a musical show, and in wartime doubly so, as even Richard Rodgers and Oscar Hammerstein had discovered.

Auditions had begun for *Green Grow the Lilacs*, which was budgeted at eighty-three thousand dollars, of which the Theatre Guild, teetering on the brink of insolvency, had not a penny.

"There wasn't anything enormously attractive about the idea," Rodgers said. "And then on top of that, Oscar and I were working together for the first time. Who knew whether this combination would work or not? They were sure in my case it wouldn't, because I'd had all those years with Hart and suddenly came the split. They knew that I had to fail."

Practically everyone Rodgers and Hammerstein knew told them the same thing Larry had told them: they were crazy to do "a cowboy musical." The big producers wouldn't touch it: apart from Rodgers, what did it have? Director Rouben Mamoulian had not done a Broadway show for seven years. The choreographer, Cecil B. De Mille's niece Agnes, might be all right for the Ballets Russes de Monte Carlo, but Broadway? And Hammerstein hadn't written anything anyone wanted to see since *Music in the Air* in 1932. Do another show with Larry Hart, Rodgers was urged. Give us another *By Jupiter*, but not, for God's sake, a musical about which of two cowboys gets to take a girl to a box social.

Reactions like these forced Rodgers and Hammerstein into what must have been the most humiliating experience of their theatrical careers. Here were two men with half a century of hits behind them, a formidable record of successes on both stage and screen, reduced to going cap in hand around the "penthouse circuit" to raise money. Theresa Helburn went to MGM, offering them 50 percent of the show for a seventy-five-thousand-dollar investment; they turned her down. Through producer Max Gordon, the Guild approached Harry Cohn, the much-hated head of Columbia Pictures. Cohn agreed to back the show, but his board of directors vetoed the proposition. Cohn anyway put in fifteen thousand dollars of his own money, which persuaded Max Gordon to do likewise, and *Away We Go!* at last became a viable project.

On New Year's Eve 1942, with the Rodgers and Hammerstein show in rehearsal, Larry and his mother went down to Florida accompanied by Teddy's wife Dorothy. Larry's favorite hotel, the Roney Plaza, had been requisitioned by the military, so they stayed at a small place. When the ladies on the porch took it for granted that Dorothy was Larry's wife, Frieda teasingly told him about it. Even with everything her son said—and did—telling her it would never happen, she seems somehow to have still been clinging to the hope he would straighten out his life and settle down, as she had done all those years ago in Hollywood when she told Sam Marx's wife Marie, "I wish my Lorry would marry an angel."

While they were in Florida, Dorothy Hart said, they talked about the future, and she realized Larry didn't know what to do with his life or himself. There had been no fight with Rodgers, but he didn't want to talk with him. The break didn't mean anything, she said; he *wanted* the break. He was interested in Teddy's career, but nothing else. A few days later, however, Larry's moodiness and heavy drinking brought the vacation to a premature end. There was a scene in a restaurant when they didn't have something on the menu he wanted. He was even becoming difficult and almost abusive to his mother, so Frieda—herself not in the best of health—decided to return to New York.

January 1943 was the month of Casablanca, when Roosevelt and Churchill hammered out the policy of "unconditional surrender" and General Dwight

David Eisenhower was appointed Supreme Commander; and of *Casablanca*, the imperishable movie melodrama starring Humphrey Bogart and Ingrid Bergman which had opened in December and made an all-time hit of Herman Hupfeld's "As Time Goes By," a forgotten song from a 1931 Shubert production, *Everybody's Welcome*. January 1943 was also the month in which the siege of Leningrad was lifted, and the German army under Field Marshal von Paulus surrendered at Stalingrad. The Warsaw ghetto was razed, and the Battle of the Coral Sea was won. The RAF launched its first daylight attack on Berlin. Guadalcanal was liberated. For the first time in the war, the possibility of victory could be discerned.

Broadway, its lights dimmed in wartime precaution against air raids that never came, was about to experience a spring season unlike any other that had gone before. It began conventionally enough when producer Mike Todd put Herbert and Dorothy Fields together with Cole Porter and Ethel Merman and came up with *Something for the Boys*, which opened at the Alvin on January 7. The legitimate stage offered a new Sidney Kingsley play, *The Patriots*; Zachary Scott and Jan Sterling in *This Rock*, a Pollyanna-ish play about England during the Blitz; and an adaptation of a Russian war story, *Counterattack*, with Richard Basehart and Karl Malden. Helen Hayes had a popular success as authoress Harriet Beecher Stowe in *Harriet*, directed by Elia Kazan. Author Irwin Shaw's *Sons and Soldiers*, directed by Max Reinhardt and starring Geraldine Fitzgerald and Gregory Peck, lasted just twenty-two performances.

On March 17, George Abbott's production of the F. Hugh Herbert comedy *Kiss and Tell*, featuring radio actor Richard Widmark, cover girl Joan Caulfield from *Beat the Band*, and the unknown Kirk Douglas, was a smash hit at the Biltmore. Then on the last day of March, Rodgers and Hammerstein brought in their new show. Only the second musical of the year, advance word on *Oklahoma!*, as it now was, had been almost universally bad. Mike Todd, who'd seen it out of town, had damned it with a phrase that stuck when Winchell put it in his column: no gags, no girls, no chance. Opening night was not a sellout.

Larry Hart attended the premiere with his mother. He, too, had seen the show in New Haven; doubtless he thought, along with everyone else, that it was a hopeless mess and did not have a chance. He cannot have been prepared, any more than anyone else in the auditorium was, for the enormous changes that had been wrought during the tryouts in Boston. The curtain went up on a bright, sunlit scene, an old woman churning butter in a farmyard. Offstage, the unaccompanied baritone voice of Alfred Drake sang of the bright golden haze on the meadow, and the corn as high as an elephant's eye. The audience's response was immediate. "It was like the light from a thousand lanterns," Oscar Hammerstein said. "You could *feel* the glow. It was that bright."

"Larry sat in a box applauding, howling with laughter, yelling bravos!" Joshua Logan remembered. "Oh, my God, it must have been painful. He had taken his mother and he had to sit there with her, this sensitive, sensitive little man, seeing a revolutionary development in the theatre, brought about by his partner but without his own participation."

Even so, Larry Hart knew he had seen something special that night at the St. James Theatre. After the show everyone went over to Sardi's to have supper and wait for the reviews. Suddenly Dick saw Larry jostling his way through the crowd, grinning from ear to ear. "You have at least another *Blossom Time*," Larry said as he threw his arms around his partner. "This show of yours will run forever."

What that cost him to say, no one will ever know; but there was never any doubt in Rodgers's mind that he meant it.

For Dick Rodgers, and for Oscar Hammerstein too, *Oklahoma!* was a rebirth. But their accomplishing what he had been reaching toward all his life and never quite achieved was, and could only be, a blow to the heart for Larry. Just how hurtful a blow is revealed in a story told by Alan Jay Lerner. He, Fritz Loewe, and Larry were together one evening when there was an air-raid alert blackout. Loewe switched on the radio; it was playing one of the tunes from *Oklahoma!* In the dark, Larry puffed angrily on his cigar. Fritz tried another station. Same thing. Even more furious puffing. Fritz tried again, with the same result. Hart's cigar glowed brighter and brighter. Eventually, they found a station playing some other tune, and the glow of the cigar subsided. When the lights came on again, Larry continued their conversation as though nothing had happened, but as Lerner later said, they knew what they had witnessed was the sight of a man made all too painfully aware of his own obsolescence.

Hard on the heels of *Oklahoma!* came an even worse blow. Early in the morning of Saturday, April 17, Frieda Hart woke up in unbearable pain. She was rushed to Doctors' Hospital for an operation, but it was not successful. A second one was scheduled, but the hospital decided against proceeding; she was too frail. On one visit, perhaps his last, Larry tearfully tried to convince his mother she would soon be home. He thought he was reassuring her, but it was clear Frieda was under no illusions at all. She hung on for a week, tended by Dr. Edward Carey of the hospital, and died at 4:50 a.m. on Easter Sunday, April 25.

Larry could not face his mother's funeral, which took place two days later. He disappeared as the family left the apartment; services had to be delayed at the Julius Steinfeld mortuary on 171 West 85th Street until he was located in a nearby bar. The cortège drove out to Queens, and Frieda was buried alongside her Max in the Mount Zion Cemetery plot, one of many his lodge had distributed over forty years earlier; a joint memorial was erected later. Nearby in the same plot were Max's mother Taubchen, brother Harry, and sisters Hattie Strauss and Sophie Herman. Among the mourners was old Uncle Willie Hart, ninety-two. The half-drunk Larry was unable to restrain his macabre sense of humor. "Look at Uncle Willie," he whispered to his brother. "There hardly seems any point in his going home."

After the funeral when she and her husband returned to Larry's apartment, Dorothy Hart found Larry drunk and incoherent. There were half a dozen of his so-called friends making themselves very much at home, as though at long last the house was theirs. Stolidly ignoring the disorder, Big Mary went sullenly about her duties, but not without first ostentatiously removing the food from the refrigerator and taking it up to her room as she had always done when Mister Lorry brought what she called "the riffraff" home.

Larry Hart had lost everything he loved. First Dick, now Frieda. His life was pointless. He was drunk much of the time. He made rash promises to collaborate with people who fortunately did not take him seriously, like Lester Cowan's wife, songwriter Anne "Willow, Weep For Me" Ronell. Acting as Larry's agent, Doc Bender—still the "half-assed entrepreneur," as Milton Pascal described him—tried to set up a collaboration between Larry and Jerome Kern, but Larry wasn't interested. Pascal recalled being at the apartment, playing pinochle with Larry and Phil Charig, when a wire arrived from Hollywood. "It said, 'When can you come out here and start a picture with Jerome Kern?' And I said, 'Not bad, from Rodgers to Kern.' Larry just looked at the telegram, then tore it up and said, 'Deal the cards.'"

In spite of this the story persists. According to Dorothy Hart, at the time of his death Larry had a thirty-eight-thousand-dollar contract to write a movie *Ziegfeld Follies*. This raises an unanswerable question: with whom? Kern had no contract with MGM; and there is no indication that producer Arthur Freed ever planned to include work by either of them in the production, which eventually became *Ziegfeld Follies of 1944* and featured songs by half a dozen different writers. The only movie Kern was to become involved in—his last—was a Universal picture which began life as *Caroline* and eventually became *Can't Help Singing*, the first of the movie responses touched off by *Oklahoma!* However, Kern didn't start on it until well after Larry was dead; his lyricist on the picture, which premiered on Christmas Day 1944, was Yip Harburg.

A little while after Frieda's death, Larry gave up his apartment at the Ardsley and moved to a smaller but nonetheless quite luxurious penthouse at the Delmonico Hotel on Park Avenue, with spectacular views north and south over the city. Soon after this, there was an unhappy ending to the huge success of *By Jupiter*, which had become the longest running of all the Rodgers and Hart shows.

Ray Bolger, for whom Rodgers and Hart had earlier in the year provided a new song, "The Girl I Love To Leave Behind," for a cameo in a United Artists movie, *Stage Door Canteen*, abruptly announced his withdrawal from the cast. His doctors had ordered him to rest; he was too exhausted even to take on the role of Jack Donahue in the film biography of Marilyn Miller he'd signed to do in Hollywood. (It was finally made in 1949 as *Look for the Silver Lining*, when he did appear in it, opposite June Haver.)

Without Bolger in the lead, producers Wiman and Rodgers had no choice but to shut down an extremely profitable show. When press stories appeared to the effect that there was nothing wrong with the dancer but that he was just loafing, Rodgers wrote a bitter verse which appeared in the *New York Daily News* on June 16:

> *By Jupiter* is shuttered so Bolger is a loafer
> I read it in your columns and it's stuff you shouldn't go fer.
> Let's read the line correctly
> Here's the way it should be uttered
> Bolger is a loafer, so *By Jupiter* is shuttered.
>
> P.S. Gross last week $21,655. Performers out of work nearly one hundred.
> Love, Richard Rodgers.

He had cause to regret his petulance some weeks later when it was revealed that Bolger had left the show in order to be free to fly out to the South Pacific war zone to entertain the troops.

———※———

During the summer, with his new partner engaged in bringing *Carmen Jones* to Broadway under the auspices of producer Billy Rose, Dick Rodgers accepted an invitation from an old friend, Jesse Lasky Jr., to see an advance print of a new Warner Bros. movie he had produced. Based on a Harold M. Sherman play, with a screenplay by Alan Le May, *The Adventures of Mark Twain* starred Fredric March and Alexis Smith. When the lights went up after the screening, sitting in front of Dick was none other than lawyer Charles Tressler Lark. Bells rang in Rodgers's head. He had been racking his brains to think of a project he might interest Larry in, but had at the same time been concerned that whatever he might come up with would be too much for Larry to cope with. What about a revival of *A Connecticut Yankee*?

Shortly thereafter he contacted Herbert Fields; together they mapped out a strategy for bringing the play—and Larry Hart—back to Broadway. First, the existing show could easily be updated to the present day; there would be some new dialogue, but it would not require a completely new book. Neither would it need a completely new score. Most important of all, if they could expand the role of Morgan Le Fay to make a juicy singing part for Vivienne Segal, Larry might become enthusiastic enough about the show to do it. And so it proved.

38

※ ※ ※ ※

What Have I Lived For?

For a brief period, Larry Hart appeared to come back to life. Sober most of the time, he seized upon the new show with great zest, working hard—and fast—on the new libretto. "He would come up to stay at our place in Connecticut," Rodgers wrote, "and we'd work regularly at reasonable hours. I don't think he took a drink the entire time. There was no question that he was making a genuine effort to rehabilitate himself and to prove that the team of Rodgers and Hart was still a going concern."

With Vivienne Segal set from the start, Rodgers and Hart found their new Sir Boss in the form of senator's son Dick (actually John Nicholas) Foran, a former band vocalist turned low-budget singing movie cowboy and journeyman actor. Their new Alisande was Julie Warren; Arthur, Robert Chisholm. Evelyn was played by Vera-Ellen, real name Vera-Ellen Westmyr Rohe, a lithe youngster who'd started dancing at age ten, performed as a Radio City Rockette, and most recently sparkled in *By Jupiter*. Rehearsals were scheduled for mid-September.

A copy still exists of Larry's handwritten draft for the second act, full, like the new lyrics, of good jokes and topical allusions. Merlin was informed that his "weather prophecy" had a low Crossley rating (the name of a radio-audience survey); the "in" place to drink was the "Connecticut Yankee Clipper"; heroine Sandy organizes the Camelot International Objectors, a women's protest group that just coincidentally has the same initials as John L. Lewis's mighty steel union.

Taking his cue from a series of forty-five-minute American Theatre Wing productions supervised by Kurt Weill that had been staged at the Todd Shipyard in Brooklyn in June 1942 (which might suggest where Larry spent at least some of his time) under the catch-all title *Lunchtime Follies*, he introduced a radio program for the factory workers of Camelot, "Ye Lunchtime Follies," which even had its own "crooner" singing deliberately, deliciously unadulterated corn to sweet "Elaine," who he loved when they met in the rain. And the girls in the factory would shriek and "swoon" like Frank Sinatra's bobby-soxers.

The song appears to have been cut from the finished score, as was another lyric that has escaped all the collectors who have scoured the archives in search of unpublished gems by Lorenz Hart; "I Won't Sing A Song" was written for Vivienne Segal as Morgan Le Fay to sing to Sir Boss in the first scene of Act Two. No melody seems to have been written, and charming though it is, the song was never used.

They retained the best numbers from the 1927 production, of course: "On A Desert Island With Thee," "I Feel At Home With You," "Thou Swell," and "My Heart Stood Still." The new songs, too, were good: "This Is My Night To Howl," "The Camelot Samba" (capitalizing on a current craze), "You Always Love The Same Girl," and the haunting, touching "Can't You Do A Friend A Favor?," which was written during rehearsals. And brilliantly demonstrating that, no matter what his problems, he had not lost his lyrical touch is the song Larry wrote especially for Vivienne Segal, in which the much-married Queen describes how she "bumped off" her many husbands "To Keep My Love Alive."

Taking her through it at rehearsal, Larry suggested that at the end of each refrain, on the word "alive," she take it a half-tone lower. For a soprano this was, to say the least, unusual. As with each refrain she sank further down the scale, Vivienne balked.

"Larry," she protested, "if I go any lower I'll turn into a man."

"If you do," he retorted darkly, "you'll be the only one in the show."

Larry was reading Fielding's *Tom Jones*, which he thought would make a great show with Teddy Hart and Hope Emerson in the leads and a score by Rodgers and Hart. But once rehearsals were over, he backslid into the alcohol. By the time they took the show up to the Forrest Theatre in Philadelphia for tryouts beginning October 28, he was drinking as heavily as ever.

The reason: Vivienne had turned him down again. "While we were in Philadelphia just before he died, he asked me to marry him," she said. "He was afraid that on account of his drinking 'they' were going to put him away. I said, no, I couldn't do that, I'd always be a good friend of his, but marriage was out of the question. I mean, I never even *kissed* Larry. And I hated cigars!" When Dorothy Hart went up to Philadelphia, where Larry was sharing a suite with Herbert Fields, Herb confided how dismayed he was about Larry's condition. Everyone knew the problem; nobody knew the cure.

When Larry and his sister-in-law ran into Dick Rodgers in the lobby at the hotel, it seemed to her that Larry was fearful, and the contrast between his own pathetic appearance and that of Dick's new partner Oscar Hammerstein seemed to penetrate even Larry's befuddled mind. In retrospect, of course, it is easy to see he ought not to have been in Philadelphia at all, but back in New York in Doctors' Hospital. Within a week he was anyway.

Wednesday, November 17, was a filthy day, bitterly cold, with freezing squalls of rain lashing down the concrete canyons of New York. The war news was suitably grim to match: stalemate in the fighting to recapture Bougainville from the Japanese, a German Panzer counter-attack on Kiev, the Allied advance up the leg of Italy halted by winter snows. Kept secret was the fact that, just a few days earlier, a live torpedo unwittingly launched during an exercise on the escort destroyer *William D. Porter* had narrowly missed the battleship *Iowa*, on board which were not only President Franklin D. Roosevelt but the highest-ranking officers of the U.S. military, en route to the Teheran conference with Churchill and Stalin scheduled for November 28.

Early that Wednesday afternoon Larry Hart called at his brother's apartment at 333 West 57th Street and delivered a dozen tickets for that night's premiere to be distributed to friends. Just the night before, he had telephoned Dorothy Hart to ask her to buy a gold cigarette case for Herb Fields and something special for Vivienne Segal. When he arrived at Teddy's he was already on his way to not being sober, which was not good news; he would be smashed by curtain-up. They knew, and Larry probably knew, too, that Dick Rodgers had given orders to the stage manager to have him removed from the theatre if he became difficult. The signs were they were in for a bad night.

From his apartment at Delmonico's, Larry telephoned Helen Ford, working as a senior hostess and mistress of ceremonies at the Stage Door Canteen, and invited her to the premiere of *A Connecticut Yankee*. She arranged to pick him up at the hotel; they could take a cab together down to the Martin Beck Theatre on 45th Street.

> I got there and went down into the dining room—there were three or four steps—
> and I saw Larry. He was with a young couple, a beautiful young man and an
> actress I recognized vaguely. He was absolutely falling-down drunk, and there was
> no food. It was the most awful thing to see. He introduced me as the most beauti-
> ful singer in America, and went on and on about how he was still the best lyric
> writer in this town. He kept on ordering more drinks until finally we got him into
> a cab and went to the theatre.

At the Martin Beck a small task force had been deployed to waylay Larry before he could get in. Dorothy Hart had instructed Larry's cousin, press agent Billy Friedberg, and a writer friend, Harry Irving, to watch out for him and head him off; if he got past them, Dick Rodgers had two men standing by ready to handle any disturbance Larry might make.

When Helen Ford and Larry Hart arrived at the theatre, the usual first-night crush of people was crowded in the lobby; the diminutive Larry immediately disappeared in the throng. Helen Ford spotted Dorothy Hart, who was telling other friends that no tickets had been set aside; Larry had simply assumed he could turn up and ask for as many as he wanted. The two women went backstage to find Dick Rodgers.

"Oh, my God!" he blurted. "Is Larry here?"

When the curtain went up and the lights went down, Larry appeared in his accustomed place in the gangway at the back of the theatre. His two "minders" stood watchfully nearby. The first act passed without incident; at intermission Larry—his overcoat still in the checkroom—went out for a drink at a nearby bar, ignoring the downpour. He came back soaked—in both senses of the word—and resumed his watch from the rear of the house. He was quiet at first, but as the second act progressed he began talking the lines, reciting the lyrics in an audible undertone that grew more feverish and agitated. When his voice became too loud to be ignored, the two watchdogs grabbed him and bundled him yelling and kicking into the foyer.

Dorothy Hart, who had heard the disturbance, came out and took him to her apartment on West 57th Street in a taxi. After a while she got him settled down on the sofa, and he fell asleep fully dressed. "When I looked in on him, several times during this night," she wrote, "he was perspiring and breathing heavily, but by morning he had left the apartment without telling anyone." And without an overcoat.

According to Alan Jay Lerner, the following evening, a cousin of Larry's—probably Billy Friedberg—rang him and Fritz Loewe to ask them to help find Hart. "Fritz found him, literally sitting on a curb in the pouring rain in a state of drunken paralysis. Fritz put him in a cab and took him to the hotel where he was staying. He made Larry promise he would go right to bed. The following day he was taken to the Doctors' Hospital suffering from pneumonia."

Without amplifying, Dorothy Hart adds that, when this happened, at five-thirty in the afternoon, an actor friend of Teddy's named Seldon Bennett was with Larry. When they got to Doctors' Hospital they were told Larry's condition was "serious, but not critical." Sulfanilomide was administered after a blood transfusion. For the next couple of days they all kept a vigil as treatment continued. Dick and Dorothy Rodgers were there much of the time. Many of Larry's friends came by to see how he was: Max Dreyfus's secretary Irene Gallagher, Milton Pascal, Irving Eisman, Philip Charig, Willie Kron, Helen Ford and her husband, and, of course, Doc Bender.

Irene Gallagher called producer John Golden, who was acquainted with Eleanor Roosevelt. Through her intercession penicillin, not yet available to the general public, was flown in; but it was no use. Later they learned Larry's heart had stopped twice and was restarted after emergency treatment. He lapsed into a coma. "I talked to a nurse up there," George Ford said, "and she told me the last thing Larry said to her was, 'What have I lived for?'"

Shortly after nine o'clock in the evening of Monday, November 22, there was an air-raid alert. As all the lights of New York went out, the last faint flicker of life left Larry Hart's body. In the darkness, a hospital doctor came out. "He's gone," he told them. It was nine-thirty exactly.

A moment later, all the lights came back on again.

Coda

*T*wo days later, on Wednesday, November 24, a funeral service attended by more than three hundred mourners was conducted by Rabbi Nathan Perilman of the Temple Emanu-El in the Universal Funeral Chapel at Lexington Avenue and 52nd Street. Among those present were Teddy Hart and his wife, Dick and Dorothy Rodgers, Dick's father and his brother Mortimer, Max Dreyfus, Moss Hart, Cole Porter, Dorothy Parker, Arthur Schwartz, John O'Hara, Dorothy Fields, Danny Kaye, Emmerich Kalmán, Willie Kron, and Doc Bender. Rites began at 12:30; a simultaneous service, attended by a large congregation that included Oscar Hammerstein, Jerome Kern, and Herb Fields, was conducted in Hollywood.

There was no music; Teddy Hart said afterwards Larry had always maintained funerals should be simple affairs without displays of emotion. Rabbi Perilman read Psalms 19, "To the chief musician," and 121, "A song of degrees," followed by a passage from John Greenleaf Whittier's poem "Snowbound" that ended, "Life is ever Lord of Death, and love can never lose its own."

"Lorenz Hart," Rabbi Perilman said, "was gifted with a fine mind, a warm understanding, and an ability to interpret the moods and hungers of our time, which he translated into poetry and song." He spoke of Larry's human qualities, his benevolent humor, and how his attitude was unchanged by success and his mind touched by genius. He ended with a short prayer and the Twenty-third Psalm.

The service took little more than twenty minutes; the cortege left immediately for Mount Zion, where Larry was interred in a grave facing that of his parents. It is a sad and empty place to spend eternity, a trash incineration plant on one side, the roaring Long Island Expressway on the other.

At the end of the year there was an unsavory squabble over Larry's will, drawn June 17, 1943, and filed for probate a week after his death. There was even a rumor to the effect that Larry had married a nurse on his deathbed, and the family had the marriage annulled. Hart's net estate was valued at $196,971. Of this sum, he left $26,828 to charity. His brother Teddy received $5,000 cash and a life interest in a trust of 70 percent of the residue. $2,500 in cash and succeeding interest in that trust went to Dorothy Hart. A life interest in 30 percent of the residue was left to William Kron of 865 West End Avenue. After agonizing over it for some time—"Can I go on the stand and say my brother was a drunkard?" he asked Milton Pascal, who told him he had no choice—Teddy Hart filed to prevent probate. On November 21, *Variety* reported Teddy had lost a second suit in pursuit of the estate, this time an insurance policy where the beneficiary had been changed to the estate.

Teddy Hart's petition, which was heard December 28, charged that because his brother "lacked mental sobriety for the last few years because he was addicted to alcohol," William Kron had exerted undue influence on Larry. "In the last three years of his life," Teddy told the *New York Times*, "he acted like a man mentally unbalanced, and one who did not know what he was doing and did not understand the nature of his acts."

Kron counter-petitioned that he had had no hand in the will. Dick Rodgers testified that Larry had turned out some of his best work in the last six months of his life. "Of course I knew he drank," he said, "but he always had his mental faculties. All you have to do is look at his record to see the amount of work he did in the last six months and you can be assured his mind was clicking properly." In view of the unassailable evidence to the contrary, this was a remarkable statement. "Dick perjured himself on the stand," Milton Pascal said. "All he wanted was to control the copyrights."

It would have been a squalid end to the story, but for the fact that the story has no end. In the years following Larry Hart's death—and to what was often Rodgers's intense annoyance—the music of Rodgers and Hart became and has gone on growing increasingly popular. The renaissance began five years after Larry's death with Arthur Freed's MGM film biography, *Words and Music,* and has continued to the present day. Even people to whom the names Rodgers and Hart mean nothing know "The Lady Is A Tramp," "My Heart Stood Still," "Where Or When," "My Funny Valentine," and many more of their songs. Few of them either know, or care, that the man who wrote the words virtually changed the craft of lyric writing singlehanded, but it is so.

When Larry Hart first teamed up with Dick Rodgers in 1919, the commercial song lyric was a thing of trite phrase and cliché, of cloying Victorian sentiment, a tired and hackneyed commodity whose only job was to "get the tune over." Hart changed all that, always avoiding the obvious, always aiming for the unexpected phrase that would twang the nerve or touch the heart. From the beginning, and throughout his all-too-short life, Larry Hart took the way people talk, the street argot he heard every day, and wove it into his lyrics: "Ten Cents A Dance," "I Didn't Know What Time It Was," "I Like To Recognize The Tune," "The Most Beautiful Girl In The World," "Try Again Tomorrow," "What's The Use Of Talking," "You Took Advantage Of Me," "A Little Birdie Told Me So"—the list could be much, much longer. Songs had never been written like that before. It is hard to believe now that songs were ever written any other way.

Hart never kept copies of what he had written. As with his attitude to money, any amateur psychologist would diagnose this as a statement of how he felt about it, and those who knew him best agree he viewed his craft as something considerably less than momentous. He once told Ted Fetter, another talented lyricist, that he couldn't believe he made so much money for what he did. "It's completely disproportionate to the amount of work I do to earn it." Later still, he would almost contemptuously dismiss his work as "sophomoric." Yet when he felt so inclined, he could discuss it in terms well above the head of most laypeople, as this 1928 interview shows:

> In the old days musical lines were written in regular measures—meter-iambic strophes; but today the introduction of the offbeat brings musical accents in the most unexpected places, so that rhymes appear not only at the end of the metric line, but at irregular intervals, due to the strange occurrence of accents in jazz music. Moreover the old love song because of the quiet seductive strains of the then popular waltz was usually a quiet exemplification—portrayal—of the innocent amatory music; but today the barbaric quality of jazz dance music demands expressions of love that are much more dynamic and physical.

As with his craft, so with his life. He loved what he was doing and hated it at the same time. Russell Bennett once asked him the source of his inspiration. "Oh, Russ, when I write a lyric, the only time I'm inspired is when I have a pencil in my hand and a piece of paper in front of me. *That's* the inspiration," Larry replied. In other words, he didn't start writing until he started writing; inspiration was likeliest to come while he and Dick were kicking around ideas. Larry had a fine musical ear and an intuitive feel for a good melody. This made it easy for him to work within the framework of Rodgers's musical thinking, while Rodgers appreciated the subtlety of Hart's lyrical demands and was able to accommodate them melodically.

Hart's lyric writing might best be compared with Bix Beiderbecke's cornet playing: effortless and, even when he wasn't really trying, brilliant. Bix was a man who couldn't—didn't know how to—play a bad chorus: Larry was a man who couldn't—didn't know how to—write a bad lyric. P. G. Wodehouse, himself no mean wordsmith, said that what he found most striking about Hart's work, even more than its brilliance, was its consistency:

316 Lorenz Hart – A Poet on Broadway

Larry Hart was always good. If there is a bad lyric of his in existence, I have not come across it. It seems to me he had everything. He could be ultra-sophisticated and simple and sincere. He could handle humor and sentiment. And his rhyming, of course, was impeccable. But the great thing about his work was that, as somebody once wrote, he was the first to make any real assault on the intelligence of the songwriting public. He brought something quite new into a rather tired business.

Henry Myers again:

While his mother lived, Larry wrote, not merely songs, but shows. During, and concurrent with, all the events that I happened to witness, as well as through all his moments away, wherever and however spent, he wrote shows, shows, shows; he thought in terms of shows. They were his unit of measurement and he turned them out *sans cesse;* few lyricists ever rhymed, rhythmed, thought, and felt their way through so many complete entertainments.

Larry Hart brought to lyric writing not just his erudition but his own wit and charm, in a way that seems always spontaneous and effortless. In spite of the fact that they almost invariably originated in one, there is never the smell of a smoke-filled room around his work. To quote Rodgers: "He didn't care where he lived, how much he earned, what the social or financial status of his friends was, or what row he sat in on opening night. He did care tremendously, however, about the turn of a phrase or the mathematical exactness of an interior rhyme."

Equally significant was his contribution to the musical theatre. While Rodgers and Hart were not alone in revolutionizing the way shows were created, they were at the forefront of the change, not only because of their songs but because of Larry Hart's seemingly instinctive feel for what could be done and what should be done. Over and over again comes the picture of him bustling down the aisle of an empty theatre, rubbing his hands, puffing joyously on that damned cigar, and calling out in his high-pitched voice, "Hey, Dick, listen, did anyone ever write a musical about a three-day bicycle race?" Or a hoofer who gets mixed up with the ballet, or a gang of kids putting on a show in a barn, or the War of the Rebellion, or the movies, or . . . ?

"Then what is it," Joshua Logan asked rhetorically, "that has caused people to forget to mention him? Is it because Oscar Hammerstein's career with Dick Rodgers was so much more successful, from a financial point of view? You see, I think Larry Hart was one of the great milestone geniuses in the history of our American musical theatre. In a sense he invented it, he started it. There just weren't that many good musicals around when he came along. He had a color, a tone, a sort of bitter beauty that no one else had or ever will have, the most sensitive, the most touching, almost Chaplinesque ability to get laughs and make you cry at the very same time. And that's why he must be saluted, he must be remembered, he must be honored, and he must be kept alive somehow, because he's something we all need."

No more heartfelt or genuine encomium has ever been spoken about Larry Hart and yet, and yet . . . it is difficult not to imagine him squirming had he been around to hear it. Sensitive, touching, Chaplinesque—these were not words Larry Hart associated with himself. Hip, sardonic, clever, witty, yes. Generous, fun to be with, dynamic, a great guy, a good sport, certainly. The best goddamn lyric writer on Broadway, unquestionably. But color, tone, a sort of bitter beauty? "Ix-nay on all that," one hears him jeer. "What're you trying to do, make this *Camille?*"

Yet the brilliant, tragic, infuriating, lovable, irrepressible, and unutterably sad bundle of contradictions that was Larry Hart was all of these things. So how, then, to best remember him? Perhaps with this story, told by Vivienne Segal, the woman he always loved but never won:

> It was at an audition for the revival of *A Connecticut Yankee*, and we were watching some girls try out. One girl danced and sang, and I thought she was awfully good, but Larry said no.
> "Larry!" I said, surprised. "What's wrong with you? She'd be great. She's beautiful."
> "She is not," he said.
> And I said, "Are you crazy? She's beautiful!"
> "No she is *not*," Larry insisted.
> "All right," I said. "You tell me *your* definition of beautiful."
> And he said, "Talent is beautiful."

APPENDIX I

Lyrics by Hart:
A Show-by-show Listing

Hitherto unpublished lyrics now located are asterisked; the many lyrics still missing are indicated thus: #. Songs which were cut or dropped from their various productions either prior to or following the opening are identified by a (c) placed before the title.

Die Tolle Dolly, 1916
   Music by Walter Kollo
      Meyer, Your Tights Are Tight *
      Hubby Dances On A String #
      Every Man Needs A Wife #
   Music by Hans Kronert
      Ticky, Ticky, Tack *
      Bummel, Bummel, Bummel *
   Music by Arthur Steinke
      The Kiss Lesson #

Take a Chance, Columbia Varsity Show, 1919
   Music by Roy Webb
      The Sandman

Camp songs, Brant Lake Camp, 1919–20
   Music by Milton Thomas [Mickey Tomaschevsky]
      Stop! Stop! I Am The Traffic Cop! #
      Oh, Mr. Postman, Don't Pass Me By!
      B-R-A
      Green and Gray

Old Brant Lake
He Lights Another Mecca
She's Camping at Red Wing
The Green Song
The Gray Song
Brant Lake Blues
I Used to Love Them All #
Tee Ta Tee #
Horicon Hop #
I've Got A Girl In Chestertown

Music by Arthur Schwartz
Down at the Lake *
B-L-C *
I Love To Lie Awake In Bed
[The?] Last Night #

Miscellaneous
Music by Arthur Schwartz
I Know My Girl by Her Perfume #

Music by Richard C. Rodgers (?)
Venus (There's Nothing Between Us) #

A Lonely Romeo, 1919
Music by Richard C. Rodgers
Any Old Place With You

You'd Be Surprised, Akron Club, 1920
Music by Richard C. Rodgers
Kid, I Love You
Don't Love Me Like Othello
The Boomerang
Breath Of Spring
Spain #
You Don't Have To Be A Toreador #
Poor Fish #
Little Girl, Little Boy #
China #
Aphrodite #
My World Of Romance #
When We Are Married #
I Hate To Talk About Myself #
Flying The Blimp #

Fly with Me, Columbia Varsity Show, 1920
   Music by Richard C. Rodgers
        Gone Are The Days
        A Penny For Your Thoughts
        Another Melody In F
        Working For The Government
        Inspiration
        Peek In Pekin
        Dreaming True
        A College On Broadway
        Call Me Andrè
        Moonlight And You
        If I Were You
        Gunga Dhin
        The Third Degree Of Love
        If Only I Were A Boy #
   Re-used:
        Don't Love Me Like Othello
        Kid, I Love You

Poor Little Ritz Girl, 1920
   Music by Richard C. Rodgers
        The Midnight Supper
        Poor Little Ritz Girl
        Lady Raffles—Behave!
        Let Me Drink In Your Eyes
        Will You Forgive Me?
        The Gown Is Mightier Than The Sword
        Call The Doctor
        Love Will Call
        You Can't Fool Your Dreams
        All You Need To Be A Star
        The Daisy And The Lark
        Love's Intense In Tents
        Souvenirs #
        The Lord Only Knows #
        What Happened Nobody Knows # (may be same as above)
        I Surrender #
   Re-used:
        The Boomerang

Say Mama, Akron Club, 1921
   Music by Richard C. Rodgers
        Chorus Girl Blues

Watch Yourself
Wake Up
Priscilla
In Caroline
Poor Little Model #
First Love #
Show Him The Way #
Under The Mistletoe #
Jack and Jill #
When The Crime Waves Roll #
Re-used:
I Surrender

You'll Never Know, Columbia Varsity Show, 1921
Music by Richard C. Rodgers
Prologue: We Take Only The Best
Opening: Don't Think That We're The Chorus
Virtue Wins The Day
I'm Broke
When I Go On The Stage.
Just A Little Lie
You'll Never Know
Your Lullaby
Something Like Me
Mr. Director
Jumping Jack
Re-used:
Will You Forgive Me?
Let Me Drink In Your Eyes
Watch Yourself
Chorus Girl Blues
Music by Ray L. Perkins
Law #

Miscellaneous songs, 1921–22
Music by Mel Shauer
My Cameo Girl *
Fool Me With Kisses *
Vixen *

Say It with Jazz, Benjamin School for Girls, 1922.
Music by Richard C. Rodgers; lyric collaborator Frank Hunter
Just Remember Coq d'Or #

See The Golden Rooster #
Working For The Institute #
Oh, Harold! #
Re-used:
    The Moon And You
    Something Like Me
    Dreaming True
    Chorus Girl Blues
    Just A Little Lie
    Don't Love Me Like Othello
    Your Lullaby

Steppe Around, Columbia Varsity Show, 1922
    Music by Richard C. Rodgers
        Kiki*
        When You're Asleep*

Vaudeville, Georgie Price, 1922
    (with Morrie Ryskind)
        Shakespeares of 1922

Weber & Fields Re-united, 1922
    Music by Clapson [F. H. Bode]; collaborator Herbert Fields
        The Pelican*
    Music by Joseph Trounstine
        Chloe, Cling to Me

Jazz à la Carte, Institute of Musical Art, 1922.
    Music by Richard C. Rodgers
    Re-used:
        Another Melody In F
        Breath Of Springtime [Breath Of Spring]
        Moonlight And You

Winkle Town (unproduced), 1922.
    Music by Richard C. Rodgers
        One A Day
        The Hollyhocks Of Hollywood
        I Know You're Too Wonderful For Me
        Old Enough To Love
        The Hermits
        Congratulations
        Darling Will Not Grow Older

    Baby Wants To Dance
    I Want A Man
    Comfort Me
    Since I Remember You
    The Three Musketeers
    If I Were King
    Manhattan
    I'll Always Be An Optimist
    We Came, We Saw, We Made 'Em!

Half Moon Inn, Columbia Varsity Show, 1923
  Music by Richard C. Rodgers
  Re-used:
    Jack and Jill #

If I Were King, Benjamin School for Girls, 1923
  Music by Richard C. Rodgers
    The Band Of The Ne'er-Do-Wells
  Re-used:
    If I Were King

A Danish Yankee in King Tut's Court, Institute of Musical Art, 1923.
  Music by Richard C. Rodgers
    If You're Single #
  Re-used:
    Will You Forgive Me?
    The Hermits
    If I Were King

Helen of Troy, New York, 1923
  Music by W. Franke Harling
    Moonlight Lane *

Temple Belles, Park Avenue Synagogue, 1924
  Music by Richard C. Rodgers
  Re-used:
    Just A Little Lie
    A Penny For Your Thoughts
    The Hermits

The Melody Man, 1924
  Music and lyrics by "Herbert Richard Lorenz"
    Moonlight Mama
    I'd Like To Poison Ivy

Miscellaneous Songs, 1923–24
   Music by Mel Shauer
      The Spanish Dancer
   Music by Richard C. Rodgers
      Candy Opening
      Daddy's Away
      Daughters Of The Evolution
      Flow On, River
      Good Bad Woman
      Good Provider
      He Was The Last Rose Of Summer
      I'm Getting Better
      Jake The Baker
      Let Me Walk
      A Lonely Traveling Man
      Madame Esther, Queen of Hester Street
      Meet My Mother
      My Daddies (probably a variation of Daddy's Away)
      On The Bahamas
      Robin Hood
      Telltale Eyes
      The Wandering Minstrel
      A Wedding Trip

Melodies (no lyrics): Richard Rodgers archive, Library of Congress
   Music by Richard C. Rodgers
      Let's Go Home And Kiss Our Mothers #
      My Hobby #
      Pierrot #
      To Him #
      Try #
      Wee Little China Rose #
      When We Are In Love #
      When the Willow Blooms #
      When Your Heart Lets You Down #
      You're Too Wonderful For Me #

Bad Habits of 1925, Evelyn Goldsmith Home, 1925
   Music by Richard C. Rodgers
      I'd Like To Take You Home To Meet My Mother
      The Merrie Merrie #
      In Gingham #

Across The Garden Wall #
Mah-Jongg Maid #
Re-used:
If I Were King
Darling Will Not Grow Older

NOTE: unless otherwise indicated, all music from this point onward is by Richard Rodgers.

Half Moon Inn (2nd edition), Columbia Varsity Show, 1925
Babbitts In Love

June Days, 1925
Anytime, Anywhere, Anyhow #

Vaudeville, Jay Velie and Renee Robert, 1925
Terpsichore And Troubadour #

The Garrick Gaieties, 1925
Soliciting Subscriptions
Gilding The Guild
April Fool
Stage Manager's Chorus (Walk Upon Your Toes)
The Joy Spreader
Ladies Of The Box Office
Do You Love Me, I Wonder?
Black And White
On With The Dance
Sentimental Me
And Thereby Hangs A Tail
It's Quite Enough To Make Me Weep
Re-used:
Manhattan
The Three Musketeers

Dearest Enemy, 1925
Heigh-ho, Lackaday
War Is War
I Beg Your Pardon
Cheerio
Full-Blown Roses

Here In My Arms
Finale, Act One
Gavotte
I'd Like To Hide It
Where The Hudson River Flows
Bye And Bye
Sweet Peter
Here's A Kiss
The Pipes Of Pansy
Girls Do Not Tempt Me
(c) Ale, Ale, Ale #
(c) Dear Me (Oh, Dear) #
(c) Dearest Enemy #
(c) How Can We Help But Miss You? #
Re-used:
The Hermits
Old Enough To Love

The Fifth Avenue Follies, 1926
In The Name Of Art
Maybe It's Me
Where's That Little Girl?
Susie
A City Flat #
Do You Notice Anything? #
Lillie, Lawrence and Jack #
Mike #
High Hats #

The Girl Friend, 1926
Hey! Hey!
The Simple Life
The Girl Friend
Goodbye, Lenny
The Blue Room
Cabarets
Why Do I?
The Damsel Who Done All The Dirt
He's A Winner
Town Hall Tonight
Good Fellow Mine
Creole Crooning Song
What Is It?

In New Orleans
(c) Sleepyhead
(c) Hum To #
(c) Two Of A Kind #
(c) Turkey In The Straw # (an early version of Red Hot Trumpet?)
Re-used:
    I'd Like to Take You Home (rewrite of I'd Like To Take You Home
      To Meet My Mother)
    (c) The Pipes of Pansy

The Garrick Gaieties (2nd edition), 1926
    Six Little Plays
    We Can't Be As Good As Last Year
    Mountain Greenery
    Keys To Heaven
    "The Rose of Arizona"
      Vacationing
      It May Rain
      Davy Crockett
      Say It With Flowers (American Beauty Rose)
      To Hell With Mexico!
    Tennis Champs (Helen, Suzanne and Bill) (rewrite of Lillie,
      Lawrence and Jack)
    Four Little Songpluggers
    What's The Use Of Talking?
    Idles Of The King
    Gigolo
    Queen Elizabeth
    Finale, Act Two (You Can't Be As Good As Last Year)
    (c) Allez-Up
    (c) A Little Souvenir
    (c) Somebody Said
Re-used:
    (c) Sleepyhead

Lido Lady, 1926
    A Cup Of Tea
    You're On The Lido Now
    Lido Lady
    A Tiny Flat Near Soho Square
    Finale, Act One (Good Old Harry)
    The Beauty Of Another Day

My Heart Is Sheba Bound
Try Again Tomorrow
Finale, Act Two (Do You Really Mean To Go?)
What's The Use? (same music as Where's That Little Girl?)
Atlantic Blues
(c) Camera Shoot
(c) I Want A Man. (new version of I Want A Man, 1922)
(c) Ever-Ready Freddie
(c) Exercise
(c) Two To Eleven
(c) Morning Is Midnight (rewrite of part of Good Old Harry)
(c) Chuck It
I Must Be Going #
Re-used:
Here In My Arms

Peggy-Ann, 1926
Hello!
A Tree In The Park
Howdy To Broadway
A Little Birdie Told Me So
Charming, Charming
Where's That Rainbow?
Finale, Act One (Wedding Procession)
We Pirates From Weehawken
(c) I'm So Humble
Havana
Give This Little Girl A Hand
Peggy, Peggy
(c) Come And Tell Me
Paris Is Really Divine
(c) Trampin' Along
(c) In His Arms #
Re-used:
(c) The Pipes Of Pansy
Chuck It
(c) Maybe It's Me

Betsy, 1926
The Kitzel Engagement
My Missus
Stonewall Moskowitz March (collaborator: Irving Caesar)
One Of Us Should Be Two

Sing
This Funny World
Bugle Blow
Finale, Act One (Come And Tell Me)
Cradle Of The Deep
If I Were You
Birds On High
Shuffle
In Our Parlor On The Third Floor Back #
Follow On #
Push Around #
(c) At The Saskatchewan
(c) Is My Girl Refined?
(c) Six Little Kitzels
(c) You're The Mother Type
(c) Burn Up
(c) Social Work
(c) Transformation
(c) Viva Italia!
(c) A Melican Man
(c) A Ladies' Home Companion
(c) Show Me How To Make Love # (early version of Tell Me I Know
   How To Love, 1933?)
(c) In Variety #
Re-used:
(c) Come And Tell Me

Lady Luck, 1927
Lady Luck
Re-used:
Sing
If I Were You

One Dam Thing After Another, 1927
The Election
My Heart Stood Still
Make Hey, Hey, While The Moon Shines
Opening, Act Two (11th Dam Thing)
I Need Some Cooling Off
My Lucky Star
One Dam Thing After Another
(c) Pretty Little Lady
(c) Idles Of The King

Re-used:
> Shuffle
> Paris Is Really Divine
> Gigolo

A Connecticut Yankee, 1927
> Thou Swell
> At The Round Table
> On A Desert Island With Thee
> Nothing's Wrong #
> Finale, Act One (Hymn To The Sun)
> I Feel At Home With You
> The Sandwich Men
> Evelyn, What Do You Say? (Morgan Le Fay)
> (c) I Blush
> (c) You're What I Need
> (c) Someone Should Tell Them
> (c) Britain's Own Ambassadors #

Re-used:
> My Heart Stood Still
> A Ladies' Home Companion

She's My Baby, 1928
> This Goes Up
> Here She Comes
> Smart People
> When I Saw Him Last
> Where Can The Baby Be?
> A Baby's Best Friend
> The Swallows #
> (c) How Was I To Know?
> (c) Whoopsie
> (c) Wasn't It Great?

Re-used:
> My Lucky Star
> You're What I Need
> When I Go On The Stage
> Try Again Tomorrow
> Camera Shoot
> A Little House In Soho
> I Need Some Cooling Off
> (c) Morning Is Midnight
> (c) The Pipes Of Pansy

Present Arms, 1928
> Tell It To The Marines
> You Took Advantage Of Me
> Do I Hear You Saying "I Love You"?
> A Kiss For Cinderella
> Is It The Uniform?
> Crazy Elbows
> Down By The Sea
> I'm A Fool, Little One
> Finaletto, Act Two (Rescue scene)
> Blue Ocean Blues (same music as Atlantic Blues)
> Hawaii
> (c) I Love You More Than Yesterday
> Kohala Welcome #
> What Price Love? #

Chee-Chee, 1928
A great many of the musical pieces in this show are very short. The principal songs are:
> Dear, Oh Dear
> I Must Love You
> Better Be Good To Me
> The Tartar Song
> Singing A Love Song
> Moon Of My Delight

The others are:
> Await Your Love
> Farewell, O Life
> Her Hair Is Black As Licorice
> I Am A Prince
> I Bow A Glad Good Day
> I Grovel To Your Cloth
> I'll Never Share You
> In A Great Big Way
> I Wake At Morning
> Joy Is Mine
> Khonghouse Song
> Living Buddha
> Oh, Gala Day, Red-Letter Day
> Owl Song
> Sleep, Weary Head
> Thank You In Advance
> The Most Majestic Of Domestic Officials

We Are The Horrors Of Deadliest Woe
We're Men Of Brains
You Are Both Agreed

Spring Is Here, 1929
    (c) A Cup Of Tea (There's Magic In The Cup)
    Spring Is Here (In Person)
    Yours Sincerely
    We're Gonna Raise Hell
    You Never Say Yes
    With A Song In My Heart
    Baby's Awake Now
    Finale, Act One (Oh, Look!)
    This Is Not Long Island
    Red Hot Trumpet
    What A Girl!
    Rich Man, Poor Man!
    Why Can't I?
    (c) A Word In Edgeways
    (c) The Color Of Her Eyes.
  Re-used:
    (c) Lady Luck

Heads Up!, 1929
  Originally *Me for You;* these songs were not in the revised show:
    (c) The Three Bears #
    (c) Have You Been True To Me? #
    (c) Kindly Nullify Your Fears
    (c) Sweetheart, You Make Me Laugh
  These songs (some from *Me for You*) were in *Heads Up!:*
    Me For You (Wouldn't You Love It?)
    My Man Is On The Make
    A Ship Without A Sail
    (c) I Can Do Wonders With You
    It Must Be Heaven
    You've Got To Surrender
    Playboy
    Mother Grows Younger
    Why Do You Suppose? (same music as How Was I To Know?)
    Ongsay And Anceday
    Knees
    (c) Sky City
    (c) As Though You Were There

(c) They Sing! They Dance! They Speak!
(c) Mind Your P's And Q's #
(c) It's A Man's World #
(c) We're An English Ship (Bootlegger's Chantey) #
(c) Now Go To Your Cabin #
(c) Harlem On The Sand

Lady Fingers, 1929
  Re-used:
    I Love You More Than Yesterday
    Sing

The Play's the Thing (unproduced and incomplete), 1929 (?)
    Opening, Act One (Ladies And Gentlemen, Good Evening!)
    Music Is Emotion
    Italy

Simple Simon, 1930
    Coney Island
    Don't Tell Your Folks
    Magic Music
    (c) I Still Believe In You (same music as Singing A Love Song)
    Send For Me (same music as I Must Love You)
    Sweetenheart
    Hunting The Fox
    Come On, Men
    Ten Cents A Dance
    Rags And Tatters
    He Was Too Good To Me
    (c) Opening, Act Two (Service and Courtesy)
    (c) Oh, So Lovely
    Peter Pan
    The Simple Simon Instep
    Prayers Of Tears And Laughter
    Hunting Song
    Come Out Of The Nursery
    Sing Glory Hallelujah
    He Dances On My Ceiling (Dancing On The Ceiling)
    (c) Hands
    (c) Steps
    (c) Say When, Stand Up, Sit Down #
    (c) Dull And Gay #
    (c) I Want That Man (I Want A Man?)

Re-used:
>   (c) I Can Do Wonders With You

Follow Through, 1929 (movie)
>   I'm Hard To Please
>   It Never Happened Before
>   Softer Than A Kitten
>   Because We're Young

Ever Green, 1930
>   Harlemania
>   Doing A Little Waltz Clog
>   Dear! Dear!
>   Nobody Looks At The Man
>   Waiting For The Leaves To Fall
>   No Place But Home
>   The Lion King
>   When The Old World Was New (Quand Notre Vieux Monde Etait Tout Neuf) [French lyrics by J. Lenoir]
>   Lovely Woman's Ever Young (La Femme à Toujours Vingt Ans) [French lyrics by J. Lenoir]
>   In The Cool Of The Evening
>   Je M'en Fiche du Sex-Appeal!
>   Impromptu Song (Tommy's Talking Song)
>   If I Give In To You
>   The Beauty Contest
Re-used:
>   Dancing On The Ceiling
>   The Color Of Her Eyes

The Hot Heiress, 1931 (movie)
>   Nobody Loves A Riveter
>   Like Ordinary People Do
>   You're The Cats
Re-used:
>   He Looks So Good To Me (rewrite of He Was Too Good To Me)

America's Sweetheart, 1931
>   Mr. Dolan Is Passing Through #
>   In Californ-i-a #
>   My Sweet #
>   I've Got Five Dollars

Sweet Geraldine
There's So Much More (rewrite of Someone Should Tell Them)
We'll Be The Same
How About It?
Innocent Chorus Girls Of Yesterday
A Lady Must Live
(c) I Want A Man (second rewrite; Milton Pascal claimed he wrote
the third refrain of this version at Larry's invitation)
Now I Believe
You Ain't Got No Savoir-Faire #
Two Unfortunate Orphans #
Tonight Or Never #
(c) Come Across #
(c) Tarts In Ermine #
(c) Tennessee Dan #
(c) God Gave Me Eyes #
(c) I'll Be A Star # (early version of You've Got That?)
(c) A Cat Can Look At A Queen #
Re-used:
(c) I'm Hard To Please

Billy Rose's Crazy Quilt, 1931
Rest Room Rose #

Love Me Tonight, 1932 (movie)
That's The Song Of Paree
Isn't It Romantic?
Lover
Mimi
A Woman Needs Something Like That
The Poor Apache
Cleaning Up The Floor With Lulu (earlier version of The Poor Apache)
Love Me Tonight
The Son-Of-A-Gun Is Nothing But A Tailor!
(c) The Man For Me (The Letter Song)
(c) Give Me Just A Moment
also "rhythmic dialogue"

Million Dollar Legs, 1932 (movie)
Music by Ralph Rainger
It's Terrific When I Get Hot (uncredited)

The Phantom President, 1932 (movie)
The Country Needs A Man

Somebody Ought To Wave A Flag
The Medicine Show (some lyrics with Jimmy Durante)
Give Her A Kiss
The Convention
also "rhythmic dialogue"

Hallelujah, I'm a Bum!, 1933 (movie)
I Gotta Get Back To New York
My Pal Bumper
Laying The Cornerstone
Bumper Found A Grand
Dear June
What Do You Want With Money?
Hallelujah, I'm A Bum! (two different songs with this title were written)
Kangaroo Court
I'd Do It Again
You Are Too Beautiful
(c) Sleeping Beauty
also "rhythmic dialogue"

Peg o' My Heart, 1933 (movie)
(c) When You're Falling In Love With The Irish
(c) Tell Me I Know How To Love

I Married an Angel, 1933 (unproduced movie)
Opening (Love Is Queen, Love Is King)
Face The Facts
Animated Objects
Why Have You Eyes?
I Married An Angel
Bath and Dressmaking sequence
also "rhythmic dialogue"
Re-used:
Tell Me I Know How To Love

Dancing Lady, 1933
That's The Rhythm Of The Day
(c) Dancing Lady (two different songs with this title were written)

Please, 1933 (London revue)
Rhythm
Re-used:
A Baby's Best Friend

Hollywood Party, 1933 (movie)
> Hollywood Party (two different songs with this title were written)
> Hello!
> Reincarnation
> (c) You Are
> (c) Black Diamond
> (c) Prayer (see also *Manhattan Melodrama*)
> (c) The Pots
> (c) I'm One Of The Boys
> (c) Burning
> (c) Baby Stars
> (c) You've Got That
> (c) Fly Away To Ioway
> (c) Give A Man A Job
> (c) I'm A Queen In My Own Domain
> (c) My Friend The Night
> (c) Keep Away From The Moonlight

Meet the Baron, 1933 (movie)
> (c) Yes, Me
> (c) The Mahster's Coming

The Merry Widow, 1934 (movie)
> Music by Franz Lehár
>> Girls, Girls, Girls!
>> Vilia
>> The Melody Of Paris In The Spring
>> Maxim's
>> The Merry Widow Waltz
>> Widows Are Gay, I've Heard Men Say
> Music by Richard Rodgers
>> (c) It Must Be Love
>> (c) A Widow Is A Lady
>> (c) Dolores

Nana, or, *Lady of the Boulevards*, 1934 (movie)
> That's Love

Manhattan Melodrama, 1934 (movie)
> (c) Manhattan Melodrama (It's Just That Kind Of A Play)
> The Bad In Ev'ry Man
> These two songs and Prayer all had the melody which became
> Blue Moon.

Mississippi, 1935
>No Bottom
>Roll, Mississippi
>Soon
>Down By The River
>It's Easy To Remember
>(c) The Notorious Colonel Blake
>(c) Pablo, You Are My Heart
>(c) The Steely Glint In My Eye
>(c) I Keep On Singing *

The Post-Depression Gaieties, 1935
>What Are You Doing In Here? *

Something Gay, 1935
>You Are So Lovely And I'm So Lonely

Let's Have Fun, 1935 (radio show)
>A Little Of You On Toast
>Please Make Me Be Good
>You Carry My Heart (early version of Please Make Me Be Good)

Miscellaneous
>Someone Must Be Getting Married Somewhere
>Little Dolores/ Muchacha (for revue with Laurence Stallings, 1934?)
>The Desert Lullaby (for Bert Lahr in *The Show Is On,* 1936?)

Jumbo, 1935
>Over And Over Again
>The Circus Is On Parade
>The Most Beautiful Girl In The World
>Laugh
>My Romance
>Little Girl Blue
>Song Of The Roustabouts
>Women
>Memories Of Madison Square Garden
>Diavolo
>(c) The More I See Of Other Girls (The Elephant Song)
>(c) There's A Small Hotel

On Your Toes, 1936
>Two-A-Day For Keith
>The Three B's

It's Got To Be Love
Too Good For The Average Man
The Heart Is Quicker Than The Eye
Quiet Night
Glad To Be Unhappy
On Your Toes
  Re-used:
There's A Small Hotel
(c) Olympic Games #

The Show Is On, 1936
Go! Go! Benuti! #
Rhythm (London lyric revised for New York)

The Dancing Pirate, 1936 (movie)
Are You My Love?
When You're Dancing The Waltz

All Points West, 1936 ("Symphonic Narrative")
All Points West

Babes in Arms, 1936
Where Or When
Babes In Arms
I Wish I Were In Love Again
All Dark People
Way Out West (On West End Avenue)
My Funny Valentine
Johnny One Note
Imagine
All At Once
The Lady Is A Tramp
You Are So Fair

Charlie McCarthy Show, 1937 (radio)
Re-used:
A Little Of You On Toast (new lyrics)
Please Make Me Be Good (new lyrics)

I'd Rather Be Right, 1937
A Homogeneous Cabinet
Have You Met Miss Jones?
Take And Take And Take

Spring In Vienna/ Milwaukee
(c) The World Is My Oyster (same music as Spring In Vienna)
A Little Bit Of Constitutional Fun
Not So Innocent Fun (Nine Young Girls And Nine Old Men) (earlier
    version of A Little Bit of Constitutional Fun)
Sweet Sixty-Five
We're Going To Balance The Budget
Labor Is The Thing
I'd Rather Be Right (two different songs with this title were written)
Off The Record
A Baby Bond
(c) We Just Sing And Dance
(c) Ev'rybody Loves You
(c) A Treaty My Sweetie With You
(c) His Chances Are Not Worth A Penny

Fools for Scandal, 1938 (movie)
There's A Boy In Harlem
Food For Scandal
How Can You Forget?
(c) Love Knows Best
(c) Once I Was Young
(c) Let's Sing A Song About Nothing
also "rhythmic dialogue"

I Married an Angel, 1938
Did You Ever Get Stung?
I'll Tell The Man In The Street
The Modiste
How To Win Friends And Influence People
Spring Is Here
Angel Without Wings
A Twinkle In Your Eye
At The Roxy Music Hall
Re-used:
I Married an Angel
(c) Women Are Women (probably Love Is Queen, Love Is King)
(c) Men Of Old Milwaukee
(c) Othello #

The Boys from Syracuse, 1938
I Had Twins.
Dear Old Syracuse.

What Can You Do With A Man?
Falling In Love With Love
The Shortest Day Of The Year
This Can't Be Love
Ladies Of The Evening
He And She
You Have Cast Your Shadow On The Sea
Come With Me (To Jail)
Big Brother
Sing For Your Supper
Oh, Diogenes!
(c) The Greeks Had No Word For It
(c) Who Are You?
(c) Cinderella (probably an earlier version of My Prince; see *Too Many Girls* below)

Too Many Girls, 1939
Heroes In The Fall (lyric by Rodgers)
Tempt Me Not
My Prince
Pottawatomie
Love Never Went To College
Spic And Spanish
I Like To Recognize The Tune
Look Out!
The Sweethearts Of The Team
She Could Shake The Maracas
I Didn't Know What Time It Was
Too Many Girls
Give It Back To The Indians
(c) The Hunted Stag

Higher and Higher, 1940
A Barking Baby Never Bites
From Another World
Morning's At Seven
Nothing But You
Disgustingly Rich
Blue Monday
Ev'ry Sunday Afternoon
Lovely Day For A Murder
How's Your Health?
It Never Entered My Mind

I'm Afraid
(c) It's Pretty In The City
(c) Life, Liberty, And The Pursuit Of You

Two Weeks with Pay, 1940 (revue)
  Now That I Know You (retitled second version of I'd Rather Be
    Right)

The Boys from Syracuse, 1940 (movie)
 Re-used:
  The Greeks Had No Word For It
  Who Are You?

Too Many Girls, 1940 (movie)
  You're Nearer

Pal Joey, 1940
  You Mustn't Kick It Around
  I Could Write A Book
  Chicago
  That Terrific Rainbow
  What Is A Man?
  Happy Hunting Horn
  Bewitched, Bothered And Bewildered
  Pal Joey (What Do I Care For A Dame?)
  The Flower Garden Of My Heart
  Zip
  Plant You Now, Dig You Later
  Den Of Iniquity
  Do It The Hard Way
  Take Him
  (c) I'm Talking To My Pal
  (c) Love Is My Friend (same melody as What Is A Man?)

They Met in Argentina, 1941 (movie)
  North America Meets South America
  I Congratulate You, Mr. Cowboy/ (c) Contrapunto
  You've Got The Best Of Me
  Carefree Carretero
  Amarillo
  Lolita
  Cutting The Cane
  Never Go To Argentina

Simpatica
(c) We're On The Track
(c) Encanto
(c) Back To Texas #
(c) Bury Me Not #

Keep 'Em Rolling, 1942 (short movie)
Keep 'Em Rolling

By Jupiter, 1942
For Jupiter And Greece
Jupiter Forbid
Life With Father
Nobody's Heart
The Gateway Of The Temple Of Minerva
Here's A Hand
No, Mother, No!
The Boy I Left Behind
Ev'rything I've Got
Bottoms Up
Careless Rhapsody
The Greeks Have Got The Girdle
(c) Wait Till You See Her
Now That I've Got My Strength
(c) Fool Meets Fool
(c) Life Was Monotonous
(c) Nothing To Do But Relax

Miscellaneous, 1942 (written for the Army Air Force Aid Society)
The Bombardier Song
Shorty The Gunner

Yankee Doodle Dandy, 1942 (movie)
Off The Record (topical extra verse of lyric)

Miss Underground, 1942
  Music by Emmerich Kalmán
Opening (Messieurs, Mesdames)
It Happened In The Dark
Mother, Look, I'm An Acrobat
The One Who Yells The Loudest Is The Captain
Do I Love You? (No, I Do Not!)
Fall In Love (You Crazy Little Things)

Alexander's Blitztime Band *
Otto's Patter Song *
Vendor's Song *
Otto's German/English Song # (same as above?)
Lucio's Victorian Family #
Otto And The Elephants #
The Bad Little Apple And The Wise Old Tree #
Get Your Man #
New York number (Miss Underground) #
Jean's Magic Song #
France Is Free #

Stage Door Canteen, 1943 (movie)
The Girl I Love To Leave Behind

A Connecticut Yankee, 1943
Here's Martin The Groom
This Is My Night To Howl
Ye Lunchtime Follies
Can't You Do A Friend A Favor?
You Always Love The Same Girl
To Keep My Love Alive
The Camelot Samba #
(c) Elaine *
(c) I Won't Sing A Song *
Re-used:
My Heart Stood Still
On A Desert Island With Thee
Thou Swell
At The Round Table
I Feel At Home With You
Finale, Act One (Hymn To The Sun)

APPENDIX II

Lyrics by Hart:
An Alphabetical Listing

Across The Garden Wall
Ale, Ale, Ale
Alexander's Blitztime Band
All At Once
All Dark People
All Points West
All You Need To Be A Star
Allez-Up
Amarillo
American Beauty Rose
And Thereby Hangs A Tail
Angel Without Wings
Animated Objects
Another Melody In F
Any Old Place With You
Anytime, Anywhere, Anyhow
Aphrodite
April Fool
Are You My Love?
As Though You Were There
At The Round Table
At The Roxy Music Hall
At The Saskatchewan
Atlantic Blues
Await Your Love

B-L-C
B-R-A

Babbitts In Love
Babes In Arms
Baby Bond, A
Baby Stars
Baby Wants To Dance
Baby's Awake Now
Baby's Best Friend, A
Back To Texas
Bad In Ev'ry Man, The
Bad Little Apple And The Wise Old Tree, The
Band Of The Ne'er-Do-Wells, The
Barking Baby Never Bites, A
Bath and Dressmaking sequence
Beauty Contest, The
Beauty Of Another Day, The
Because We're Young
Better Be Good To Me
Bewitched, Bothered and Bewildered
Big Brother
Birds On High
Black And White
Black Diamond
Blue Monday
Blue Moon
Blue Ocean Blues
Blue Room, The
Bombardier Song, The

Boomerang, The
Bootlegger's Chantey, The
Bottoms Up
Boy I Left Behind, The
Brant Lake Blues
Breath Of Spring
Breath Of Springtime
Britain's Own Ambassadors
Bugle Blow
Bumper Found A Grand
Burn Up
Burning
Bury Me Not
Bye And Bye

Cabarets
Call Me André
Call The Doctor
Camelot Samba, The
Camera Shoot
Candy Opening
Can't You Do A Friend A Favor?
Carefree Carretero
Careless Rhapsody
Cat Can Look At A Queen, A
Charming, Charming
Cheerio
Chicago
China
Chloe, Cling To Me
Chorus Girl Blues
Chuck It
Cinderella
Circus Is On Parade, The
City Flat, A
Cleaning Up The Floor With Lulu
College On Broadway, A
Color Of Her Eyes, The
Come Across
Come And Tell Me
Come On, Men
Come Out Of The Nursery
Come With Me (To Jail)

Comfort Me
Coney Island
Congratulations
Contrapunto
Convention, The
Country Needs A Man, The
Cradle Of The Deep
Crazy Elbows
Creole Crooning Song
Cup of Tea, A
Cup Of Tea (There's Magic In The Cup), A
Cutting The Cane

Daddy's Away
Daisy And The Lark, The
Damsel Who Done All The Dirt, The
Dancing Lady (2)
Dancing On The Ceiling
Darling Will Not Grow Older
Daughters Of The Evolution
Davy Crockett
Dear! Dear!
Dear June
Dear Me
Dear, Oh Dear
Dear Old Syracuse
Dearest Enemy
Den Of Iniquity
Desert Lullaby, The
Diavolo
Did You Ever Get Stung?
Disgustingly Rich
Do I Hear You Saying "I Love You"?
Do I Love You?
Do It The Hard Way
Do You Love Me, I Wonder?
Do You Notice Anything?
Do You Really Mean To Go?
Doing A Little Waltz Clog
Dolores
Don't Love Me Like Othello

Don't Tell Your Folks
Don't Think That We're The
  Chorus
Down At The Lake
Down By The River
Down By The Sea
Dreaming True
Dull And Gay

Elaine
Election, The
Elephant Song, The
Encanto
Evelyn, What Do You Say?
Ever-Ready Freddie
Every Man Needs A Wife
Ev'ry Sunday Afternoon
Ev'rybody Loves You
Ev'rything I've Got
Exercise

Face The Facts
Fall In Love (You Crazy Little
  Things)
Falling In Love With Love
Farewell, O Life
First Love
Flow On, River
Flower Garden Of My Heart, The
Fly Away To Ioway
Flying The Blimp
Follow On
Food For Scandal
Fool Me With Kisses
Fool Meets Fool
For Jupiter And Greece
Four Little Songpluggers
France Is Free
From Another World
Full-Blown Roses

Gateway Of The Temple Of
  Minerva, The

Gavotte
Get Your Man
Gigolo
Gilding The Guild
Girl Friend, The
Girl I Love To Leave Behind, The
Girls Do Not Tempt Me
Girls, Girls, Girls!
Give A Man A Job
Give Her A Kiss
Give It Back To The Indians
Give Me Just A Moment
Give This Little Girl A Hand
Glad To Be Unhappy
Go! Go! Benuti!
God Gave Me Eyes
Gone Are The Days
Good Bad Woman
Good Fellow Mine
Good Old Harry
Good Provider
Goodbye, Lenny
Gown Is Mightier Than The
  Sword, The
Gray Song, The
Greeks Had No Word For It, The
Greeks Have Got The Girdle, The
Green And Gray
Green Song, The
Gunga Dhin

Hallelujah, I'm A Bum!
Hands
Happy Hunting Horn
Harlem On The Sand
Harlemania
Havana
Have You Been True To Me?
Have You Met Miss Jones?
Hawaii
He And She
He Dances On My Ceiling
He Lights Another Mecca

He Looks So Good To Me
He Was The Last Rose Of Summer
He Was Too Good To Me
Heart Is Quicker Than The Eye, The
Heigh-ho, Lackaday
Helen, Suzanne and Bill
Hello!
Her Hair Is Black As Licorice
Here In My Arms
Here She Comes
Here's A Hand
Here's A Kiss
Here's Martin The Groom
Hermits, The
Heroes In The Fall (lyric by
  Rodgers)
He's a Winner
Hey! Hey!
High Hats
His Chances Are Not Worth
  A Penny
Hollyhocks Of Hollywood, The
Hollywood Party (2)
Homogeneous Cabinet, A
Horicon Hop
How About It?
How Can We Help But Miss You?
How Can You Forget?
How To Win Friends And Influence
  People
How Was I To Know?
Howdy To Broadway
How's Your Health?
Hubby Dances On A String
Hum To
Hunted Stag, The
Hunting Song
Hunting The Fox
Hymn To The Sun

I Am A Prince
I Beg Your Pardon

I Blush
I Bow A Glad Good Day
I Can Do Wonders With You
I Congratulate You, Mr Cowboy
I Could Write A Book
I Didn't Know What Time It Was
I Feel At Home With You
I Gotta Get Back To New York
I Grovel To Your Cloth
I Had Twins
I Hate To Talk About Myself
I Keep On Singing
I Know My Girl By Her Perfume
I Know You're Too Wonderful For
  Me
I Like To Recognize The Tune
I Love To Lie Awake In Bed
I Love You More Than Yesterday
I Married An Angel
I Must Be Going
I Must Love You
I Need Some Cooling Off
I Still Believe In You
I Surrender
I Used To Love Them All
I Wake At Morning
I Want A Man
I Want That Man
I Wish I Were In Love Again
I Won't Sing A Song
I'd Do It Again
I'd Like To Hide It
I'd Like To Poison Ivy
I'd Like To Take You Home
I'd Like To Take You Home To
  Meet My Mother
I'd Rather Be Right (2)
Idles Of The King
If I Give In To You
If I Were King
If I Were You
If Only I Were A Boy

If You're Single
I'll Always Be an Optimist
I'll Be A Star
I'll Never Share You
I'll Tell The Man In The Street
I'm A Fool, Little One
I'm A Queen In My Own Domain
I'm Afraid
I'm Broke
I'm Getting Better
I'm Hard To Please
I'm One Of The Boys
I'm So Humble
I'm Talking To My Pal
Imagine
Impromptu Song
In A Great Big Way
In Californ-i-a
In Caroline
In Gingham
In His Arms
In New Orleans
In Our Parlor On The Third Floor
  Back
In The Cool Of The Evening
In The Name Of Art
In Variety
Innocent Chorus Girls Of Yesterday
Inspiration
Is It The Uniform?
Is My Girl Refined?
Isn't It Romantic?
It Happened In The Dark
It May Rain
It Must Be Heaven
It Must Be Love
It Never Entered My Mind
It Never Happened Before
Italy
It's A Man's World
It's Easy To Remember
It's Got To Be Love

It's Just That Kind Of A Play
It's Pretty In The City
It's Quite Enough To Make Me
  Weep
It's Terrific When I Get Hot
I've Got A Girl in Chestertown
I've Got Five Dollars

Jack And Jill
Jake The Baker
Je M'en Fiche du Sex-Appeal!
Jean's Magic Song
Johnny One Note
Joy Is Mine
Joy Spreader, The
Jumping Jack
Jupiter Forbid
Just A Little Lie
Just Remember Coq d'Or

Kangaroo Court
Keep Away From The Moonlight
Keep 'Em Rolling
Keys To Heaven
Khonghouse Song
Kid, I Love You
Kiki
Kindly Nullify Your Fears
Kiss For Cinderella, A
Kiss Lesson, The
Kitzel Engagement, The
Knees
Kohala Welcome

Labor Is The Thing
Ladies and Gentlemen, Good
  Evening!
Ladies' Home Companion, A
Ladies Of The Box Office
Ladies Of The Evening
Lady Is A Tramp, The
Lady Luck

Lady Must Live, A
Lady Raffles—Behave!
Last Night
Laugh
Law
Laying The Cornerstone
Let Me Drink In Your Eyes
Let Me Walk
Let's Go Home And Kiss Our
    Mothers
Let's Sing A Song About Nothing
Letter Song, The
Lido Lady
Life, Liberty, And The Pursuit Of
    You
Life Was Monotonous
Life With Father
Like Ordinary People Do
Lillie, Lawrence and Jack
Lion King, The
Little Birdie Told Me So, A
Little Bit Of Constitutional Fun, A
Little Dolores
Little Girl Blue
Little Girl, Little Boy
Little House In Soho, A
Little Of You On Toast, A
Little Souvenir, A
Living Buddha
Lolita
Lonely Traveling Man, A
Look Out!
Lord Only Knows, The
Love Is My Friend
Love Is Queen, Love Is King
Love Knows Best
Love Me Tonight
Love Never Went To College
Love Will Call
Lovely Day For A Murder
Lovely Woman's Ever Young
Lover

Love's Intense In Tents
Lucio's Victorian Family

Madame Esther, Queen of Hester
    Street
Magic Music
Mah-Jongg Maid
Mahster's Coming, The
Make Hey, Hey, While The Moon
    Shines
Man For Me, The
Manhattan
Manhattan Melodrama
Maxim's
Maybe It's Me
Me For You (Wouldn't You
    Love It?)
Medicine Show, The
Meet My Mother
Melican Man, A
Melody Of Paris In The Spring, The
Memories Of Madison Square
    Garden
Men Of Old Milwaukee
Merrie Merrie, The
Merry Widow Waltz, The
Messieurs, Mesdames
Meyer, Your Tights Are Tight
Midnight Supper, The
Mike
Mimi
Mind Your P's And Q's
Modiste, The
Moon And You, The
Moon Of My Delight
Moonlight And You
Moonlight Lane
Moonlight Mama
More I See Of Other Girls, The
Morgan Le Fay
Morning Is Midnight
Morning's At Seven

Most Beautiful Girl In The
  World, The
Most Majestic Of Domestic
  Officials, The
Mother Grows Younger
Mother, Look, I'm An Acrobat
Mountain Greenery
Mr. Dolan Is Passing Through
Mr. Director
Muchacha
Music Is Emotion
My Cameo Girl
My Daddies
My Friend The Night
My Funny Valentine
My Heart Is Sheba Bound
My Heart Stood Still
My Hobby
My Lucky Star
My Man Is On The Make
My Missus
My Pal Bumper
My Prince
My Romance
My Sweet
My World Of Romance

Never Go To Argentina
New York number (Miss
  Underground)
Night Was Made For Dancing, The
Nine Young Girls And Nine Old
  Men
No Bottom
No, Mother, No!
No Place But Home
Nobody Looks At The Man
Nobody Loves A Riveter
Nobody's Heart
North America Meets South
  America
Not So Innocent Fun

Nothing But You
Nothing To Do But Relax
Nothing's Wrong
Notorious Colonel Blake, The
Now Go To Your Cabin
Now I Believe
Now That I Know You
Now That I've Got My Strength

Off The Record
Oh, Dear
Oh, Diogenes!
Oh, Gala Day, Red-Letter Day
Oh, Harold!
Oh, Look!
Oh, Mr. Postman, Don't Pass Me By!
Oh, So Lovely
Old Brant Lake
Old Enough To Love
Olympic Games
On A Desert Island With Thee
On The Bahamas
On With The Dance
On Your Toes
Once I Was Young
One A Day
One Dam Thing After Another
One Of Us Should Be Two
One Who Yells The Loudest Is The
  Captain, The
Ongsay And Anceday
Othello
Otto And The Elephants
Otto's German/English Song
Otto's Patter Song
Over And Over Again
Owl Song

Pablo, You Are My Heart
Pal Joey
Paris Is Really Divine
Peek In Pekin

Peggy, Peggy
Pelican, The
Penny For Your Thoughts, A
Peter Pan
Pierrot
Pipes Of Pansy, The
Plant You Now, Dig You Later
Playboy
Please Make Me Be Good
Poor Apache, The
Poor Fish
Poor Little Model
Poor Little Ritz Girl
Pots, The
Pottawatomie
Prayer
Prayers Of Tears And Laughter
Pretty Little Lady
Priscilla
Push Around

Queen Elizabeth
Quiet Night

Rags And Tatters
Red Hot Trumpet
Reincarnation
Rest Room Rose
Rhythm
Rich Man, Poor Man!
Robin Hood
Roll, Mississippi

Sandman, The
Sandwich Men, The
Say It With Flowers
Say When, Stand Up, Sit Down
See The Golden Rooster
Send For Me
Sentimental Me
Service and Courtesy
Shakespeares
She Could Shake The Maracas

She's Camping At Red Wing
Ship Without A Sail, A
Shortest Day Of The Year, The
Shorty The Gunner
Show Him The Way
Show Me How To Make Love
Shuffle
Simpatica
Simple Life, The
Simple Simon Instep, The
Since I Remember You
Sing
Sing For Your Supper
Sing Glory Hallelujah
Singing A Love Song
Six Little Kitzels
Six Little Plays
Sky City
Sleep, Weary Head
Sleeping Beauty
Sleepyhead
Smart People
Social Work
Softer Than A Kitten
Soliciting Subscriptions
Somebody Ought To Wave A Flag
Somebody Said
Someone Must Be Getting Married
Somewhere
Someone Should Tell Them
Something Like Me
Son-Of-A-Gun Is Nothing But A
   Tailor!, The
Song of the Roustabouts
Soon
Souvenirs
Spain
Spanish Dancer, The
Spic And Spanish
Spring in Vienna/ Milwaukee
Spring Is Here (In Person)
Spring Is Here
Stage Manager's Chorus

Steely Glint In My Eye, The
Steps
Stonewall Moskowitz March
Stop! Stop! I Am The Traffic Cop!
Susie
Swallows, The
Sweet Geraldine
Sweet Peter
Sweet Sixty-Five
Sweetenheart
Sweetheart, You Make Me Laugh
Sweethearts Of The Team, The

Take And Take And Take
Take Him
Tartar Song, The
Tarts In Ermine
Tee Ta Tee
Tell It To The Marines
Tell Me I Know How To Love
Telltale Eyes
Tempt Me Not
Ten Cents A Dance
Tennessee Dan
Tennis Champs
Terpsichore And Troubadour
Thank You In Advance
That Terrific Rainbow
That's Love
That's The Rhythm Of The Day
That's The Song of Paree
There's A Boy In Harlem
There's A Small Hotel
There's Magic In The Cup
There's No Fool Like A Young Fool
There's So Much More
They Sing! They Dance! They
   Speak!
Third Degree Of Love, The
This Can't Be Love
This Funny World
This Goes Up
This Is My Night To Howl

This Is Not Long Island
Thou Swell
Three Bears, The
Three B's, The
Three Musketeers, The
Ticky, Ticky, Tack
Tiny Flat Near Soho Square, A
To Hell With Mexico!
To Him
To Keep My Love Alive
Tommy's Talking Song
Tonight Or Never
Too Good For The Average Man
Too Many Girls
Town Hall Tonight
Trampin' Along
Transformation
Treaty My Sweetie With You, A
Tree In The Park, A
Try
Try Again Tomorrow
Turkey In The Straw
Twinkle In Your Eye, A
Two-A-Day For Keith
Two Of A Kind
Two To Eleven
Two Unfortunate Orphans

Under The Mistletoe

Vacationing
Vendor's Song
Venus (There's Nothing Between
   Us)
Vilia
Virtue Wins The Day
Viva Italia!
Vixen

Wait Till You See Her
Waiting For The Leaves To Fall
Wake Up
Walk Upon Your Toes

Wandering Minstrel, The
War Is War
Wasn't It Great?
Watch Yourself
Way Out West (On West End Avenue)
We Are The Horrors of Deadliest Woe
We Came, We Saw, We Made 'Em!
We Can't Be As Good As Last Year
We Just Sing And Dance
We Pirates From Weehawken
We Take Only The Best
Wedding Trip, A
Wee Little China Rose
We'll Be The Same
We're An English Ship
We're Going To Balance The Budget
We're Gonna Raise Hell
We're Men Of Brains
We're On The Track
What A Girl!
What Are You Doing In Here?
What Do I Care For A Dame?
What Can You Do With A Man?
What Do You Want With Money?
What Happened Nobody Knows
What Is A Man?
What Is It?
What Price Love?
What's The Use?
What's The Use Of Talking?
When I Go On The Stage
When I Saw Him Last
When The Crime Waves Roll
When The Old World Was New
When The Willow Blooms
When We Are In Love
When We Are Married
When Your Heart Lets You Down
When You're Asleep
When You're Dancing The Waltz

When You're Falling In Love With The Irish
Where Can The Baby Be?
Where Or When
Where The Hudson River Flows
Where's That Little Girl?
Where's That Rainbow?
Who Are You?
Whoopsie
Why Can't I?
Why Do I?
Why Do You Suppose?
Why Have You Eyes?
Widow Is A Lady, A
Widows Are Gay, I've Heard Men Say
Will You Forgive Me?
With A Song In My Heart
Woman Needs Something Like That, A
Women Are Women
Women
Word In Edgeways, A
Working For The Government
Working For the Institute
World Is My Oyster, The

Ye Lunchtime Follies
Yes, Me
You Ain't Got No Savoir-Faire
You Always Love The Same Girl
You Are
You Are Both Agreed
You Are So Fair
You Are So Lovely And I'm So Lonely
You Are Too Beautiful
You Can't Be As Good As Last Year
You Can't Fool Your Dreams
You Carry My Heart
You Don't Have To Be a Toreador
You Have Cast Your Shadow On The Sea

You Mustn't Kick It Around
You Never Say Yes
You Took Advantage Of Me
You'll Never Know
Your Lullaby
You're Nearer
You're On The Lido Now
You're The Cats
You're The Mother Type

You're Too Wonderful For Me
You're What I Need
Yours Sincerely
You've Got That
You've Got The Best Of Me
You've Got To Surrender

Zip

BIBLIOGRAPHY

Abbott, George. *"Mr. Abbott."* New York: Random House, 1963.

Anderson, Jervis. *Harlem: The Great Black Way.* London: Orbis, 1981.

Arnaz, Desi. *A Book.* New York: Morrow, 1976.

Astaire, Fred. *Steps in Time.* New York: Harper & Bros., 1959.

Behlmer, Rudy. *Memo from David O. Selznick.* New York: Viking, 1972.

Berg, A. Scott. *Goldwyn: A Biography.* New York: Knopf, 1989.

Bergreen, Laurence. *As Thousands Cheer: The Life of Irving Berlin.* New York: Viking Penguin, 1990.

Blum, Daniel. *A Pictorial History of the American Theatre 1900–56.* New York: Greenberg, 1950.

Bordman, Gerald. *American Musical Theatre: A Chronicle.* New York: Oxford University Press, 1978.

———. *Jerome Kern: His Life and Music.* New York: Oxford University Press, 1980.

Bronner, Edwin J. *The Encyclopedia of the American Theatre, 1900–75.* New York: Barnes, 1980.

Burns, George, with David Fisher. *All My Best Friends.* New York: Putnam, 1989.

Cerf, Bennett. *At Random.* New York: Random House, 1977.

Cochran, Charles. *Cock-a-Doodle-Doo.* London: Dent, 1941.

Conrad, Earl. *Billy Rose: Manhattan Primitive.* Cleveland: World, 1968.

Csida, Joseph, and June Bundy. *American Entertainment.* New York: Watson Guptill, 1978.

Dietz, Howard. *Dancing in the Dark: An Autobiography.* New York: Quadrangle/New York Times, 1974.

Eames, John Douglas. *The MGM Story.* New York: Crown, 1975.

Ewen, David. *The Life and Death of Tin Pan Alley.* New York: Funk & Wagnalls, 1964.

———. *Richard Rodgers.* New York: Holt, 1957.

Ferber, Edna. *A Peculiar Treasure.* Garden City, N.Y.: Doubleday Doran, 1939.

Fordin, Hugh. *Getting to Know Him: A Biography of Oscar Hammerstein II.* New York: Random House, 1977.

Fowler, Gene. *Schnozzola.* New York: Viking, 1951.

Furia, Philip. *The Poets of Tin Pan Alley.* New York: Oxford University Press, 1990.

Geist, Kenneth L. *Pictures Will Talk.* New York: Scribner's, 1978.

Goldman, Herbert. *Fanny Brice: The Original Funny Girl.* New York: Oxford University Press, 1992.

———. *Jolson: The Legend Comes To Life.* New York: Oxford University Press, 1988.

Goldstein, Malcolm. *George S. Kaufman, His Life, His Theatre.* New York: Oxford University Press, 1979.

Green, Stanley. *Ring Bells! Sing Songs!* New York: Arlington House, 1971.

———. *Rodgers & Hammerstein Fact Book.* New York: Lynn Farnol Group, 1980.

———. *The World of Musical Comedy.* South Brunswick, N.J.: Barnes, 1968.

Harriman, Margaret Case. *Take Them Up Tenderly: A Collection of Profiles.* New York: Knopf, 1944.

Hart, Dorothy, *Thou Swell, Thou Witty: The Life and Lyrics of Lorenz Hart.* New York: Harper, 1976.

Hart, Dorothy, and Robert Kimball, eds. *The Complete Lyrics of Lorenz Hart.* New York: Knopf, 1986.

Hart, Moss. *Act One: An Autobiography.* New York: Random House, 1959.

Haver, Ronald. *David O. Selznick's Hollywood.* New York: Knopf, 1980.

Havoc, June. *Early Havoc.* New York: Harper & Row, 1980.

Hemming, Roy. *The Melody Lingers On: The Great Songwriters and Their Movie Musicals.* New York: Newmarket, 1986.

Hepburn, Katharine. *Me: Stories of My Life.* New York: Knopf, 1991.

Higham, Charles. *Ziegfeld.* Chicago: Regnery, 1972.

Hirschhorn, Clive. *The Hollywood Musical.* London: Octopus, 1981.

———. *The Warner Bros. Story.* New York: Crown, 1979.

Hulbert, Jack. *The Little Lady Is Always Right.* London: W. H. Allen, 1975.

Jablonski, Edward. *Gershwin: A Biography.* Garden City, N.Y.: Doubleday, 1987.

Jablonski, Edward, and Lawrence D. Stewart. *The Gershwin Years.* Garden City, N.Y.: Doubleday, 1973.

Katz, Ephraim. *The International Film Encyclopedia.* New York: Perigee, 1979.

Kaufman, George S., Moss Hart, and Lorenz Hart. *I'd Rather Be Right: Libretto.* New York: Random House, 1937.

Keats, John. *You Might as Well Live: The Life and Times of Dorothy Parker.* New York: Simon & Schuster, 1970.

Kreuger, Miles. *Show Boat.* New York: Oxford University Press, 1977.

Lerner, Alan Jay. *The Street Where I Live.* New York: Norton, 1978.

Logan, Joshua. *Josh—My Up and Down, In and Out Life.* New York: Delacorte, 1976.

MacShane, Frank. *The Life of John O'Hara.* New York: Dutton, 1980.

Marx, Samuel. *Mayer and Thalberg: The Make-Believe Saints.* New York: Random House, 1975.

Marx, Samuel, and Jan Clayton. *Rodgers and Hart: Bewitched, Bothered and Bedevilled.* New York: Putnam, 1976.

Matthews, Jessie. *Over My Shoulder.* London: W. H. Allen, 1974.

Meredith, Scott. *George S. Kaufman and His Friends.* Garden City, N.Y.: Doubleday, 1974.

Merman, Ethel, with George Eels. *Merman: An Autobiography.* New York: Simon & Schuster, 1978.

Mordden, Ethan. *Broadway Babies.* New York: Oxford University Press, 1983.

Morehouse, Ward. *George M. Cohan: Prince of the American Theater.* Philadelphia: Lippincott, 1943.

Nolan, Frederick. *The Sound of Their Music.* New York: Walker, 1978.

O'Hara, John. *Pal Joey: The Libretto.* New York: Random House, 1952.

Rodgers, Richard. *Musical Stages.* New York: Random House, 1975.

Rodgers, Richard, and Oscar Hammerstein II, eds. *The Rodgers and Hart Songbook.* New York: Simon & Schuster, 1951.

Sann, Paul. *The Lawless Decade.* New York: Crown, 1957.

Schulberg, Budd. *Moving Pictures.* New York: Stein & Day, 1981.

Schwartz, Charles. *Cole Porter: A Biography.* New York: Dial, 1977.

———. *Gershwin: His Life and Music.* Indianapolis: Bobbs-Merrill, 1973.

Sheean, Vincent. *Oscar Hammerstein I: The Life and Exploits of an Impresario.* New York: Simon & Schuster, 1956.

Sillman, Leonard. *Here Lies Leonard Sillman.* New York: Citadel Press, 1959.

Simon, George T. *The Big Bands.* New York: Macmillan, 1967.

Stagg, Jerry. *The Brothers Shubert.* New York: Random House, 1968.

Stott, William, with Jane Stott. *On Broadway.* Austin: University of Texas Press, 1978.

Suskin, Stephen. *Show Tunes, 1905–85.* New York: Dodd, Mead, 1986.

Wilder, Alec, and James T. Maher. *American Popular Song: The Great Innovators, 1900–50.* New York: Oxford University Press, 1972.

Wilk, Max. *They're Playing Our Song.* New York: Atheneum, 1974.

Wilkerson, Tichi, and Marcia Borie. *The Hollywood Reporter.* New York: Coward McCann, 1984.

Wodehouse, P. G., and Guy Bolton. *Bring On the Girls.* New York: Simon & Schuster, 1953.

WPA Guide to New York City [1939]. Facsimile edition. New York: Pantheon, 1982.

INDEX I

Rodgers and Hart Shows and Movies

Amateur Shows

Bad Habits of 1925, 58, 75
Danish Yankee in King Tut's Court, A, 45, 58, 104
Fly with Me, 31–32
If I Were King, 44, 62
Say It With Jazz, 38, 39
Say Mama, 35
Temple Belles, 56
Twinkling Eyes, 23
Up Stage and Down, 23, 61
You'd Be Surprised, 31, 33
You'll Never Know, 38, 98

Stage Musicals

All's Fair. See By Jupiter.
America's Sweetheart, 2, 149–51, 153, 196
Arabian Nights, The (projected), 256
Babes in Arms, 2, 215, 217, 218, 223, 235, 236, 250, 255, 257, 262
Bad Man, The (projected), 117
Betsy, 2, 88, 90, 91, 93, 94, 97
Betsy Kitzel. See Betsy.
Boys from Syracuse, The, 2, 252–54, 259, 264
Buy Buy Betty. See Betsy.
By Jupiter, 294–96, 298, 299, 302, 308
Came The Dawn. See America's Sweetheart.
Chee-Chee, 2, 119–22, 138
Come Across. See America's Sweetheart.
Connecticut Yankee, A, 2, 103, 106–10, 115 (revival), 306, 308–10
Dear Enemy. See Dearest Enemy.
Dearest Enemy, 2, 60, 68–71, 74, 75, 79, 81, 83, 88, 90, 255

Ever Green, 2, 136, 141, 142, 144–46, 153
Fifth Avenue Follies, The, 74, 82, 83, 88, 93
Garrick Gaieties (First), 2, 61, 64–66, 68, 88, 118, 193, 208, 270
Garrick Gaieties (2nd), 75, 79, 80, 81, 84, 246.
Girl Friend, The, 2, 75, 76, 80, 81, 88, 96, 105, 110
Heads Up!, 132, 133, 135, 136, 141, 179, 203, 225
Henky. See Melody Man, The.
Higher and Higher, 264, 267, 268–70, 272
Hotel Splendide (projected), 287
I Married an Angel, 2, 224, 235, 241, 242, 246–50, 256, 267
I'd Rather be Right, 2, 224, 229, 230–36, 250, 255, 259, 291
Jazz King, The. See Melody Man, The.
Jumbo, 2, 183, 203–06, 207, 209, 210, 211, 213, 216, 249
Lady Fingers, 123, 124
Lady Luck, 96
Lido Lady, 74, 79, 82, 83, 92, 96, 113, 131
Love Champion, The. See Lido Lady.
Loving Ann. See Spring Is Here.
Me For You, 131, 137. See also *Heads Up!*
Melody Man, The, 57, 73, 106
Messer Marco Polo (projected), 263
Muchacha (projected), 300
On Your Toes, 2, 199, 201, 203, 204, 207, 210, 212–14, 223, 237, 241, 250, 255
One Dam Thing After Another, 97–100, 141
Pal Joey, 2, 272–82, 291, 295
Peggy. See Peggy-Ann.
Peggy-Ann, 2, 87–90, 92, 93, 96, 97, 105, 108, 120, 147

Play's The Thing, The (projected), 130, 131
Poor Little Ritz Girl, 2, 32, 33, 36, 88, 298
Present Arms, 116, 117, 119, 142, 229
Saratoga Trunk (projected), 287
She's My Baby, 38, 112–14, 133
Simple Simon, 137–140
Spring Is Here, 2, 125–27, 131
Sweet Rebel. See *Dearest Enemy*.
Terpsichore and Troubadour, 58
Three Men On A Horse (projected), 263
Too Many Girls, 253, 256, 258, 260–62, 264, 265, 272
Winkle Town (projected), 54, 83

Movies

Big Liar, The. See *Meet The Baron*.
Dancing Lady, 186–88
Dancing Pirate, The, 214
East Side. See *Manhattan Melodrama*.
Evergreen, 201
Follow Through, 139, 142
Fools For Scandal, 225, 226, 229, 241, 247
Good Bad Girl, The. See *Hot Heiress, The*.
Hallelujah, I'm A Bum!, 170–73, 177
Happy Go Lucky. See *Hallelujah, I'm A Bum!*
Heads Up!, 142

Heart of New York. See *Hallelujah, I'm A Bum!*
Hollywood Party, 181–84, 186, 201, 205, 209
Hollywood Revue of 1933. See *Hollywood Party*.
Hot Heiress, The, 143, 144, 151
I Married An Angel, 174, 175, 183
Jumbo, 206
Leathernecking, 142
Love Me Tonight, 158–61, 163, 169, 172–74, 190, 198, 199, 231, 247, 256
Love of Michael (projected), 151, 152
Makers of Melody, 127–29, 155, 156
Manhattan Melodrama, 176, 183, 193
Meet The Baron, 182, 183
Merry Widow, The, 189–90, 193
Mississippi, 194, 198, 199, 200–202, 203, 217
Nana, 178–180, 191, 201, 272
New Yorker, The. See *Hallelujah, I'm A Bum!*
On Your Toes, 268
Optimist, The. See *Hallelujah, I'm A Bum!*
Pal Joey, 282
Peg o' My Heart, 175, 186
Phantom President, The, 164–66, 169, 172, 229
Spring Is Here, 142
They Met in Argentina, 284
Too Many Girls, 284
Words and Music, 314

INDEX II

Rodgers and Hart Songs

"Ale, Ale, Ale," 69
"All Dark People," 218
"All Points West," 216, 217, 235
"And Thereby Hangs A Tail," 66
"Another Melody In F," 32
"Any Old Place With You," 26, 27
"Anytime, Anywhere, Anyhow," 67, 82
"April Fool," 65
"Are You My Love?," 214
"As Though You Were There," 133

"Babbitts In Love," 45
"Baby Stars," 183
"Baby Wants To Dance," 55
"Baby's Best Friend Is His Mother, A," 113
"Bad In Ev'ry Man, The," 176, 193
"Band Of The Ne'er Do Wells, The," 44
"Barking Baby Never Bites, A," 269
"Better Be Good To Me," 120
"Bewitched, Bothered And Bewildered," 274, 280, 282
"Black And White," 65, 66
"Black Diamond," 181, 183
"Blue Day," 205
"Blue Moon," 176, 183, 184, 193–94, 201, 202
"Blue Room, The," 75, 92, 105, 127, 128
"Britain's Own Ambassadors," 108
"Bugle Blow," 94
"Burning," 183

"Camelot Samba, The," 309
"Camera Shoot," 92
"Can't You Do A Friend A Favor?," 309

"Careless Rhapsody," 297
"Cat Can Look At A Queen, A," 149
"Cinderella," 253
"Circus Is On Parade, The," 205
"Circus Wedding, The," 206
"College Baby," 58
"College On Broadway, A," 32
"Color Of Her Eyes, The," 141
"Come And Tell Me," 92
"Come With Me," 253
"Comfort Me," 55
"Country Needs A Man, The," 165
"Cradle Of The Deep," 94
"Cup Of Tea, A," 82

"Damsel Who Done All The Dirt, The," 75
"Dancing Lady," 187
"Dancing On The Ceiling," 138, 141, 145
"Dear, Dear," 141
"Dear, Oh, Dear," 120
"Dear Me," 69
"Dearest Enemy," 69
"Den Of Iniquity," 281, 282
"Desert Lullaby, The," 216
"Diavolo," 205
"Disgustingly Rich," 269
"Do I Hear You Saying 'I Love You'?," 116
"Do You Love Me, I Wonder?," 65
"Do You Notice Anything?," 74
"Dolores," 190
"Don't Love Me Like Othello," 32
"Don't Tell Your Folks," 138
"Down By The River," 198
"Dreaming True," 32

"Elaine," 309
"Election, The," 98
"Evelyn, What Do You Say?," 108
"Ever-Ready Freddy," 83
"Ev'ry Sunday Afternoon," 270
"Ev'rybody Loves You," 230, 233
"Ev'rything I've Got," 297

"Falling In Love With Love," 253, 254
"Flower Garden Of My Heart," 274, 275
"Fly Away To Ioway," 183
"Follow On," 93
"For Jupiter and Greece," 297
"From Another World," 269

"Gateway Of The Temple of Minerva," 297
"Gigolo," 98
"Girl Friend, The," 75, 105, 127, 128
"Girl I Love To Leave Behind, The," 306
"Give A Man A Job," 183
"Give Her A Kiss," 165
"Give It Back To The Indians," 260
"Give Me Just a Moment," 159
"Give This Little Girl A Hand," 93
"Glad To Be Unhappy," 211
"Go, Go, Benuti," 216
"Gone Are The Days," 32
"Good Fellow Mine," 75
"Good Provider," 101
"Greeks Have No Word For It, The," 254
"Guild Gilded, The," 65

"Hallelujah, I'm A Bum!," 171
"Hallelujah, I'm A Tramp!," 172
"Hands," 139
"Havana," 93
"Have You Met Miss Jones?," 230, 233
"He and She," 253
"He Looks So Good To Me," 144
"He Was Too Good To Me," 138, 144
"Heart Is Quicker Than The Eye, The," 211
"Heigh-ho, Lackaday," 61, 71
"Hello," 183
"Here In My Arms," 71, 79, 83, 127, 128, 218
"Hermits, The," 45, 55, 71
"Heroes In The Fall," 261
"Hey, Hey," 75
"His Chances Are Not Worth A Penny," 234
"Hollyhocks Of Hollywood, The," 55
"How About It?," 199
"How Can We Help But Miss You?," 69

"How Can You Forget?," 226
"How To Win Friends and Influence People," 246
"How Was I To Know?," 134
"Howdy To Broadway," 92

"I Blush," 108
"I Can Do Wonders With You," 137, 138
"I Could Write A Book," 274, 282
"I Didn't Know What Time It Was," 260, 262, 273, 315
"I Feel At Home With You," 309
"I Gotta Get Back To New York," 171
"I Keep On Singing," 199
"I Know You're Too Wonderful For Me," 55
"I Like To Recognize The Tune," 261, 315
"I Love You More Than Yesterday," 116, 124
"I Married An Angel," 174, 177
"I Must Love You," 120, 138
"I Need Some Cooling Off," 98
"I Still Believe In You," 138, 140
"I Surrender," 36
"I Want A Man," 55, 83, 149, 150
"I Want That Man," 138
"I Wish I Were In Love Again," 217, 218
"I Won't Sing A Song," 309
"I'd Do It Again," 171
"I'd Like To Hide It," 70
"I'd Like To Poison Ivy," 57
"I'd Like To Take You Home," 75, 105
"I'd Like To Take You Home To Meet My Mother," 59
"Idles Of The King," 98
"If I Give In To You," 141
"If I Were King," 44, 58
"If I Were You," 94, 96
"If You're Single," 45
"I'll Be A Star," 149
"I'll Tell The Man In The Street," 246
"I'm A Queen In My Own Domain," 183
"I'm Afraid," 269
"I'm Hard To Please," 139
"I'm One Of The Boys," 183
"I'm Talking to My Pal," 278
"Imagine," 218
"In Our Parlor On The Third Floor Back," 93
"In The Cool of the Evening," 141
"In The Name of Art," 74
"Innocent Chorus Girls of Yesterday," 150
"Inspiration," 32
"Isn't it Romantic?," 2, 159–61

"It May Rain (When The Sun Stops Shining)," 81
"It Must Be Love," 190
"It Never Entered My Mind," 269, 270
"Italy," 131
"It's Easy to Remember," 2, 200, 203, 251
"It's Got To Be Love," 211
"It's Just That Kind Of A Play," 176, 193
"I've Got Five Dollars," 149, 151, 216, 229

"Johnny One Note," 199, 218, 253
"Joy Spreader, The," 65
"Just Remember Coq d'Or," 38

"Keep 'em Rolling," 292
"Keep Away From The Moonlight," 183
"Keys To Heaven," 80
"Kiki," 41
"Kiss For Cinderella, A," 116

"Ladies' Home Companion, A," 108
"Ladies Of The Box Office," 65
"Lady Is A Tramp, The," 218, 257, 314
"Lady Must Live, A," 150
"Laugh," 205
"Life Was Monotonous," 297
"Life With Father," 297
"Life, Liberty, And The Pursuit of You," 269
"Like Ordinary People Do," 144
"Lillie, Lawrence And Jack," 74, 77, 80
"Little Birdie Told Me So, A," 92, 315
"Little Girl Blue," 205
"Little Of You On Toast, A," 208, 227
"Lolita," 284
"Look Out!," 260
"Love Knows Best," 225
"Love Me Tonight," 159
"Love Never Went To College," 260
"Love Will Call," 33
"Lovely Woman's Always Young," 141
"Lover," 159, 161, 173, 180, 191
"Love's Intense In Tents," 33

"Mahster's Coming, The," 183
"Make Hey! Hey! While The Moon Shines," 98
"Make Me A Star," 182
"Man For Me, The," 159
"Manhattan," 55, 63, 64–67, 127, 128, 257
"Maybe It's Me," 82, 93
"Memories Of Madison Square," 205
"Men Of Old Milwaukee," 246
"Mike," 74

"Mimi," 159
"Moon Of My Delight," 120
"Moonlight Mama," 57
"More I See Of Other Girls, The," 206
"Morning Is Midnight," 83
"Most Beautiful Girl In The World, The," 206, 217, 315
"Mountain Greenery," 80, 105
"Muchacha," 201, 214
"Music Is Emotion," 131
"My Friend The Night," 183
"My Funny Valentine," 191, 217, 314
"My Heart Stood Still," 2, 99, 100, 107, 108, 110, 136, 141, 239, 257, 309, 314
"My Lucky Star," 98
"My Man Is On The Make," 132
"My Prince," 253
"My Romance," 206

"Never Go To Argentina," 284
"Night Was Made For Dancing, The," 179
"No Bottom," 198, 199
"No Place But Home," 141
"Nobody Looks At The Man," 141
"Nobody Loves A Riveter," 144
"Nobody's Heart," 296, 297
"Nothing But You," 269
"Nothing's Wrong," 108
"Notorious Colonel Blake, The," 199
"Now That I've Got My Strength," 297

"Off The Record," 232, 233
"Oh, Harold!," 38
"Old Enough To Love," 55
"Olympic Games," 211
"On A Desert Island With Thee," 108, 309
"On With The Dance," 66
"On Your Toes," 216
"Once I Was Young," 226
"One A Day," 55
"One Dam Thing After Another," 98
"Ongsay And Anceday," 225
"Over And Over Again," 205

"Pablo, You Are My Heart," 199, 217
"Paris Is Really Divine," 98, 100
"Party Waltz," 183, 205
"Peek In Pekin," 32
"Pipes of Pansy, The," 69, 75, 92, 113
"Plant You Now, Dig You Later," 276
"Please Make Me Be Good," 208, 227
"Pots, The," 183
"Prayer," 183, 184, 193

"Prayers Of Tears And Laughter," 138
"Princess Of The Willow Tree," 33
"Push Around," 94

"Questions And Answers," 211
"Quiet Night," 211

"Reincarnation," 183
"Rest Room Rose," 152
"Rhythm," 201, 216
"Roll, Mississippi," 199
"Rose Of Arizona," 80, 81, 96
"Roxy Music Hall, The," 245, 248

"Say When—Stand Up—Drink Down," 138
"See The Golden Rooster," 38
"Send For Me," 138
"Sentimental Me," 66, 67
"She Could Shake The Maracas," 260
"Ship Without A Sail, A," 133
"Shuffle," 98
"Silver Threads," 55
"Simpatica," 284
"Simple Life, The," 75
"Since I Remember You," 55
"Sing," 96, 124
"Singing A Love Song," 120, 138
"Sleepyhead," 75, 80
"Soliciting Subscriptions," 64
"Somebody Ought To Wave A Flag," 165
"Someone Should Tell Them," 108, 149
"Song Of The Roustabouts, The," 205
"Soon," 2, 199
"Spring In Vienna," 234
"Spring Is Here," 249
"Spring Song," 205, 249
"Steely Glint In My Eye, The," 199
"Susie," 74
"Sweet Peter," 71
"Sweet Sixty-five," 230
"Sweetenheart," 138

"Take Him," 280
"Tartar Song, The," 120
"Tell Me I Know How To Love," 175
"Ten Cents A Dance," 139, 140, 183, 315
"Terpsichore And Troubadour," 132
"That Terrific Rainbow," 274, 275
"That's Love," 179, 191, 201
"That's The Rhythm Of The Day," 187, 188
"That's The Song of Paree," 15, 9
"There's A Small Hotel," 211, 206, 213, 251
"There's So Much More," 149

"They Sing! They Dance! They Speak!," 133
"This Can't Be Love," 253
"This Funny World," 93
"This Is My Night To Howl," 309
"Thou Swell," 108, 309
"Three Musketeers, The," 55, 64, 65
"To Hell With Burgundy," 245
"To Hell With Mexico!" 80, 81, 246
"To Keep My Love Alive," 141, 309
"Too Good For The Average Man," 211
"Treaty, My Sweetie, With You, A," 233
"Tree In The Park, A," 92
"Try Again Tomorrow," 315
"Tune Up, Bluebird," 233
"Twinkle In Your Eye, A," 246
"Two A Day For Keith," 211

"Venus," 23

"Wait Till You See Her," 296, 297
"Wasn't It Great?," 113
"Way Out West," 217
"We'll Be The Same," 151
"We're Going To Balance The Budget," 233
"What a Prince," 253
"What Are You Doing In Here?," 202
"What Can You Do With a Man?," 253
"What Do You Want With Money?," 171
"What's The Use?" 83
"What's The Use Of Talking?," 315
"When I Go On The Stage," 38, 98, 113
"When The Old World Was New," 141
"When You're Asleep," 41
"When You're Dancing The Waltz," 214
"When You're Falling In Love With The
    Irish," 175
"Where Or When," 217, 218, 314
"Where's That Rainbow?," 92
"Who Are You?," 255
"Why Can't I?," 126
"Why Do You Suppose?," 133, 134
"Widow Is A Lady, A," 190
"Will You Forgive Me?," 33
"With A Song In My Heart," 2, 125, 257
"Women Are Women," 246
"Women!," 205
"Working For The Government," 32, 38
"Working For The Institute," 38
"World Is My Oyster, The," 233

"Ye Lunchtime Follies," 308
"Yes, Me," 183
"You Always Love The Same Girl," 309

"You Are So Lovely," 205
"You Are Too Beautiful," 171
"You Are," 183
"You Can't Fool Your Dreams," 33, 88
"You Took Advantage Of Me," 116, 257, 315
"You Who Are Lovely," 205
"You're So Lovely And I'm So Lonely," 203

"You're The Cats," 144
"You're What I Need," 108
"Yours Sincerely," 126
"You've Got That," 149, 183, 209

"Zip," 274, 278, 308

INDEX III

General

Aarons, Alex, 111, 115, 123, 125, 131, 133, 134, 148, 149
Aarons, Mrs. Ella, 123
Abbott, George, 86, 125, 204, 207, 211–14, 252, 254, 256, 258–62, 272, 274–77, 279, 281, 287, 289–91, 298, 303
Abie's Irish Rose, 106, 169
Actors' strike of 1919, 29, 30
Adams, Franklin P., 13, 28
Adding Machine, The, 40,
Adler, Larry, 291
Adler, Polly, 6, 262, 296
Admirable Crichton, The, 200
Adventures of Mark Twain, The, 306
Ah, Wilderness, 195
Ainley, Richard, 297
Air Eagles, 157
Akron Club, The, 17, 18, 23, 31, 35, 61
Albert, Eddie, 253, 268
Alexander, John, 285
"Alexander's Blitztime Band," 300
"Alexander's Ragtime Band" (parody), 9
All Aboard, 26
"All Alone Monday," 107
"All The Things You Are," 265, 286
Allgood, Sara, 105
Allyson, June, 269
Alton, Robert, 274, 275, 295, 298
"Always," 77, 78
Always You, 31, 37
Amberg, Gustave, 7, 12, 15, 23
Ameche, Don, 225–227
American, The, 47
"American In Paris, An," 123, 216

American Jubilee, 252
American Society of Composers, Authors, and Publishers (ASCAP), 71, 236, 282
Americana, 86
Anderson, John Murray, 60, 69, 207
Andrews, Lyle, 105, 124
Androcles and the Lion, 80
Angelus, Muriel, 253
Animal Crackers, 127, 152
"Annie Doesn't Live Here Any More," 267
Anything Goes, 294
Applause, 158
Appointment in Samarra (novel), 271
"April Showers," 86
Aquarium Club, 71
Arms and The Man, 80
Arnaz, Desi, 258–62
Arno, Peter, 62, 123, 262
Arsenic and Old Lace, 285
Artists and Models, 182
Arzner, Dorothy, 179
As Thousands Cheer, 180, 196
"As Time Goes By," 126, 303
As You Desire Me, 179
As You Were, 113
Astaire, Adele, 100, 111, 152
Astaire, Fred, 100, 111, 148, 152, 171, 188, 199
At Home Abroad, 207
"At Long Last Love," 255
At 9:45, 28
Atkinson, Brooks, 150, 249, 262, 281, 282
Atwell, Roy, 33
Aubert, Jeanne, 150
Away We Go! See Rodgers & Hammerstein.

Axelrod, George, 37
Axelrod, Herman, 37

Back Stage Club, The, 73, 86
Bad Man, The, 143, 214
Baddeley, Hermione, 82
Badger, Clarence, 143
Baker, Belle, 89, 90, 93, 94
Baker, Edythe, 98–100
Baker, Josephine, 101
Baker, Kenny, 225
Balaban, Barney, 168, 200
Balanchine, George, 204, 212, 213, 215,
    246–48, 272, 294, 301,
Ballets Russes de Monte Carlo, Les, 204, 266,
    301
Ballyhoo, 148
"Baltimore, Md., You're The Only Doctor
    For Me," 86
Band Wagon, The, 152
Bandbox Club, The, 71
Banjo Eyes, 294
Bankhead, Tallulah, 192, 203, 266
Barbour, Joyce, 116–18, 141
Bargy, Roy, 255
"Barney Google," 50
Bartholomae, Philip, 105
Basie, Count, 240
Beat The Band, 303
Beebe, Lucius, 234
Beery, Wallace, 198
Beggar's Opera, The, 283
"Begin The Beguine," 209
Behrman, S. N., 170
Beiderbecke, Bix, 215, 315
Belasco, David, 41, 44, 86, 124
Bellamy, Ralph, 225, 226
Belmore, Bertha, 297
Bemelmans, Ludwig, 287, 290
Benchley, Robert, 14, 41, 72, 111, 113, 151, 188
Bender, Milton "Doc," 30, 31, 38, 49, 102,
    148, 192, 201, 204, 213, 221, 236–38,
    241, 256, 258, 259, 265, 272, 277, 284,
    295, 300, 301, 305, 311, 313
Benjamin School for Girls, The, 42, 44, 56
Bennett, Joan, 143, 198, 203
Bennett, Robert Russell, 134, 315
Bennett, Seldon, 311
Benny, Jack, 85, 167, 234
Bergen, Edgar, and Charlie McCarthy, 226,
    249
Berger, Ludwig, 156
Berkeley, Busby, 106, 115, 152, 177, 184

Berle, Milton, 239, 256
Berlin, Irving, 14, 60, 77, 78, 89, 91, 94, 107,
    139, 143, 153, 168, 180, 196, 227, 234, 297
Besser, Joseph (Besser & Amy), 53
Best Foot Forward, 284, 289–91, 293
Between The Devil, 241
"Beyond The Blue Horizon," 86, 151
Bickel, Fred. See March, Fredric.
Big Boy, 170
Big Broadcast, The, 207
Big Pond, The, 156
"Bill," 112
Billboard Magazine, 117, 127
Billy Duffy's Club, 71
Billy Rose's Crazy Quilt, 152
Binyon, Claude, 198
"Birth of the Blues, The," 98
Bitter Sweet, 136, 141, 225 (movie)
"B.L.C.," 52
"Black Bottom, The," 85
Blackbirds, 196
Blackbirds of 1928, 116
Blane, Ralph, 289
Blind Alley, 212
Blinn, Holbrook, 28, 30, 130
Blond Beast, The, 40, 43, 44
Blossom Time, 148, 304
Blue Paradise, The, 246
"Blue Skies," 94, 108
"Blueberry Hill," 282
Bode, F. D., 41
"Body and Soul," 148, 295
Bogart, Humphrey, 79, 127, 196, 226, 266, 303
Bolger, Ray, 85, 131, 202, 203, 212, 294, 306
Bolton, Guy, 39, 43, 74, 79, 87, 91, 107, 114,
    131, 137, 256
Boom Boom, 127
Bordoni, Irene, 113
Borem and "Cherry," 99
Boy Meets Girl, 214
Bracken, Eddie, 259–62
Brady, "Diamond Jim," 7
Brant Lake Camp, 10, 15, 20, 23, 31, 38, 47,
    49, 51, 126, 264
Bread, Butter and Rhythm, see *Happy Landing.*
Brent, Romney, 62, 64, 65, 80, 208, 293, 294
Brewster's Millions, 155
Brice, Fanny, 75, 86, 90, 96, 147, 152, 192,
    225
Broadcast Music, Inc., 282
Broadway, 86
Broadway Melody, 127
Broadway Melody of 1938, The, 226

Broadway to Hollywood, 180
Bromfield, Louis, 143, 148
Brooke, Clifford, 197
Broun, Heywood, 41, 153
Brown, John Mason, 150, 232
Brown, Les, 240
Brown, Lew, 83, 85, 139
Brown, Louise, 105
Brown, Nacio Herb, 175, 184, 187
Brown, Porter Emerson, 214
Bryan, William Jennings, 69
Bryant, Nana, 106
Buchanan, Jack, 74, 77, 96, 117, 135, 241
"Bummel, Bummel, Bummel," 12
Bunker, Ralph, 32
Burdon, Albert, 141
Burke, Billie (Mrs. Florenz Ziegfeld), 91, 196, 202
Burke, Joe, 143
"But Not For Me," 148
"But Not Today" 83
"Butcher, The Baker, And The Candlestick Maker, The," 65
BUtterfield-8 (novel), 271
Butterworth, Charles, 161, 181, 183, 184
"By Myself," 241
By The Way, 77, 83
Byng, Douglas, 98

Caesar, Irving, 28, 50, 89, 91, 115, 170, 171
Cagney, James, 40, 86, 159, 165
Caldwell, Erskine, 195
Calling All Stars, 206
Camille, 150, 153, 317
Camp Paradox, 9, 15, 24, 155
Campbell, Mary "Big Mary," 101, 160, 237, 305
Campbell's Funeral Church, 85
"Can't Help Loving Dat Man," 112
Can't Help Singing, 305
"Can't We Be Friends?," 126
Cantor, Eddie, 125, 148, 168, 180, 294
Caravan, Inc., 40, 44
Carey, Dr. Edward, 304
"Carioca, The," 188
Carlisle, Kitty, 196, 200, 234
Carmen Jones, 298, 306
Carnegie Hall, 123, 173, 301
Caroline. See *Can't Help Singing*.
Carpenter, Constance, 106, 144
Carrillo, Leo, 198, 258
Carry On, 125
Carson, Doris, 212
Carter, Desmond, 96

Casablanca, 303
Casino de Paree Club, 195
Casto, Jean, 273, 278, 279, 281, 291
Cat and the Fiddle, The, 153, 212 movie, 246
Caviar, 196, 197
CBS Radio Network, 117, 202, 208, 226, 256
Cerf, Bennett, 20, 36, 123
Challenge, The, 28
Champagne and Orchids (unproduced), 190
Champagne Sec, 196
Chaplin, Charles, 104, 143
Charig, Phil, 50, 86, 91, 305, 311
"Charlie McCarthy Show, The" (radio), 227
Charlie's Aunt, 279
Charlot, André, 201
Charlot's Masquerade, 144
Charlot's Revue of 1924, 74
Charlot's Revue of 1926, 77, 98
Charm School, The, 67
"Chase and Sanborn Hour, The" (radio). *See* "Charlie McCarthy Show, The."
Cheat, The, 179
Chevalier, Maurice, 153, 154, 156, 159, 160, 163, 189, 190, 191, 201, 231
Chief Thing, The, 80
Childs, Gilbert, 83
China Seas, 189
Chinese Lanterns, 42
Chisholm, Robert, 308
"Chloe, Cling To Me," 42
Christie, Audrey, 195, 245, 248, 250
Ciro's Club, 113
City Lights, 104
City Streets, 158
Claire, Ina, 223, 234
Clamhouse Club, 71
Clansman, The, 28
"Clap Yo' Hands," 87
Clapson, see Bode, F. D.
Clayton, Herbert, 96, 105
Clement, Joan, 114
Clover Club, 191
Cloy, Robert, 127, 128
Clyde, Andy, 162
Clyde, June, 181
Cochran, Charles B., 97–100, 107, 136, 141, 142, 144–46, 179, 239
Cochrane, June, 64, 65, 80
Cochran's Revue of 1926, 82
"Cocktails For Two," 200
Cocoanuts, The, 77, 78
Cody, Lew, 193
Cohan, George M., 30, 164, 165, 195, 228–34
"Cohen Owes Me Twenty-Seven Dollars," 89

Cohn, Harry, 175, 302
Colbert, Claudette, 117, 124, 127, 156, 164, 165, 172
Colefax, Lady Sybil, 239
Coleman, Emil, orchestra, 137
Coleman, Robert, 93
College Rhythm, 198
Colman, Ronald, 124, 178, 226
Columbia Grammar School, 1, 12
Columbia University, 2, 3, 17, 22, 28, 41
  Varsity Shows, 2
  contributors to, 14
  Fly With Me, 31–32
  Half Moon Inn, 45
  On Your Way, 15
  Peace Pirates, The, 15
  Steppe Around, 41
  You'll Never Know, 38, 98
"Come On And Pet Me" (Sometimes I'm Happy), 115
Comedy of Errors, The, 241, 250, 254
Connecticut Yankee in King Arthur's Court, A (movie), 104
Connelly, Marc, 14, 45, 47, 125, 202, 266
"Conning Tower, The," 13, 14, 28
Conrad, Con, 50, 83, 105
Conried, Richard, 31
Considine, John, 205
Conway, Harold, 145
Conway, Jack, 176
Cooper, Merian C., 175, 188, 213
Co-Optimists, The, 83, 96
Corned Beef and Roses, 147
Corrigan, Lloyd, 214
Costello, Diosa, 258–61, 265, 284
Costello, Matty, 238
Count of Luxembourg, The, 197
Counterattack, 303
Courtneidge, Cicely, 74, 77
Courtney, Inez, 124, 125, 127, 128, 142, 143, 150, 163
Covarrubias, Miguel, 65
Coward, Noël, 81, 124, 130, 136, 144, 145, 148, 207, 232
Cradle Snatchers, 79
Craig's Wife, 79
Crawford, Joan, 170, 172, 175, 180, 181, 183, 186–89, 234
Crawford, Mimi, 98
Criss Cross, 95
Criterion Club, 73
Crosby, Bing, 183, 200, 202, 215, 225, 251
Crowther, Dorothy, 38, 45

Cucaracha, La, 214
Cukor, George, 125, 156, 158, 175, 237
Cullman, Howard, 293
Cunningham, Jack, 198
"Cup of Coffee, A Sandwich, And You, A," 51, 77
Curtain Going Up! (unproduced), 223
Cyrano de Bricabrac, 25

Daffy Dill, 43
Dali, Salvador, 246, 278
Dana, Robert, 284
"Dance With A Dolly With A Hole In Her Stocking," 28
Dancing Time, 74, 79
Dare, Phyllis, 82, 96
Dark Victory, 267
"Darn That Dream," 265
Darnton, Charles, 151
D' Arrast, Harry d'Abbadie, 170, 171
Davies, Marion, 149, 175, 183
Davis, Bette, 127, 159, 267
Davis, Owen, 28, 79, 123–25, 131, 253, 285, 286
Davis, Owen, Jr., 125
De Mille, Agnes, 301
De Mille, Cecil B., 154, 155, 301
De Mille, William, 14, 154
De Witt Clinton High School, 8, 12, 22
Dear Sir, 58
Deep River, 86
Del Ruth, Roy, 225
Dempsey, Jack, 28, 86
"Der Führer's Face," 301
D' Erlanger, Gerald, 100
Desert Song, The, 87, 100, 246
DeSylva, Buddy, 50, 83, 85, 139
Diaghilev, Sergei, 204, 211
Diamond, I.A.L., 14
Diamond, Mannie, 40
Diamond Lil, 116
Dickson, Dorothy, 97, 105
Die Fledermaus, 196
Die Tolle Dolly, 12
Die Walküre, 123
Dietrich, Marlene, 172, 179, 183, 224, 273
Dietz, Howard, 14, 52, 58, 87, 118, 126, 148, 152, 181, 184, 208, 241
Digges, Dudley 64, 243
Dillingham, Charles, 95, 106, 107, 112–14, 137, 148, 156, 196
Dinner at Eight, 175
Dixon, Peter, 208
Dizzy Club, 71

"Do I Love You? (No, I Do Not!)," 301
Dr. Jekyll and Mr. Hyde, 158
Doctors' Hospital, 1, 288, 295, 300, 304, 310,
    311
Dodsworth (novel), 195
Dolly Sisters, The, 67, 192
Don Q., Jr., 79
Donald Duck in Nutzi Land, 301
Donaldson, Walter, 125, 140, 184
Donen, Stanley, 273
Donohue, Jim, 137
"Dorothy and Dick" (radio), 260
Dorsey, Jimmy, 215, 239
Dorsey, Tommy, 215, 240
"Don't Be That Way," 240
"Don't Believe," 91
"Down At The Lake," 52
Down Beat Magazine, 240
Drake, Alfred, 217, 303
Dream Boy, 51–53
Dreyfus, Louis, 55, 97, 117
Dreyfus, Max, 50, 55, 64, 71, 91, 154, 270,
    290, 299, 311, 313
Dry Rot. See *White Cargo*.
Du Barry Was a Lady, 265
Du Bois, Raoul Pène, 265
Dubin, Al, 77, 78, 143, 177, 214
Duel in the Sun, 186
Dugan, Tom, 143
Duke, Vernon, 149, 294
Dulcy, 45, 204
Duna, Steffi, 214
Durant, Jack, 273, 276, 281, 291
Durante, Mrs. Jean Olsen, 185
Durante, Jimmy, 164, 181, 183–85, 204–6
Dwan, Allan, 181

Eames, Clare, 40
Earl Carroll Vanities (1928), 98,
Earl Carroll Vanities (1930), 153
Earl Carroll's Sketch Book, 208
Earth Between, The, 127
East Wind, 153
"Easter Parade," 196
Easy Come, Easy Go, 79, 123, 124
Eddy, Nelson, 188, 190, 226
Ederle, Gertrude "Trudy," 85
Edward, Prince of Wales, 97, 100
Edwards, Jack, 80
Eggerth, Marta, 268, 269
Eglevsky, André, 247
Ein Waltzertraum, 156
Eisenhower, Gen. Dwight D., 303

Eisman, Irving, 101, 192, 266–68, 287, 311
Eliscu, Edward, 124, 149
"Embraceable You," 148
Englesman, Ralph, 17
Equity, 29, 30, 44
Ernst, Leila, 259, 273
Erskine, Chester, 171
Ervine, St. John, 121
Etting, Ruth, 125, 139, 140
Evans, Wilbur, 301
Evelyn Goldsmith Home, 58
"Every Girlie Wants To Be A Sally," 39, 42
"Every Man Needs A Wife," 12
Everybody's Welcome, 303
"Everything I Have Is Yours," 187
"Everything's Gonna Be All Right," 93
Excess Baggage, 111

Fabray, Nanette, 292
Fadiman, Clifton, 292
Fanny, 86
Farewell To Arms, A (novel), 126
Farmer Takes a Wife, The, 196
Farnol, Lynn, 178
Farnum, Dustin, 29, 111, 155
Farnum, William, 155
Fata Morgana, 65
Fatal Wedding, The, 56
"Feelin' High," 184
Feiner, Ben, 84
Feiner, Dorothy, 84, 110, 135, 137, 140. See
    also Rodgers, Dorothy.
Felix, Seymour, 74, 88, 92, 181
Felix's Club, 71
Ferber, Edna, 14, 95, 111, 112, 234, 252, 286,
    287, 299
Fetter, Theodore, 315
Fielding, Henry, 309
Fields, Dorothy, 23, 26, 31, 35, 42, 44, 56, 62,
    116, 119, 187, 292, 303, 313
Fields, Herbert, 23, 26, 31, 33, 35, 41, 44, 55,
    56, 60, 62, 74, 76, 77, 79, 82, 87, 93,
    102–105, 110, 115, 119, 121, 135, 140,
    142, 143, 146, 148, 151, 198, 224, 287,
    292, 303, 307, 309, 310, 313
Fields, Herbert, amateur shows of:
    Danish Yankee in King Tut's Court, A, 104
    Fly With Me, 31–32
    If I Were King, 44, 62
    Prisoner of Zenda, The, 56
    Say Mama, 35
    Temple Belles, 56
    You'd Be Surprised, 31, 33

Fields, Herbert, librettos:
  America's Sweetheart, 149, 150, 151, 153
  Came The Dawn, 149
  Chee-Chee, 119–122
  Come Across, 149
  Connecticut Yankee, A, 103, 105–10, 115,
    306, 308–11
  Dear Enemy, 75
  Dearest Enemy, 74, 75, 88, 90, 255
  Du Barry Was A Lady, 265
  Fifty Million Frenchmen, 135
  Girl Friend, The, 75, 76, 88, 110
  Henky, 56
  Hit The Deck, 87, 115
  Jazz King, The, 55
  Let's Face It, 292
  New Yorkers, The, 147
  Peggy, 87, 88, 90
  Peggy-Ann, 92, 105, 108, 120
  Present Arms, 116, 117, 119, 142, 229
  Something For the Boys, 303
  Star Dust, 153
  Sweet Rebel, 60
Fields, Herbert, movie scripts:
  Fighting Coward, The, 198
  Fools for Scandal, 225, 226
  Good Bad Girl, The, 143
  Hot Heiress, The, 144, 151
  Love of Michael, 151, 152
Fields, Joseph, 224, 226
Fields, Lew, 14, 23, 26, 27, 31–34, 39, 41, 42,
  55, 56, 60, 74–76, 88, 92, 103, 105–10,
  106, 107, 108, 110, 115, 116, 119–122,
  124, 147, 196
Fields, W. C., 161, 162, 198, 199, 203
Fifth Avenue Club, 86
Fifty Million Frenchmen, 135
Firebrand, The, 58, 106
First Fifty Years, The, 39, 41
Fitzgerald, F. Scott, 79, 125, 191, 266
Fitzgerald, Geraldine, 303
Five Million, The, 28
Florodora, 17
Flying Down to Rio, 284
Follow A Star, 144
Follow Through, 139, 142
Fontanne, Lynn, 64, 116
"Fool Me With Kisses," 36
For Whom The Bell Tolls (novel), 268
Foran, Dick, 308
Ford, Dick, 209
Ford, George, 37, 68–70, 119, 311, 312
Ford, Harry, 69, 70, 119

Ford, Helen, 37, 43, 45, 47, 54, 60, 68–70, 73,
  84, 88, 90, 92, 119–21, 196, 311
Forman Sisters, The, 150
42nd Street, 177, 183,
Frank and Jack's Club, 71
Frazier, Brenda, 258, 262, 278
Frederici, Blanche, 161
Freed, Arthur, 175, 184, 187, 314
Freedley, Vinton, 111, 123, 125, 131, 133,
  134, 148, 149, 256
Freedman, David, 89, 91
Friedberg, Billy, 311
"Friendship," 265
Friml, Rudolf, 45, 70, 71, 81, 87, 111, 116, 147
Froman, Jane, 197, 292
Front Page, The, 204
Funny Face, 89, 111, 114

Gable, Clark, 176, 188, 189, 193, 226
Gabriel, Gilbert, 41, 150
Gallagher, Irene, 311, 312
Gallico, Paul, 300
Gang's All Here, The, 151
Garbo, Greta, 179, 181, 189, 191
Garden of Allah, The, 243
Gardiner, Reginald, as "Al Fleegle," 216
Garrett, Oliver H.P., 176
Garrick Gaieties, The (3rd), 149, 217
Gaxton, William, 106, 214
Gay, Maisie, 105
Gear, Luella, 212
Gelbfisz, Shmuel. See Goldwyn, Samuel.
Gensler, Lewis, 151
Gentle Grafters, 125
George White Scandals of 1926, 85, 98
George White Scandals of 1939–40, 265
Gerber, Alex, 34
Gershwin, George, 28, 50, 55, 73, 74, 87, 89,
  114, 123, 124, 126, 139, 152, 208, 216,
  223, 266
  death of, 227, 228
Gershwin, George, musical works of
  "An American in Paris," 123, 216
  "Rhapsody in Blue," 152, 216
  Song of the Flame, 77
Gershwin, George, songs of
  "Heaven on Earth," 87
  "Swanee" 28
  "That Certain Something You've Got," 87
Gershwin, Ira, 55, 72, 86, 87, 114, 123, 141,
  147, 149, 224, 231, 290
Gershwins, The (George & Ira), 59, 100, 120,
  136, 140, 153, 223

Gershwins, The (George & Ira), movie
    musicals of
  Goldwyn Follies, The, 223, 227, 241
Gershwins, The (George & Ira), musicals of
  Curtain Going Up! (projected), 223
  Funny Face, 89, 111, 114
  Girl Crazy, 148, 149, 216
  Lady, Be Good!, 73, 100
  Of Thee I Sing, 231
  Oh, Kay!, 87, 106, 131
  Porgy and Bess, 207, 208
  Rosalie, 114, 116
  Show Girl, 137, 140
  Strike Up The Band, 120
  Tip Toes, 72, 73, 77
  Treasure Girl, 123
Gershwins, The (George & Ira), songs of
  " 'S Wonderful," 111
  "But Not For Me," 148
  "Clap Yo' Hands," 87
  "Embraceable You," 148
  "He Loves And She Loves," 111
  "High Hat," 111
  "I Got Rhythm," 80, 148, 216
  "Looking For A Boy," 73
  "Love Walked In," 240
  "Man I Love, The," 114
  "My One and Only," 111
  "Oh, Kay, You're Okay With Me," 87
  "Our Love Is Here To Stay," 141
  "Someone To Watch Over Me," 87
  "That Certain Feeling," 73
Gerstenzang, Robert, 15, 38
"Get Happy," 139
Geva, Tamara, 148, 211, 212
Gibbons, Cedric, 189
"Gibson Family, The" (radio), 208
Gilbert, W. S, and Arthur Sullivan, 59, 150
Gingham Girl, The, 43, 54, 69
Girl Crazy, 148, 149, 216
Givot, George, 184
Gizi, Mizi, 12
Glaenzer, Jules, 123, 130, 143, 230
Glaenzer, Mrs. Kendall, 123, 130
Glass Slipper, The, 80
Glazer, Benjamin, 36, 200
Glimpse of the Great White Way, A, 26
Go West, Young Man, 102
Goat Song, 80
Goetschius, Percy, 38
Goetz, Ray, 153
Goetz, William, 189
Going Hollywood, 200

Going Places and Doing Sings, 203
Gold Diggers of 1933, 225
Golden, John, 312
Golden Dawn, 111
  Movie, 142
Goldfish, Samuel. See Goldwyn, Samuel.
Goldman, Edwin Frank, orchestra, 117
Goldwyn, Samuel, 154, 155, 164, 178–180,
    191, 198, 227, 234
Goldwyn Follies, The, 227, 241
Goldwyn Girls, The, 152
Gon, Jeni le, 226
Gone With The Wind, 88, 226, 267
Good Boy, 161
Good Fellow, The, 75
Goodman, Benny, 148, 239, 240
Gordon, Max, 148, 153, 263, 265, 302
Gould, Allan, 127, 128
Goulding, Edmund, 181, 182, 189
Grable, Betty, 152, 162, 265
Grafton, Gloria, 207
Graham, Ronald, 253, 297
Grand Hotel, 172
Grand Street Follies, The, 62
Gravet, Fernand, 224, 225, 229
Gray, Lawrence, 142
Great God Brown, The, 79
Great Waltz, The, 202
Green, John, 148, 295, 297, 298
Green, Mitzi, 217, 218
Green, Morris, 60, 151
Green Grow the Lilacs, 299–301. *See also*
    Rodgers & Hammerstein.
Greene, Charles and Henry, 157, 158
Greenwich Village Follies of 1924, The, 2, 3, 67, 77
Grey, Clifford, 116
Groody, Louise, 88, 113, 115
Guardsman, The, 65, 66
Guilford, Nanette, 196, 221, 300
Guinan, Texas, 68, 71, 93
Guys and Dolls, 148

Hale, Alan, 26
Hale, Sonnie, 98, 141, 201
Haley, Jack, 85, 268, 269
"Hallelujah!," 87, 115
Halliday, Hildegarde, 64, 65, 80
Halliday, Robert, 87
Hamilton, Nancy, 224
Hamlet, Prince of Denmark, 57
Hammerstein I, Oscar, 8, 28, 54, 111, 121
Hammerstein II, Oscar, 8, 14, 15, 17, 31, 36,
    38, 54, 58, 70, 86, 87, 95, 111, 115, 135,

Hammerstein II, Oscar, *(continued)*,142, 143, 148, 151, 153, 159, 161, 190, 198, 231, 241, 253, 263, 265, 287, 290, 298, 299, 301, 303, 310, 313, 316
Hammerstein II, Oscar, movie musicals of
  Champagne and Orchids (unproduced), 190
  Golden Dawn, 142
  Sunny, 143, 152
Hammerstein II, Oscar, musicals of
  Always You, 37
  American Jubilee, 252
  Ballyhoo, 148
  Camille (projected), 153
  Carmen Jones, 298, 306
  Daffy Dill, 43
  Desert Song, The, 87, 100
  Golden Dawn, 111
  Good Boy, 161
  Mary Jane McKane, 54, 115
  Music In The Air, 159, 247, 301
  New Moon, The, 89
  Rose-Marie, 54, 58, 87, 244
  Sally, 81
  Show Boat, 96, 111, 112, 116, 124, 168, 198, 200, 229, 231, 286
  Song Of The Flame, 77
  Sunny, 70, 71, 77, 96, 113
  Sweet Adeline, 135
  Very Warm For May, 265, 286
  Wild Rose, 87
  Wildflower, 54
Hammerstein II, Oscar, plays of
  Gypsy Jim, 54
  New Toys, 54
Hammerstein, Arthur, 37, 91, 95, 111, 147, 148
Hammerstein, Willie, 7
Hammond, Percy, 41
Hanemann, Henry W. 37
Happiness Ahead, 151
Happy Days, 30
Happy Landing, 225
Harbach, Otto, 70, 87, 95, 105, 107, 111, 153, 161, 196
Harburg, E. Y. "Yip," 149, 256, 305
Harling, W. Franke, 46, 47, 86, 152
Harlow, Jean, 172, 182–84, 186, 189
Harriet, 303
Harriman, Margaret Case, 251
Harris, Sam H., 29, 228–30, 232, 233
Hart, Dorothy, 4, 9, 30, 36, 237, 257, 258, 266, 267, 286, 288, 302, 305, 309, 310, 311, 313, 314

Hart, Frieda, 2, 3, 9, 10, 30, 36, 101, 102, 122, 158, 236, 237, 238, 255, 257, 262, 267, 302
  death of, 304, 305
Hart, James, 4
Hart, Lorenz Milton
  birth of, 4
  characterized, 48, 63, 92, 99, 110, 145, 150, 156, 157, 212, 220–22, 237, 243, 244, 248, 249, 253, 254, 261, 273, 277, 278, 288, 290, 291, 297, 310, 316, 317
  death of Frieda Hart and, 305
  death of Max Hart and, 122
  death of, 312
  "disappearances" of, 253, 261, 269, 288, 293, 295, 296
  drinking of, 72, 191, 248, 254, 258, 264, 266, 268, 274, 277, 286, 287, 289–92, 295, 296, 301, 302, 305, 309–311
  generosity of, 10, 66, 73, 192, 310
  hospitalizations of, 254, 255, 288, 295, 300, 310
  income of, 236, 251
  lifestyle of, 72, 102, 157, 158, 163, 172, 191–93, 220, 221, 236, 238, 257, 258, 289, 290
  low self-esteem of, 219, 288, 289, 291, 299, 310
  meets Richard Rodgers, 18
  obituary of, 2
  physical appearance of, 16, 243, 251
  Rodgers, Richard on, 18, 22, 129, 130, 211, 239, 254, 288, 308
  sexual ambivalence of, 20, 38, 39, 172, 238, 258, 296
  working methods of, 75, 76, 82, 118, 136, 139, 158, 159, 218, 244, 261, 278, 315
Hart, Lorenz, early works of
  "Alexander's Ragtime Band" (parody), 9
  "B.L.C.," 52
  "Bummel, Bummel, Bummel," 12
  "Chloe, Cling To Me," 42
  "Down At The Lake," 52
  Dream Boy, 51–53
  "Every Man Needs a Wife," 12
  "Fool Me With Kisses," 36
  "Hubby Dances on a String," 12
  "I Know My Girl By Her Perfume," 53
  "I Love To Lie Awake In Bed," 52, 126
  Inky Ike, the Ashbarrel Detective, 9
  "I've Got A Girl In Chestertown," 52
  "Kiki," 41
  "Kiss Lesson, The," 12

La Belle Helene, 36
"Last Night," 52
Liliom, 36
Mexico, 16
"Meyer, Your Tights Are Tight," 12
"Moonlight Lane," 47
"My Cameo Girl," 36
New Brooms, 8
"Oh, Mr. Postman, Don't Pass Me By," 52
"Pass It Along To Father," 9
"Pelican, The," 41
"Rock of Refuge, The," 12
"Stop! Stop! I Am The Traffic Cop!," 52
"Ticky, Ticky, Tack," 12
"Vixen," 41
[That] Lady in Ermine, 36
Hart, Lorenz, produces
    Blond Beast, The, 40, 43, 44
    First Fifty Years, The, 39, 41
Hart, Margie, 278
Hart, Max, 2–6, 40, 56, 7, 9, 11, 48, 49, 101,
    117, 262, 305
    death of, 122
Hart, Moss, 14, 101, 173, 180, 196, 202, 207,
    223, 224, 229, 230, 233, 255, 272, 287,
    313
Hart, Theodore Van Wyck "Teddy," 4, 6, 10,
    122, 158, 162, 192, 204, 237, 241, 253,
    255, 256, 286, 288, 302, 305, 309–11,
    313, 314
Hart family
    lifestyle of, 6, 7, 10, 101, 102, 237, 267
Hartwig, Brigitta. *See* Zorina, Vera.
Havoc, June, 273, 275, 276, 281, 291
Hayden, Rita (Heiden), 97
Hayes, Helen, 217, 303
Hayes, Rich, the Eccentric Juggler, 99
"He Loves and She Loves," 111
Healy, Peggy, 195
Healy, Ted, 151
Healy, Ted, and his Stooges, 181, 183, 188
Hearst, William Randolph, 90, 123, 176,
    233
Heart of Annie Wood, The, 37
"Heat Wave," 196
Heatherton, Ray, 217, 257
"Heaven on Earth" 87
Hecht, Ben, 170, 204, 214
"Heigh Ho, The Gang's All Here," 187
Helburn, Theresa, 64, 66, 80, 299, 302
Helen of Troy, New York, 45–47, 68
Hell-Bent fer Heaven, 13, 204
Hello, Paris, 147

Henderson, Ray, 50, 83, 85, 98, 139
Henson, Leslie, 96
Hepburn, Katharine, 287, 292, 293
Herbert, Victor, 21, 59
Here Comes Happiness, 151
Here Is My Heart, 200
Here Today, 179
Herman, Sophie, 305
Herman, Woody, 240
Hertz, Harry, 122, 305
Hertz, Meyer, 3
Hertz, Taubchen "Tessie," 3, 305
Heyman, Edward, 197
"High Hat," 111
High Kickers, 292,
him, 117
"Hindoo Moon," 38
Hines, Elizabeth, 82, 91, 96
History Is Made at Night, 243
Hit The Deck, 87, 115
Hitchy-Koo of 1922, 98
Hitler, Adolf, 210, 234, 267, 291
Hodges, Joy, 230
Hodkinson, W. W., 155
Hoffman, Dr. Richard, 288
Hoity-Toity, 25
Hold Everything!, 143
    movie, 142
Hold On To Your Hats, 256
Hold On To Your Hats, Boys!, 224
Holloway, Sterling, 64, 65, 80, 193
Hollywood Club, 195
Hollywood Music Box Revue, 96
Hollywood Revue, The, 181
Holm, John Cecil, 287, 289
Holman, Elizabeth (Libby), 64, 65, 126, 148,
    234, 255
Holmes, Taylor, 230
"Hooray For Captain Spaulding," 152
Hooray For What!, 240
Hoover, Herbert, 35, 123, 126, 167, 177
Hope of Heaven (novel), 271
Hopkins, Miriam, 117, 163
Hopper, De Wolf, 21
Hornblow, Arthur, Jr., 163, 198, 200
"Hot Choc'late Soldiers," 184
Hot-Cha!, 168, 181
Hotsy-Totsy, 71
"How Can I Ever Be Alone?," 252
How to Win Friends and Influence People (book),
    246
Howard, Sidney, 40, 65, 195
Howe, James Wong, 181

"Hubby Dances On A String," 12
Hughes, Hatcher, 13
Hulbert, Jack, 74, 77, 79, 82, 83, 96, 144
Hunchback of Notre Dame, The, 157
Hunter, Frank, 38, 39, 42
Hunter, Glenn, 125, 126
Hurlbut, Gladys, 268, 269
Huston, Walter, 195, 214, 231, 255
Hyman, Dr. Albert S., 122
Hyman, Dr. Harold, 295

"I Can't Give You Anything But Love, Baby,"
    116
"I Found A Million Dollar Baby In A Five
    And Ten Cent Store," 152
"I Got Rhythm," 80, 148, 216
"I Guess I'll Have To Change My Plan," 52,
    126
"I Hold Her Hand And She Holds Mine," 50
"I Know My Girl By Her Perfume," 53
"I Let A Song Go Out of My Heart," 240
"I Like A Big Town," 47
"I Love To Lie Awake In Bed," 52
"I Love Louisa," 152
"I See Your Face Before Me," 241
"I Want To Be Happy," 70
I Was An Adventuress, 268, 271
"I Wonder Who's Dancing With You
    Tonight?," 51
"If I Had A Girl Like You," 51
"If I Knew," 38
If I Were King, 44
Ileana, Princess, of Rumania, 85, 89, 114
"I'll Be Loving You, Mona," 78
"I'll Remember Only You" 175
"I'm Always Chasing Rainbows," 92
"I'm Crazy 'bout The Charleston," 82
"I'm In Love," 105
"I'm In Love Again," 67
"I'm So Shy," 23
In Caliente, 214
In Old Chicago, 227
Inky Ike, The Ashbarrel Detective, 9
Innocents of Paris, 156
Institute of Musical Art, 38, 42, 45, 104
"It All Depends On You," 83
"It Happened In The Dark," 301
It Happened One Night, 188
"It Was Meant To Be," 47
"It's Nobody's Fault But Mine," 51
"It's Terrific When I Get Hot," 162
"I've Got A Girl In Chestertown," 52
"I've Had My Moments" 183, 184

Jackson, Robert, 68
Janney, Russell, 44, 45
Jazz à la Carte, 42
Jazz Singer, The, 79
    movie, 108
Jeans, Isabel, 225
Jeans, Ronald, 79, 97
Jessel, George, 45, 46, 68, 79, 292
Johnson, Van, 259, 273
Johnston, Arthur, 77
Jolson, Al, 28, 86, 108, 142, 147, 164, 169,
    170, 171, 177, 256
Jones, A. L., 60
Juilliard School of Music, 38
June Days, 67, 82
"June in January," 200

Kahn, Gus, 105, 140, 184, 193
Kahn, Otto, 45, 123, 205
Kalmán, Emmerich, 111, 142, 300, 301, 313
Kalmán, Yvonne, 301
Kalmar, Bert, 45, 47, 50, 60, 79, 107, 113,
    125, 143, 161, 163, 292
Kane, Helen, 132
Karloff, Boris, 285
Katz, Sam, 168, 200
Kaufman, Beatrice, 234
Kaufman, George S., 14, 45, 46, 68, 72, 75,
    77, 78, 101, 120, 125, 126, 152, 153, 172,
    173, 223, 224, 229–33, 238, 255, 270
Kaufman, Harry, 202, 203, 256
Kaufman, S. Jay, 41, 127
Kaye, Benjamin, 61, 62, 64, 80
Kaye, Danny, 292, 313
Keeler, Ruby (Mrs. Al Jolson), 137, 147, 151,
    169, 177
Kelly, Gene, 255, 273, 274, 277–280, 281, 291
Kelly, George, 79
Kern, Jerome, 14, 19, 21, 24, 55, 58, 67, 70,
    81, 86, 93, 107, 111, 124, 125, 135, 153,
    159, 190, 196, 198, 212, 246, 252, 263,
    265, 299, 305, 313
Kern, Jerome, movie musicals of
    Can't Help Singing, 305
    Caroline. See Can't Help Singing.
    Cat and the Fiddle, The, 246
    Champagne and Orchids, 190
    Look For the Silver Lining, 306
    Sally, 142
    Sunny, 143, 152
Kern, Jerome, musicals of
    Camille, 153
    Cat and the Fiddle, The, 153, 212

Criss Cross, 95
Dear Sir, 58
Good Boy, 161
Lucky, 107, 125
Miss 1917, 24, 26, 246
Music In The Air, 159, 247, 301
Roberta, 196
Sally, 39, 81, 97, 138
Show Boat, 86, 96, 111, 112, 116, 124, 168,
   198, 208, 229, 231, 286
Sunny, 71, 70, 77, 96, 113
Sweet Adeline, 135
Very Good Eddie, 21
Very Warm For May, 265, 286,
Kern, Jerome, songs of
   "All The Things You Are," 265, 286
   "Bill," 112
   "Can't Help Lovin' Dat Man," 112
   "Look For The Silver Lining," 39, 93, 306
   "Make Believe," 111, 112
   "Ol' Man River," 112
   "Smoke Gets In Your Eyes," 196
   "Till The Clouds Roll By," 81
   "Two Little Bluebirds," 71
   "Who?," 71, 152
   "Why Do I Love You?," 112
   "Why Was I Born?" 135
   "You Are Love," 112
Kerry, Norman, 157, 158
Kesselring, Joseph, 285
Kibitzer, 127
Kid Millions, 180
Kiki, 41
"Kiki," 41
Kilgallen, Dorothy, 260, 262
King, Charles, 113, 115, 117, 127
King, Dennis, 70, 116, 244, 245
King and the Chorus Girl, The, 225
King of Jazz, 152, 200
Kiss and Tell, 303
"Kiss Lesson, The," 12
Kitty's Kisses, 105
Knickerbocker Holiday, 255, 293
Kohlmar, Fred, 179
Kollmar, Richard, 259–62, 293
Kollo, Walter, 12
Kostelanetz, André, 257, 292
Krimsky, John, 252
Kron, William, 251, 252, 258, 311, 313, 314
Kronert, Hans, 12

La Belle Hélène, 36
La Bohème, 218
La Conga Club, 258, 262

La Guardia, Fiorello, 228, 233
Lady Be Good!, 73, 100
Lady Comes Across, The, 294, 297
Lady Fingers, 123, 124
Lady in the Dark, 224, 284, 290
Lady in Ermine, The, 36, 41
Lady in Ermine, That (movie), 142
Laemmle, Carl, 127
Lahr, Bert, 90, 168, 216, 265
Laid in Mexico. See *Hot-Cha!*
Lake, Harriette. *See* Sothern, Ann.
Lambs Club, 154, 290
Lamour, Dorothy, 227
"Land Where The Camp Songs Go, The,"
   24
Lane, Burton, 187, 188, 256
Langdon, Harry, 170
Langner, Lawrence, 64, 80, 299
Lardner, Ring, 14, 148
Lark, Charles Tressler, 104, 105, 306
Lasky, Jesse, 127, 130, 143, 154–56, 166, 168,
   306
"Last Night (The)," 52
Laurel, Stan, and Oliver Hardy, 181
Lawrence, Gertrude, 74, 77, 87, 123, 124,
   145, 179, 234
Laye, Evelyn, 98, 136, 141, 241
Lazybones, 125
Le Sueur, Lucille. See Crawford, Joan.
Leave It To Me, 255
Leavitt, Philip, 3, 17, 18, 23, 59
LeBreton, Flora, 116, 117
Lee, Gypsy Rose, 203, 273, 274, 278, 279
Leftwich, Alexander, 105, 106, 115, 119, 121,
   126, 131
LeMaire, Rufus, 45, 46, 68
Leonard, Robert Z., 186
Lerner, Alan Jay, 39, 131, 286, 291, 304, 311
LeRoy, Hal, 260, 261
LeRoy, Mervyn, 177, 224, 225, 229
Let's Face It, 292
"Let's Not Talk About Love," 292
Levy, Benn, 141
Levy, Jacob, 21, 118
Levy, Rachel, 21
Levy, William Auerbach, 251
Lewis, Sinclair, 14, 45, 195
Life Begins at 8:40, 202
"Life Is A Merry-Go-Round," 187
Lightnin', 28
Liliom, 36, 131
Lillie, Beatrice, 74, 77, 107, 112–14, 124, 144,
   201, 216, 234, 274
Lippmann, Walter, 278

Little Caesar, 225
Little Johnny Jones, 164
Little Nemo, 21
Little Pal. See *Say It With Songs*.
Little Show, The, 52, 126, 148
Lockridge, Francis and Richard, 285
Loeb, Philip, 64–66, 80
Loewe, Frederick, 304, 311
Logan, Ella, 206, 207
Logan, Joshua, 238, 241, 243–46, 247, 249, 255, 256, 267–70, 274, 293, 295–98, 301, 304, 316, 317
Lombard, Carole, 200, 225, 230
Lonely Romeo, A, 26, 27
Long, Lois, 208, 209
Look For The Silver Lining, 306
"Look For The Silver Lining," 93
Look Homeward, Angel (novel), 127
"Looking For A Boy," 73
"Lorenz, Herbert Richard," 56, 57
Losch, Tilly, 135, 152
Lost Horizon, 226
Louis Bergen's, 72
"Louise," 156
Louis's 21 West 43rd Street ("21" Club), 71, 273
"Love In Bloom," 200
"Love Is Just Around the Corner," 200
"Love Me or Leave Me," 125, 140
"Love Nest," 92
Love Parade, The, 156, 162, 189
"Love Walked In," 240
Loving Ann, 123, 125
Low and Behold, 192
Loy, Myrna, 160, 161, 163, 176, 198
Luana, 147
Lubitsch, Ernst, 156, 163, 189, 190
Luce, Clare Booth, 223, 268
Lucky, 107, 125
"Lucky Kentucky," 51
Luddy, Barbara, 227
Lunceford, Jimmy, 240
Lunchtime Follies, 308
Lynn, Eve, 26
Lyon, Ben, 143, 144
Lyons, Leonard, 262

"Ma, He's Making Eyes at Me," 50
MacArthur, Charles, 204, 205, 207, 214
MacDonald, Jeanette, 73, 127, 151, 156, 160, 163, 173, 189, 190, 201, 226, 246, 247
Macfarlane, Bruce, 195
Mackay, Ellin, 78
Magnolia, 198

"Make Believe," 111, 112
Make Me A Star, 183
"Makin' Whoopee," 125
Malin, Jean, 192, 193
Mammy, 142, 169
Mamoulian, Rouben, 158, 159, 301
"Man I Love, The," 114
Mana-Zucca, Madame. See Zuccaman, Augusta.
Mandel, Frank, 54, 87, 149, 153
Maney, Richard, 111, 206, 207
Manhattan Madness, 157
Mankiewicz, Herman, 14, 15, 75, 161, 163
Mankiewicz, Joseph, 161–63, 176, 181–83
Manson, Frances, 191
Mantle, Burns, 41, 86, 113
March of Time, The, 173, 180
March, Fredric, 57, 158, 163, 225, 226, 234, 306
Marie, Queen, of Rumania, 85, 89, 114, 181
Marion, Frances, 180
Marion, George, Jr., 256, 258, 260, 262, 298
Martin, Francis, 198
Martin, Hugh, 290
Martin, Quinn, 57
Marx, Groucho, 163, 223
Marx, Harpo, 14
Marx, Marie (Mrs. Samuel Marx), 302
Marx, Samuel, 174, 181, 182, 185, 290, 302
Marx Brothers, The, 77, 78, 127, 152
Mary Jane McKane, 115
"Mary, Queen of Scots," 33
Mata Hari, 179
Matthews, Jessie, 98–100, 135, 136, 141, 142, 145, 146, 201, 294
"Maxim's," 193
Maxwell, Elsa, 130, 234
May, Ada, 142
Mayer, Edwin Justus, 40, 58, 106, 200
Mayer, Louis B., 168, 172, 175, 177, 184, 185, 188, 189, 247
Maytime, 226
McConnell, Lulu, 33, 88, 93, 147
McDermott's (club), 71
McDonald, Grace, 218
McGowan, Jack, 132
McGuire, William Anthony, 89, 90, 114, 116, 170
McHale, Duke, 217
McHugh, Jimmy, 78, 79, 116, 187
McNally, Terence, 14
Me, 79, 181
"Me And My Shadow," 51
Meiser, Edith, 61–66, 72, 75, 80, 88, 193, 238, 292

Melody in Spring, 198
"Melody of Spring," 193
Mercer, Johnny, 149, 215
Merchants of Glory, 80
Merman, Ethel, 148, 200, 216, 224, 265, 294,
   303
Merry Widow, The, 301,
   movie, 190, 191, 193, 201,
Merton of the Movies, 125
   movie, 183
Messer Marco Polo, 263
"Messieurs, Mesdames," 301
Metronome Magazine, 240
Mexico, 16
"Mexiconga, The," 265
Meyer, Joseph, 50, 72, 73, 77, 124, 143
"Meyer, Your Tights Are Tight," 12
Mickey Mouse, 181, 184, 279
Middleton, Charles, 165
Midsummer Night's Dream, A, 265
Mielziner, Jo, 275, 281, 295
Milestone, Lewis, 170, 171
Miller, Glenn, 148, 240
Miller, Marilyn, 39, 70, 91, 113, 114, 116,
   142, 143, 148, 152, 196, 210, 211, 306
Million Dollar Legs, 161, 162
Miner, Worthington, 212
Minnevitch, Borrah, 93, 94
Miranda, Carmen, 274
Miss 1917, 24, 26, 246
Miss Underground, 300, 301
Mississippi Flood Relief concert, 217
Mr. and Mrs. North, 285
Mr. Smith Goes to Washington, 267
"Moanin' Low," 126
Molnar, Ferenc, 36, 130, 131, 144
Monaco, Jimmy, 50, 183
Monte Carlo, 152
"Moonlight Lane," 47
Moore, Clint, 238
Moore, Constance, 294, 297
Moore, Grace, 130, 190, 234
Moore, Victor, 79, 131, 132
Moran, Polly, 181, 183, 184
Morgan, Frank, 171, 180
Morgan, Helen, 86, 112, 135, 153, 208, 209
Morning's at Seven, 268
Morris, Davy, 237
Morse, Lee, 139
Mount Sinai Hospital, 254
Mount Zion Cemetery, 122, 305, 313
Movie studios
   Columbia, 174, 175, 188, 191, 302
   Cosmopolitan Pictures, 176

Disney, 240
Famous Players-Lasky, 155, 156
First National, 131, 142
Fox, 127, 174, 188, 225
Gaumont-British, 201
Lasky Feature Play Company, 155
Metro-Goldwyn-Mayer (MGM), 127, 130,
   168, 169, 172–77, 179, 181–83, 185, 189,
   190, 193, 195, 196, 200, 223, 236, 246,
   302, 305, 314
Paramount, 10, 117, 123, 124, 127, 130,
   142, 154–56, 161, 162, 164, 168, 169,
   198, 199–201, 207, 213, 230, 294
Paramount-Famous Players-Lasky, 127
Pioneer Pictures, 213
RKO Radio Pictures, 127, 142, 175, 176,
   182, 188, 199, 213, 284
20th Century, 175, 189
20th Century-Fox, 189, 227, 271, 293,
   294
United Artists, 117, 169, 170, 174, 179,
   306
Universal, 127, 292, 294
Warner Bros., 108, 142, 146, 149, 151, 152,
   169, 175, 223, 224, 247, 268, 287, 306
Mowbray, Alan, 165, 175
Muni, Paul, 85, 234
Munson, Ona, 88, 143, 144
Murphy, Owen, 67, 151
Murray, Ken, 208, 209, 227
Murray, Paul, 77, 79, 82, 96
Murray, Wynn, 217, 253
Music Box Revue, The, 60, 98
"Music By Gershwin" (radio), 208,
Music In The Air, 159, 247, 301
Music Master, The, 55
Music publishers
   Bourne, 50
   Chappell & Co., 299
   Edward B. Marks, 66
   Famous Music, 162
   Feist, Leo, 50
   Harms, T. B., 50, 71, 97, 117, 142
   Mills and Witmark, 50
   Remicks, 50, 142
   Rodart Music Publishing Company, 154,
     201
   Shapiro, Bernstein & von Tilzer, 50, 52
   Witmark, 142
Mussolini, Benito, 210, 291
Mutiny on the Bounty, 193
"My Cameo Girl," 36
"My Heart Belongs To Daddy," 255
"My Ideal," 156

My Man Godfrey, 225
"My One And Only," 111
Myers, Harry, 104
Myers, Henry, 13, 39, 40, 42–44, 79, 122,
   126, 149, 161, 162, 181, 220, 316
Myers, Mrs. "Muzzie," 43

Nana, 178, 190–91, 201, 272
Nathan, George Jean, 57, 113, 151, 234, 262
NBC Radio Network, 117, 167, 226, 255
Negri, Pola, 85, 179
Nervous Wreck, The, 123, 125
New Faces, 196, 215, 224
New Moon, The, 89
New York Metropolitan Opera, 196
New Yorker (club), 191, 192
New Yorker, The (magazine), 62, 151, 210,
   250, 251, 271, 279, 281, 287
New Yorkers, The, 147, 149
Newberry, Barbara, 134
Newman, Alfred, 179
Newman, Greatrex, 96
Nicholas Brothers, The, 217, 218
Night in Paris, A, 182
Night Out, A, 115
9:15 Revue, The, 139
No Other Girl, 60
No, No, Nanette, 70, 71, 77, 88
   movie, 142, 143
"Nobody Knows The Trouble I've Seen,"
   218
"Nobody Wants Me," 86
Norths Meet Murder, The (novel), 285
Nothing Sacred, 225
Novis, Donald, 207
Nymph Errant, 179

Oakie, Jack, 162, 198
Of Thee I Sing, 231
"Oh, Johnny!" 267
Oh, Kay!, 87, 106, 131
"Oh, Kay, You're Okay With Me," 87
"Oh, Mr. Postman, Don't Pass Me By," 52
O'Hara, John, 228, 271, 273, 275–77, 278,
   280, 281, 288, 313
O'Keefe, Walter, 257
Oklahoma!, 301–5
"Ol' Man River," 112
"Old Fashioned Girl, An," 63, 65
"Old Folks At Home, The," 200
On Borrowed Time, 243
On The Line, 256
On Your Marks, 161
Once In A Lifetime, 173

One Hour With You, 156, 189
One Minute, Please, 17
"One Who Yells The Loudest Is The
   Captain, The," 301
O'Neill, Eugene, 40, 79, 116, 195
"Only A Rose," 71
Only Saps Work, 124
Oppenheimer, George, 179, 272
Osborn, Paul, 268
"Otto's Patter," 301
"Our Love Is Here To Stay," 141
Outside Looking In, 86
Owen, Catherine Dale, 131

Pal Joey (novel), 271
Pancho's Club, 258
Paradise Club, 195
Paramount on Parade, 156
Paris, 143
Park Avenue Synagogue, 56
Parker, Dorothy, 14, 138, 150, 151, 179, 287,
   313
Parsons, Donovan, 82
Parsons, Louella, 143, 173
Pascal, Milton, 7, 124, 134, 236, 266, 305,
   311, 314
"Pass It Along To Father," 9
Patriots, The, 303
Patterson, Elizabeth, 161
Pearl, Jack, 168, 175, 181–183
Peerce, Jan, 292
"Pelican, The," 41, 42
Penner, Joe, 168, 198, 255
Perilman, Rabbi Nathan, 313
Personal Appearance, 102
Petit, Charles, 118
Petrified Forest, The, 226
Phantom of the Opera, 157
Phantom President, The (novel), 163
Philadelphia Story, The (movie), 287
Pickert, Rolly, 218
Pidgeon, Walter, 143, 203
Pied Piper, The, 21
Pincus, Irving, 253, 264
Pincus, Norman, 264
Pinto, Effingham A., 44
Pitkin, Walter B., 14
Pitts, Zasu, 183
Platt, Marc, 266
"Play Me A Tune," 98
Playboy of Paris, 156
Play's The Thing, The, 130
Please, 201, 216
"Please Be Kind," 240

Poe, Aileen, 33
"Polka Dots And Moonbeams," 282
Polly With a Past, 91
Popular Songs (magazine), 219
Porgy, 158, 160
Porgy and Bess, 207, 208
Porter, Cole, 81, 98, 112, 135, 143, 147, 149,
   153, 179, 207, 232, 237, 238, 255, 265,
   292, 296, 303, 313
Porter, Cole, movie musicals of
  Gay Divorcée, The, 199
  Paris, 143
Porter, Cole, musicals of
  Anything Goes, 294
  Du Barry Was A Lady, 265
  Fifty Million Frenchmen, 135
  Hitchy-Koo of 1922, 98
  Jubilee, 207
  Leave It To Me, 255
  Let's Face It, 292
  New Yorkers, The, 147, 149
  Nymph Errant, 179
  See America First, 112, 113
  Something For the Boys, 303
  Star Dust, 153
  Wake Up and Dream, 135, 141
  You Never Know, 255
Porter, Cole, songs of
  "At Long Last Love," 255
  "Begin The Beguine," 209
  "Friendship" 265
  "I'm In Love Again" 67
  "Let's Not Talk About Love," 292
  "My Heart Belongs To Daddy," 255
  "Play Me A Tune," 98
  "Well, Did You Evah?," 265
  "What Is This Thing Called Love?," 135
  "You Do Something To Me," 135
Porter, Linda (Mrs. Cole Porter), 81, 232
Possessed, 188
Post-Depression Gaieties, The, 202
Powell, William, 176, 225
Power, Tyrone, 192, 227, 279
Powers, Tom, 40, 116
Price, Georgie, 239
Princess Flavia, 77
"Prisms, Plums and Prunes," 61
Prisoner of Zenda, The, (novel), 77
Prisoner of Zenda, The (amateur show), 56
Private Lives, 136, 145
Producers, The, 301
Puck, Eva, 57, 74, 75
Purcell, Charles, 33

Quillan, Eddie, 181
Quo Vass Iss?, 25

Rachmann (producer) 12
Radio shows
  "Amos 'n' Andy," 117
  "Charlie McCarthy Show," 227
  "Family Party," 117
  "Gibson Family, The," 208
  "Let's Have Fun," 208, 209
  "Tune-up Time," 257
  "Dorothy and Dick," 260
  "Music by Gershwin," 208
  "Ziegfeld Follies of the Air," 182
Rahe, Rudi, 12
Rain, 171, 186
  movie, 170
Rainger, Ralph, 126, 162
Ralph's (bar), 72, 253
Rambeaux, Marjorie, 293
Ramblers, The, 107
Rambova, Natacha, 106
Ramona, 195
Rand, Sally, 279
Randolph, Elsie, 105
Rapf, Harry, 173, 180, 182–85
Raphaelson, Samson, 189
Raulston, Judge John T., 69
Red Dawn, 28
Red-Headed Woman, 172
Reece, Kathryn, 127, 128
Reinhardt, Max, 18, 303
Replogle, Leonard, 89
Return Engagement. See *Food for Scandal.*
"Rhapsody in Blue," 152, 216
Rich, Buddy, 240
Rich, Freddie, orchestra, 208
Richardson, Louise, 65
Richmond (club), 71
Right This Way, 241
Rio Rita, 81, 96
Riptide, 189
Ritchard, Cyril, 96
Ritz-Carlton Nights, 131
River of Romance, 198
Rivkin, Allen, 187
Road to Rome, The, 272
Road To Singapore, The, 183
Robbins, J. N., 16
Robbins, Jack, 193, 194
Robert, Renée, 58
Roberta, 196
Roberti, Lyda, 162, 198

Robinson, Edward G., 127, 159, 237
"Rock Of Refuge, The," 12
"Rockabye Your Baby With A Dixie
    Melody," 86
Rocky, Paal and Leif, 192, 193, 196
Rodgers, Dorothy, 142, 166, 169, 173, 201,
    224, 232, 236, 239, 257, 262, 284, 292,
    311, 313
Rodgers, Linda, 203
Rodgers, Mamie, 21
Rodgers, Mary, 150, 255
Rodgers, Mortimer, 17, 20–22, 36, 140, 203,
    313
Rodgers, Richard
    characterized, 22, 36
    Lorenz Hart on, 130
    letters to wife of, 171, 199
    lifestyle of, 130
    meets Lorenz Hart, 18
    partnership with Oscar Hammerstein II,
        286, 287, 290, 298, 301
    social aspirations of, 266
Rodgers, Richard, amateur shows
    Chinese Lanterns, 42
    Jazz à la Carte, 42
    One Minute, Please, 17
    Prisoner of Zenda, The, 56
Rodgers, Richard, early songs
    "Every Girlie Wants To Be A Sally," 39, 42
    "I'm So Shy," 23
    "Land Where The Camp Songs Go, The,"
        24
    "Old-Fashioned Girl, An," 63, 65
    "Prisms, Plums And Prunès," 61
    "Twinkling Eyes," 32
Rodgers, Richard, musical works
    "Danse Grotesque à la Nègre," 98
    "Princesse Zenobia, La," 211
    "Slaughter on Tenth Avenue," 211, 212,
        215
    "Ghost Town," 266
    "Nursery Ballet," 255
Rodgers, Dr. William, 17, 20, 21, 62, 123,
    196, 313
Rodgers & Hammerstein, 178, 218, 301–305
Rodgers & Hammerstein, early songs of
    "Can It," 17
    "There's Always Room For One More,"
        17, 32
    "Weaknesses," 17, 32
Rodgers & Hart,
    arguments of, 64, 103, 219, 244, 274
    break-up of, 299

divergent lifestyles of, 219
working methods of, 31, 273
Rodman, Gene, 254, 266, 274
Rogers, Buddy, 139, 168, 198
Rogers, Ginger, 148, 188, 199, 225
Rogers, Will, 82, 91, 168
Rolando, Rose, 65
Roly Poly, 26
Romberg, Sigmund, 34, 41, 77, 87, 89, 114,
    147, 153, 190
Ronell, Anne, 305
Room Service, 204, 253
Roosevelt, Eleanor, 312
Roosevelt, Franklin D., 167, 174, 177, 211,
    224, 229, 232, 234, 246, 291, 294, 302,
    310
    "Good Neighbor" Policy of, 213, 258, 284
Rosalie, 89, 114, 116
Rose-Marie, 87, 244
"Rose of Washington Square," 216
Rose, Billy, 49, 50, 51, 56, 57, 73, 77, 83, 86,
    147, 149, 152, 195, 203, 206, 207, 210,
    239, 299, 306
Rosen, Max, 197
Rosett, Marshall, 52
Ross, Lanny, 198–200
Ross, Shirley, 176, 181, 269
Rothafel, Samuel L. "Roxy," 250
Rowland, Roy, 181
Royston, Roy, 105
R.S.V.P., 83
Ruby, Harry, 45, 47, 50, 60, 79, 107, 113,
    125, 143, 161, 292
Ruggles, Charles, 160, 161
Runyon, Damon, 78, 106
Russell, Lillian, 7
Ryskind, Morrie, 11, 14, 65, 66, 77, 86, 239

"'S Wonderful," 111
Sailor Beware!, 195
Sally, 39, 81, 97, 138
    movie, 142
Sandalwood, 125
Saratoga Trunk (novel), 286, 287, 299
Sardi's (restaurant), 247, 262, 273, 304
Savo, Jimmy, 241, 252, 253
Say It With Songs, 169
Schenck, Joseph, 117, 169–72, 175, 189
Schenck, Nicholas, 127, 157, 169, 189
Schildkraut, Joseph, 36
Schloss, Herbert, 40
Schulberg, B. P., 155, 161, 168
Schulberg, Budd, 155, 277

Schwab, Laurence, 54, 55, 64, 149, 153
Schwartz, Arthur, 9, 47, 48, 51–53, 86, 88, 99,
      100, 126, 148, 152, 208, 216, 241, 252,
      253, 270, 313
Schwartz, Arthur, musicals of
      American Jubilee, 252
      Virginia, 253
Schwartz, Arthur, songs of
      "Baltimore, Md., You're The Only Doctor
          For Me," 86
      "How Can I Ever Be Alone?," 252
      "I Guess I'll Have to Change My Plan,"
          126
      "I Know My Girl By Her Perfume," 53
Schwartz, Arthur, and Howard Dietz,
      musicals of
      At Home Abroad, 207
      Between The Devil, 241
      Little Show, The, 126, 148
      Three's a Crowd, 148, 212
Schwartz & Dietz, songs of
      "By Myself," 241
      "Dancing In The Dark" 152
      "Hammacher, Schlemmer, I Love You,"
          126
      "I Love Louisa," 152
      "I See Your Face Before Me," 241
      "Something To Remember You By," 148
"Second Hand Rose," 152
Second Little Show, The, 207
See America First, 112, 113
See My Lawyer, 256
Segal, Vivienne, 87, 116, 244, 245–47, 250,
      273, 274, 277, 281, 295, 301, 307–10,
      317
Selwyn, Edgar, 157, 169
Selwyn, Mrs. Ruth, 157
Selwyn brothers, 155
Selznick, David, 143, 163, 176, 175, 182, 183,
      186–88, 189, 213, 226, 234, 243
Selznick, Irene (Mrs. David Selznick), 143, 163
Selznick, Myron, 181
Sennett, Mack, 200, 225
"September Song," 255
Set A Thief, 105
Seven Year Itch, The, 37
Shaler, Eleanor, 65
Shanghai Gesture, The, 79
Shapiro, Elliott, 52, 64
Sharaff, Irene, 295
Shauer, Mel, 10, 36, 39, 41, 154, 156, 157,
      158, 163, 190, 191, 197, 255
Shaw, Artie, 239, 240

Shawn, Ted, 245
She Loves Me Not, 200
Shean, Al, 93
Shearer, Norma, 172, 189
Shotgun Wedding, 125
Show Boat, 86, 96, 111, 112, 116, 124, 168,
      198, 208, 229, 231, 286
Show Girl, 137, 140
Show of Shows, The, 149
Showgirl in Hollywood, 142
Shubert, J. J., 15, 124, 252, 279
Shubert, Lee, 15, 29, 196, 202–4, 252, 279
Shubert Organization, 2, 15, 26, 36, 39, 41,
      77, 147, 148, 153, 170, 182, 196, 207,
      240, 245, 246, 256, 274, 303
"Shuffle Off to Buffalo," 177, 183
Sillman, Leonard, 191, 192, 217, 241
Simon, Robert, 31, 58, 151
"Sing, You Sinners," 86
Sing Out The News, 255
Skyrocket, 127
Slezak, Walter, 244, 245, 248
Smiles, 148
Smiling Lieutenant, The, 156
Smith, Paul Gerard, 132
Smith, Queenie, 47, 73
"Smoke Gets In Your Eyes," 196
Snapshots of 1921, 39, 42
Snark Was a Boojum, The, 286
      novel 285
Snow White and the Seven Dwarfs, 240
Snyder, Moe "the Gimp," 125, 139
Sobel, Larry, 100
Sobel, Louis, 262
Soma, Tony. See Tony's.
"Someday," 71
"Someone To Watch Over Me," 87
Something For The Boys, 303
Something Gay, 203
"Something To Remember You By," 148
Sometime, 37
"Sometimes I'm Happy," 87, 115
Son of the Grand Eunuch, The (novel), 118
Son of the Sheik, 179
Sondheim, Herbert, 8
Song of the Flame, 77
"Sonny Boy," 169, 216
Sons o' Guns, 170
Sons and Soldiers, 303
Sothern, Ann, 149, 153, 198
Spigelgass, Leonard, 187, 236, 237
"Springtime for Hitler," 301
Squall, The, 110, 111

Squaw Man, The, 155
St. Cyr, Lili, 279
St. Valentine's Day Massacre, The, 126
Stage Door Canteen, 306
Stagecoach, 86
Stallings, Laurence, 86, 201–3, 253
Stanislavsky, Konstantin, 178
Star Dust, 153
Star Is Born, A, 226
Starbuck, Betty, 64, 65, 80, 88, 119, 132
Steinfeld, Roselee, 59
Steinke, Arthur, 12
Sten, Anna, 178, 179, 191
"Step On The Blues," 105
Step This Way, 26
Stepping Stones, 95
Stern, G. B., (Lady Aberconway), 239
Sternberg, Josef, 179
Stevens, Onslow, 175
Stickney, Dorothy, 243
"Stop! Stop! I Am The Traffic Cop!," 52
Stork Club, 72, 273
Stothart, Herbert, 37, 43, 91, 111, 142, 175
Strange Interlude, 40, 116
Strauss, Hattie, 305
Strauss, Johann, 120
Strike Up The Band, 120
Strouse, Irving, 39, 40, 58
Sullivan, Arthur, 65
Sullivan, Ed, 262
Sun Also Rises, The (novel), 86
Sunny, 70, 71, 77, 96, 113
"Sunny Disposish," 86
Sunshine Boys, The, 26
Swan, The, 144
"Swanee," 89
Sweet Adeline, 135
Sweet Little Devil, 55
Sweet and Low, 147, 152
Sweethearts and Wives, 143
Swift, Kay, 126, 149
Swingin' The Dream, 265
Swope, Herbert Bayard, 14, 123

Taiz, Lillian, 125, 126
Take Them Up Tenderly (book), 251
"Tales of Hoffman," 91
Talmadge, Norma, 22, 138, 169
Tapps, Georgie, 291
Tarzan of the Apes, 172
Tarzan the Untamed (novel), 185
Taurog, Norman, 164, 165
"Tea For Two," 70, 89

Teagarden, Charlie, 215
Teagarden, Jack, 148, 215
"Tell Me, Pretty Maiden," 17
"Temptation," 200
"Tennessee Fish Fry," 253
Tester, Ruth, 127, 128
Texas Guinan's Club, 71
Thalberg, Irving, 172, 173, 175, 181, 189, 190, 195, 215
"That Certain Feeling," 73
"That Certain Something You've Got," 87
That Lady in Ermine. See *Lady in Ermine*.
"That Old Gang Of Mine," 50
Theatre Guild, The, 36, 40, 61, 63, 65, 66, 72, 80, 88, 149, 158, 299, 301
Theatres (New York unless otherwise stated)
  Adelphi, London, 146
  Alvin, 111, 133, 134, 235, 264, 303
  Apollo, 8, 43, 75
  Baker Theater, Rochester, 124
  Belasco, 285
  Belmont, 86
  Biltmore, 275, 303
  Broadhurst, 150
  Cambridge, London, 144
  Carlton, London, 97
  Casino, 26, 87
  Center Theatre, 202
  Central, 33, 37
  Colonial, Boston, 138, 231
  Colonial Theatre, Akron, Ohio, 69
  Columbia, 28
  Detroit Opera House, 39
  Detroit Shubert, 132
  Deutsches Theatre, 12
  Ethel Barrymore, 279, 291
  Fifth Avenue, 72
  Ford's Opera House, Baltimore, 69
  Ford's Theatre, Baltimore, 233
  Forrest, 197
  Forrest, Philadelphia, 120, 277, 309
  44th Street Theatre, 23, 26
  46th Street Theatre, 77, 147
  48th Street Theatre, 110
  Fulton, 156, 285
  Gaiety, 28, 77
  Gaiety, London, 96
  Garrick, 63–65, 80
  Globe, 70, 95, 113
  Grand Opera House, Chicago, 256
  Hammerstein, 111
  Hammerstein's Olympia, 25
  Hammerstein's Victoria, 8

Harlem Opera House, 7
Harry Miner's Bowery Theatre, 25
Henry Miller, 131
Hippodrome, 30, 203, 209, 210, 300
Hurtig & Seamon's Music Hall, 7, 8
Imperial, 151
Imperial Music Hall, 25
Irving Place Theatre, 8, 12
Knickerbocker, 70
La Salle Theatre, Chicago, 56
Liberty, 72, 116
London Empire, 100
London Pavilion, 82, 97, 100
Longacre, 275
Lyric, 77, 116, 135
Lyric, London, 144
Majestic, 21
Manhattan Opera House, 207
Mansfield, 116, 120
Martin Beck, 310
Maxine Elliott, 44
Metropolitan Opera House, 266
Music Box, 69, 126, 196
National Theatre, Philadelphia, 111
National Theatre, Washington, 91
New, London, 144
New Amsterdam, 21, 39, 70, 93, 116
New Orpheum, 7
Palace, 28, 37, 41, 58, 72, 113, 127, 131
Palace, London, 105
Palace, Manchester, 82
Pasadena Playhouse, 192
Phoenix, London, 145
Playhouse, 105
Poli's Theatre, Washington, 37
Princess, 39, 41, 79
Provincetown Playhouse, 127
Radio City Music Hall, 195, 217, 249
Royale, 117
Savoy, London, 144
Selwyn, 47, 77
Shubert, 37, 241, 249, 270, 291
Shubert, Boston, 217, 269, 297
Shubert, Newark, 150
Shubert, Philadelphia, 126, 132
Shubert-Wilbur Theatre, Boston, 33
St. James, 295, 304
St. James's, London, 144
Theatre Royal, Drury Lane, 100
Vanderbilt, 75, 91, 92, 105, 108, 124, 147
Walnut Street Theatre, Philadelphia, 108
Weber & Fields (Imperial), 25
Weber and Fields Music Hall, 26

Winter Garden, London, 100, 144
Ziegfeld, 90, 91, 96, 140
"There's A New Star In Heaven Tonight,"
   85
They Knew What They Wanted, 65
Thin Man, The, 176
This Rock, 303
Three Men on a Horse, 204, 263, 286, 294
Three Musketeers, The, 116, 246
Three Waltzes, 241
Three's A Crowd, 148, 212
"Ticky, Ticky, Tack," 12
"Till The Clouds Roll By," 81
Tillie's Nightmare, 87
Time Magazine, 206, 250, 251
Time of Your Life, The, 273, 278
"Time On My Hands" 148
Tiomkin, Dimitri, 186, 187
Tip Toes, 72, 73, 77
"To Hell With Burgundy!" (Friml), 81, 245
Tobacco Road, 195, 230
Todd, Mike, 303
Toler, Sidney, 165
Tom Jones (novel), 309
Tombes, Andrew, 33
Tone, Franchot, 188, 234, 299
Tonight at Twelve, 125
"Tonight Will Teach Me To Forget," 193
Tony's Club, 71, 72, 289
Tony's West 49th, 71
"Too Romantic," 183
Townsend Harris Hall, 22
Tracy, Lee, 86, 117
Trail of the Lonesome Pine, The, 213
Treasure Girl, 123
Tree, Dolly, 176
Trounstine, Joseph H., 42
Tucker, Sophie, 75, 144, 292
Twentieth Century, 204
   movie, 225
"Twinkle In Your Eye, The," 43
"Two Little Bluebirds," 71
Type and Print Club, 71

Ugast, Eddie, 53
Ulric, Leonore, 41
Under Two Flags, 124
Up Stage and Down, 17, 32
Urban, Joseph, 111, 112

Vagabond King, The, 45, 70, 71, 77, 80, 81,
   100, 244, 245
"Vagabond Song, The," 80

Vajda, Ernst, 189
Valentino, Rudolph, 85, 106, 179
Vallee, Rudy, 168, 195, 217
Van Dyke, W. S. "Woody," 176
Vanderbilt, Cornelius, 4
Vanderbilt Revue, The, 147
Vaszary, Janos, 173, 223
Velez, Lupe, 168, 181, 183, 184, 255, 284
Velie, Janet, 132
Velie, Jay, 58
Venuta, Benay, 294, 297
Venuti, Joe, 215
Vera-Ellen, 269, 308
Very Warm For May, 265, 286
Villa, Pancho, 16, 214
"Vixen," 41
Vogue Magazine 123, 235
Von Tilzer, Albert, 43, 216
Vreeland, Frank, 109

Wake Up And Dream, 135, 141
Wake Up, Jonathan, 13
Walker, Danton, 262
Walker, Jimmy, 123, 147, 228
Wall, Max, 98
Waller, Jack, 96, 105
Walsh, Mary Jane, 259–62, 292
Walter, Eugene, 28
Walters, Charles, 265
Wanger, Walter, 123, 124, 127, 130, 213, 243
Warburg, Felix, 38
Warburg, Gerald, 38, 123
Warner, Jack, 143, 151, 175, 234
Warner, Lewis, 143, 151
Warner, Ruth, 97
Warren, Harry, 51, 147, 177, 214
Warren, Julie, 308
Warrior's Husband, The, 292–94, 297
  movie, 293
Watts, Richard, 254
Webb, Chick, 240
Webb, Clifton, 112, 126, 148, 196, 223, 232, 255
Webb, Roy, 88, 115, 119, 121
Weber, Joe, 196
Weber, L. Lawrence, 29
Weber & Fields, 24–26, 42
Weber and Fields Re-United, 41
Weill, Kurt, 255, 308
Weingart Institute, 8
Welchman, Harry, 77, 100
"Well, Did You Evah?," 265
Wendling, Pete, 50

"We're Gonna Hang Out the Washing on
  the Siegfried Line," 267
"We're Having a Baby (My Baby and Me),"
  294
"We're In The Money" 225
We're Not Dressing, 200, 225
West, Mae, 102, 116, 257, 265
West 44th Street Club, 71
"West Point Forever," 47
Westcott, Marcy, 253, 259–262
What A Life, 259
"What Is This Thing Called Love?," 135
What Price Glory? 86
"When A Pansy Was A Flower," 147
"When Chloe Sings Her Song," 42
"When Hearts Are Young," 41
"When My Sugar Walks Down The Street,"
  116
"When The Kid Who Came From The East
  Side Found A Sweet Society Rose," 79
"When The Red, Red Robin Comes Bob,
  Bob, Bobbin' Along," 86
"When You Wish Upon A Star," 282
Whirl-i-gig, 42
Whistler, Rex, 279
White, Sammy, 57, 74, 75
White Cargo, 46
White Eagle, The, 111
White Heat, 40
Whiteman, Paul, 173, 195, 200, 204–6, 215,
  216, 220, 237, 255
Whiting, Jack, 113, 131, 132, 134, 149, 241,
  298
Whiting, Richard, 27, 143, 156
Whitney, John Hay "Jock," 203, 207, 210,
  213, 215
"Who?," 71, 152
Whoopee, 125, 148, 294
  movie, 152
Who's Who, 241
"Why Do I Love You?," 112
"Why Was I Born?," 135
Whyte, Jerome, 275
Wild Rose, 87
Wilder, Alec, 290
Wildflower, 43, 44, 54
Wiley, Lee, 133
Williams, Frances, 181
"Willow, Weep For Me" 305
Wilson, Earl, 262
Wiman, Dwight Deere, 196, 204, 212, 214,
  218, 223, 241, 243, 256, 264, 267–69,
  272, 293, 295, 306

Winchell, Walter, 93, 142, 262, 303
Winninger, Charles, 229
Wise, Rabbi Stephen S., 140, 228
Wizard of Oz, The, 268
Wodehouse, P. G., 87, 114, 116, 130, 131, 315, 316
Woman Disputed, The, 138
Wonder Bar, The, 170
Wong, Anna May, 62
Wood, Peggy, 196
Woods, Audrey, 293
Woollcott, Alexander, 14, 41, 57, 72, 78, 93, 95, 113
Woolley, Monty, 149, 212, 238
Wright, Cobina, 278
Wycherly, Margaret, 40, 106
Wynn, Ed, 135, 137, 167, 181, 182, 240
Wynn, Milton, 40

Yankee Doodle Dandy, 165
Yokel, Alex, 215, 263, 285, 286
"You Are Love," 112
You Can't Have Everything, 226
You Can't Take It With You, 101, 223
  movie, 224
"You Do Something To Me," 135
"You Gotta See Momma Ev'ry Night," 51

"You Made Me Love You" 226
You Never Know, 255
"You Oughta Be In Pictures," 197
Youmans, Vincent, 43, 59, 74, 87, 115, 143, 148, 216
Young, Roland, 170, 171
"Young and Healthy," 177
Young America Magazine, 218
"You're Getting To Be A Habit With Me," 177

Zanuck, Darryl F., 175, 177, 189
Ziegfeld, Florenz, 81, 82, 88–91, 93, 94, 96, 97, 107, 112, 114, 123, 125, 135, 136–39, 153
  death of, 168
Ziegfeld, Florenz Patricia, 90, 91
Ziegfeld Follies, 80, 89, 91, 107, 153, 182, 196, 197
"Ziegfeld Follies of the Air" (radio show), 182
Ziegfeld Follies of 1934 (Shubert production), 202
Ziegfeld Follies of 1944 (movie), 305
Zorina, Vera, 239, 241, 244, 245, 248, 256, 264, 268, 271, 278
Zuccaman, Augusta, 65
Zukor, Adolph, 10, 155, 168, 200
Zukor, Eugene, 10, 155